Appropriate Technologies for Environmental Protection in the Developing World

Ernest K. Yanful
Editor

Appropriate Technologies for Environmental Protection in the Developing World

Selected Papers from ERTEP 2007,
July 17-19 2007, Ghana, Africa

 Springer

Editor
Ernest K. Yanful
University of Western Ontario
Canada

Cover photos (from top left):

Photo 1: Regular manual street sweeping in Kumasi, Ghana, helps maintain a clean environment whilst creating employment opportunities for the urban poor: June 2008. Courtesy of Anthony Mensah, Kumasi Metropolitan Assembly

Photo 2: Preparing the base course for surface dressing of access road to Kumasi Sanitary Landfill site: July 2003. Courtesy of Anthony Mensah, Kumasi Metropolitan Assembly

Photo 3: Small mud house (kitchen) constructed with bamboo frames at Enyan Abaasa, Central Region Ghana. Courtesy of Ernest K. Yanful

Photo 4: Foreground: Kumasi Sanitary Landfill & Septage Treatment Facility. Courtesy of Anthony Mensah, Kumasi Metropolitan Assembly

ISBN 978-1-4020-9138-4   e-ISBN 978-1-4020-9139-1

Library of Congress Control Number: 2008936885

© 2009 Springer Science + Business Media B.V.
No part of this work may be reproduced, stored in a retrieval system, or transmitted in any form or by any means, electronic, mechanical, photocopying, microfilming, recording or otherwise, without written permission from the Publisher, with the exception of any material supplied specifically for the purpose of being entered and executed on a computer system, for exclusive use by the purchaser of the work.

Printed on acid-free paper

springer.com

Preface

This book is the first edited compilation of selected, refereed papers submitted to ERTEP 2007. The selected papers either dealt with technologies or scientific work and policy findings that address specific environmental problems affecting humanity in general, but more specifically, people and ecosystems in developing countries. It was not necessary for the work to have been done in a developing country, but the findings and results must be appropriate or applicable to a developing country setting. It is acknowledged that environmental research, technology applications and policy implementation have been demonstrated to improve environmental sustainability and protection in several developed economies. The main argument of the book is that similar gains can be achieved in developing economies and economies in transition.

The book is organized into six chapters along some of the key themes discussed at the conference: Environmental Health Management, Sustainable Energy and Fuel, Water Treatment, Purification and Protection, Mining and Environment, Soil Stabilization, and Environmental Monitoring. It is hoped that the contents of the book will provide an insight into some of the environmental and health management challenges confronting the developing world and the steps being taken to address them.

Acknowledgements

The Editor wishes to thank the authors of the papers for their contribution to this volume. Publications coordinator, Robyn Gaebel, did an excellent job handling the selected manuscripts and communicating with the authors. Alex Dolson, Shahenda Abou-Aly, Cindy Quintus and Erin Cullen provided editorial assistance in the compilation of papers for the original conference proceedings. Funding for the conference was provided by the Geotechnical Research Centre and Research Western at The University of Western Ontario, Duke University Pratt School of Engineering, the Canadian International Development Agency, Ministry of Local Government, Rural Development and Environment, Ghana, Goldfields Ghana Limited, Newmont Ghana Gold, AngloGold Ashanti, Tema Oil Refinery, Shell Ghana Limited, Volta River Authority, Ghana, Zoomlion Environmental Limited, New Times Corporation, Graphic Communications Group Limited and Ghana Telecom.

Introduction

Global industrialization and trade affecting developing countries have occasioned increasing human and economic activities that tend to impact negatively on the environment. Such impacts have created huge environmental problems of air, water and land pollution that confront governments and require mitigating solutions. Recent world summits have highlighted the need to develop environmental technologies and policies to protect fragile ecosystems. The purpose of the First International Conference on Environmental Research, Technology and Policy, ERTEP 2007, was to discuss grass-root environmental issues, assess efforts by government machinery and identify what communities and corporate entities can do as a social responsibility to mainstream and maintain environmental protection and integrity for sustainable development. The three-day conference attracted some 250 people delegates from 18 countries. Invited plenary lectures on policy were presented by high ranking officials from the Ghana Government, including the sector Ministers for Local Government, Rural Development and Environment (Honorable Stephen Asamoah-Boateng), Lands, Forestry and Mines (Honorable Professor Dominic Fobih), and Women and Children's Affairs (Honorable Hajia Alima Mahama). Osagyefuo Amoatia Ofori Panin, the Okyenhene (Ghana) opened the conference, while Honorable Joseph Henry Mensah, Senior Ghana Government Minister delivered a plenary address. Other plenary and keynote speakers included Dr. Ulf Jaeckel, Federal Ministry for the Environment, Berlin, Germany, Ms. Joyce Aryee, Ghana Chamber of Mines, Mr. Charles Darku, Volta River Authority, Ghana, Mr. Lars-Ake Lindahl of the Swedish Mining Association, Dr. Wanda Günther Risso, University of Sao Paulo, Brazil, and Mr. Peter Steblin, City Engineer, City of London, Professor George Nakhla, University of Western Ontario, Canada, and Dr. Clement Dorm-Adzobu, and Mr. Philip Acquah, Ghana. The plenary lectures were followed each day by technical breakout sessions during which more than 100 papers were presented under the seven themes of the conference: Environment, Health and Safety, Oil and Gas Extraction and Environment, Forestry and Environment, Mining and Environment, State-of-the-Art Technologies for Environmental Performance and Protection, Integration of Gender in Environmental Management, Environmental Monitoring Institutions and Policy Development, and Sustainability, Corporate Investment and Social Responsibility.

Nearly all technical papers presented at ERTEP 2007 were reviewed by a theme of international referees selected on the basis of their expertise in the subject area. Each paper was reviewed by at least two referees and written comments were sent to authors for the preparation of revised papers. Following the conference, the Editor of the current volume and ERTEP 2007 Conference Chair, Professor Ernest Yanful, and three editorial assistants, Robyn Gaebel, Cindy Quintus and Alex Dolson selected a number of refereed papers dealing with technologies and policy interventions appropriate for use or application in the developing world for publication. Authors of the selected were informed of the process and additional revisions were requested by the Editorial Team to produce this final volume.

Editor
Professor Ernest K. Yanful
Editorial Assistants
Robyn Gaebel, Cindy Quintus, Alex Dolson
and Long Vu

Contents

Preface v

Acknowledgements vii

Introduction ix

Part I Environmental Health Management

A Comparative Case Study of Detection of Radiation in Vegetable
Leaves in a Coastal Oil Producing and Hinterland Non-oil
Producing Regions in Akwa Ibom State 3
G.T. Akpabio and B.E. Bassey

Influence of Human Activities and Land Use on Heavy Metal
Concentrations in Irrigated Vegetables in Ghana
and Their Health Implications 9
E. Mensah, N. Kyei-Baffour, E. Ofori, and G. Obeng

An Economical Solution for the Environmental Problem
Resulting from the Disposal of Rice Straw 15
A.A. El Damatty and I. Hussain

Reversing Africa's Deforestation for Sustainable Development 25
K. Tutu and C. Akol

Life Cycle Assessment of Chocolate Produced in Ghana 35
A. Ntiamoah and G. Afrane

Sustainable Production of Traditional Medicines in Africa 43
O.A. Osunderu

Microbial Risk Assessment: Application and Stages
to Evaluate Water Quality 53
M.T.P. Razzolini, W.M.R. Günther, and A.C. Nardocci

Benefits and Dangers of Nanotechnology: Health and Terrorism 59
Y.A. Owusu, H. Chapman, T.N. Dargan, and C. Mundoma

Part II Sustainable Energy and Fuel

**Environment Friendly Biodiesel from *Jatropha curcas*:
Possibilities and Challenges** 75
C. Baroi, E.K. Yanful, M.F. Rahman, and M.A. Bergougnou

**Thermal Utilization of Solid Recovered Fuels in Pulverized
Coal Power Plants and Industrial Furnaces as Part
of an Integrated Waste Management Concept** 83
G. Dunnu, J. Maier, and A. Gerhardt

**Biogas Production from Organic Waste in Akwa
IBOM State of Nigeria** 93
E.E. Ituen, N.M. John, and B.E. Bassey

**West African Gas Pipeline (WAGP) Project: Associated Problems
and Possible Remedies** 101
E.O. Obanijesu and S.R.A. Macaulay

Part III Water Treatment, Purification and Protection

**Activated Carbon for Water Treatment in Nigeria: Problems
and Prospects** 115
I.K. Adewumi

**Impact of Feedwater Salinity on Energy Requirements
of a Small-Scale Membrane Filtration System** 123
B.S. Richards, L. Masson, and A.I. Schäfer

**Modern Technology for Wastewater Treatment
and Its Application in Africa** 139
D. Thierno and S. Asplund

**Ultrafiltration to Supply Drinking Water in International
Development: A Review of Opportunities** 151
J. Davey and A.I. Schäfer

**Groundwater Pollution in Shallow Wells in Southern Malawi
and a Potential Indigenous Method of Water Purification** 169
M. Pritchard, T. Mkandawire, and J.G. O'Neill

Photoelectrocatalytic Removal of Color from Water Using TiO_2 and TiO_2/Cu_2O Thin Film Electrodes Under Low Light Intensity 181
Z. Feleke, R. van de Krol, and P.W. Appel

Part IV Mining and Environment

Management of Acid Mine Drainage at Tarkwa, Ghana 199
V.E. Asamoah, E.K. Asiam, and J.S. Kuma

A Multi-disciplinary Approach to Reclamation Research in the Oil Sands Region of Canada... 205
C.J. Kelln, S.L. Barbour, B. Purdy, and C. Qualizza

Intelligent Machine Monitoring and Sensing for Safe Surface Mining Operations.. 217
S. Frimpong, Y. Li, and N. Aouad

Neutralization Potential of Reclaimed Limestone Residual (RLR)........... 229
H. Keith Moo-Young

Quantification of the Impact of Irrigating with Coalmine Waters on the Underlying Aquifers.. 235
D. Vermeulen and B. Usher

Application of Coal Fly Ash to Replace Lime in the Management of Reactive Mine Tailings 247
H. Wang, J. Shang, Y. Xu, M. Yeheyis, and E. Yanful

Part V Soil Stabilization

Consolidation and Strength Characteristics of Biofilm Amended Barrier Soils 257
J.L. Daniels, R. Cherukuri, and V.O. Ogunro

Bagasse Ash Stabilization of Lateritic Soil 271
K.J. Osinubi, V. Bafyau, and A.O. Eberemu

Strength and Leaching Patterns of Heavy Metals from Ash-Amended Flowable Fill Monoliths 281
R. Gaddam, H.I. Inyang, V.O. Ogunro, R. Janardhanam, and F.F. Udoeyo

Part VI Environmental Monitoring

Geoelectrical Resistivity Imaging in Environmental Studies..................... 297
A.P. Aizebeokhai

Nitrogen Management for Maximum Potato Yield, Tuber Quality, and Environmental Conservation 307
S.Y.C. Essah and J.A. Delgado

In Vitro* Analysis of Enhanced Phenanthrene Emulsification and Biodegradation Using Rhamnolipid Biosurfactants and *Acinetobacter calcoaceticus 317
N.D. Henry and M. Abazinge

Digital Elevation Models and GIS for Watershed Modelling and Flood Prediction – A Case Study of Accra Ghana 325
D.D. Konadu and C. Fosu

Trace Metal Pollution Study on Cassava Flour's Roadside Drying Technique in Nigeria 333
E.O. Obanijesu and J.O. Olajide

Impact of Industrial Activities on the Physico-chemistry and Mycoflora of the New Calabar River in Nigeria 341
O. Obire and W.N. Barade

Plants as Environmental Biosensors: Non-invasive Monitoring Techniques 349
A.G. Volkov, M.I. Volkova-Gugeshashvili, and Albert J. Osei

Index 357

Part I
Environmental Health Management

A Comparative Case Study of Detection of Radiation in Vegetable Leaves in a Coastal Oil Producing and Hinterland Non-oil Producing Regions in Akwa Ibom State

G.T. Akpabio and B.E. Bassey

Abstract The purpose of this work is to determine how safe vegetable leaves are in an oil producing area, using a case study of the Akwa Ibom State. Radioactive radiation levels were detected for five samples of vegetable leaves namely Waterleaf (Talinum triangulare), Bitter leaf (Veronia amygdalina), Fluted pumpkin (telfairia occidentalis), Editan, (Lasientera Africana) and Afang (Gnetum africanum). The vegetable leaves were collected form Uyo (hinterland region) and Ibeno (coastal region) in Akwa Ibom. Radioactivity levels in each of these samples were determined. In Uyo, waterleaf had the least radioactive level of 0.00079 Bq/g while Editan recorded the highest level of 0.0019 Bq/g for Ibeno, its Fluted pumpkin showed the least of 0.0037 Bq/g whereas waterleaf records the highest radioactive level 0.0070 Bq/g. The higher radioactive level observed in Ibeno is attributed to the presence of radioactive materials in the environment due to oil drilling activities in the area.

Keywords Radioactivity · radiation · vegetable leaves · NORM

1 Introduction

Radiation is a form of energy that originates from a source and travels through some material or through space. The term radiation in its wavelike form emits particles, such as electrons, neutrons, or alpha particles as well as electromagnetic radiations. Nuclei that are not stable are radioactive (Tippler 1991). Radioactivity can be dangerous to human health. Naturally occurring radioactive materials (NORM), produced along with oil and gas production, are often concentrated at some points in the hydrocarbon production process. Concentrations of NORM are

---

G.T. Akpabio(✉) and B.E. Bassey
Department of Physics, University of UYO, UYO, Akwa Ibom State, Nigeria
e-mail: gtakpabio@yahoo.com

found in oil and gas production equipment and waste. The radioactive component is the major environmental concern, especially in places like Akwa Ibom State and other oil producing regions.

Many processes occur in a leaf, but the distinctive and most important is the process of food manufacture. Green plants posses the ability to manufacture food from raw material derived from the soil and air. It is on these activities that not only the life of plants but also the life of all animals, including man depends (Wilson et. al. 1971). All leaves are metabolic factories equipped with photosynthetic cells, but that vary enormously in size, shape and texture (Starr and Targgart 1995). Vegetables leaves are plant or part of plant (leaves) that is eaten as food (Hornby 2001).

There are a number of devices that can be used to detect the particles and photons emitted when a radioactive nucleus decays; such devices detect the ionization that these particles and photon cause as they pass through matter (Cutnell and Johnson 1998). The presence of radioactive substance is easily detected with a Geiger Muller (GM) counter, based on the ionization produced by the radioactive emanations, (Eno 1998). One of the most common radiation detectors is the Geiger counter. These radiation detectors are based upon the ionization or excitation of atoms by the passage of energetic particles through matter (Wilson and Buffer 2000). The GM counter is non-energy dissipative, hence effectively useful for environmental radiation measurements, (Sigalo and Briggs-Kamara 2004).

The rate at which radioactive materials disintegrate or decay is almost independent of all other physical and chemical conditions (Akpabio and Ituen 2006). The activity of a radioactive sample can be expressed in terms of the rate of decay, that is, the number of disintegration per second in the sample (Howill ands Sylvester 1976). Unstable nuclei are radioactive and decay by emitting α particles (4He nuclei), β particles (electrons or protons), or γ rays (photons) (Tipler 1991). Radioactivity, or the emission of α- or β-particles and γ rays, is due to the disintegrating nuclei of atoms. All radioactivities are statistical in nature and follow an exponential decay law, Equation (4).

The number of atoms disintegrating per seconds, dN/dt, is directly proportional to the number of atoms, N, present at the instant (Nelkon and Parker 1977). Hence;

$$\frac{dN}{dt} = -\lambda N \qquad (1)$$

Where λ is a constant characteristic of the atom concerned called the radioactivity decay constant. Thus, if N_o is the number of radioactive atoms present at a time t = 0, and N is the number at the end of a time t is

$$N = N_o e^{-\lambda t} \qquad (2)$$

The time it takes for the number of nuclei or the decay rate to decrease by half is called the half-life $T_{1/2}$. Equation (2) shows that:

$$\frac{N_0}{2} = N_0 e^{-\lambda t}$$
(3)

Therefore,

$$T_{1/2} = \frac{1}{\lambda} \ln 2 = \frac{0.693}{\lambda}$$
(4)

$$T_{1/2} = \frac{0.693}{\lambda}$$
(5)

1.1 Material and Method

Waterleaf, Bitter leaf, Fluted pumpkin leaf, Afang leaf (Gnetum africanum), Editan leaf (Lasienthera africana) were each collected form Ibeno (Oil producing area) and Uyo (non-oil producing area) of Akwa Ibom State. The fresh samples of the vegetables were carefully plucked and mesh using the pestle and mortar. The meshed samples were now transferred carefully into the beaker and were placed directly under the Geiger–Muller tube. The experiment was repeated in each case in order to determine the average value and the distance between the Geiger–Muller tube and the beaker was kept constant throughout the experiment. Reports of environmental monitoring of radiation levels in which Geiger–Muller Counters have been used include Sigalo and Briggs-Kamara (2004) among others.

2 Results and Discussion

Results of the radiation levels in each of the vegetable samples for the two locations are presented in Tables 1 and 2.

At Ibeno (Table 1), waterleaf has the highest radiation level of 0.0070 Bq/g and Afang leaf the minimum of 0.0034 Bq/g. Table 2 shows that in Uyo, Editan had the highest level of radioactivity, followed by bitter leaf and the least in waterleaf. The histogram representing radioactivity in vegetables as shown in Fig. 2, give the comparison between Ibeno and Uyo. Ibeno generally has a higher level of radioactivity in the vegetable leaves when compared to Uyo. This is of great concern because according to Mgbenu et al. (1995), the exposure of human being to nuclear radiation is very harmful. Its damage to the human body depends on the absorbed dose, the exposure rate and the part of the body exposed. If the permissible level is greatly exceeded, the individual can suffer effects, which may be (a) somatic and (b) Genetic (Mgbenu et al. 1995). Similar results were found for roots of plants within the same environment and were obtained by Akapbio and Ituen (2006). Oil drilling sites and production facilities have many radioactive materials associated with them.

Table 1 Experimental results for samples obtained from Ibeno LGA

Samples	Mass	Background count for 10 min	Background + sample count for 10 min	Sample count for 10 min	Corrected count for 1 s	Counts per gram (Bq/g)
Water leaf	6.029	109.0	134.0	25.0	0.042	0.0070
Bitter leaf	9.093	119.3	142.7	23.4	0.039	0.0039
Pumpkin leaf	9.0881	118.0	135.0	17.3	0.029	0.0037
Editan leaf (Lasienther africanan)	5.644	112.3	131.3	19.0	0.032	0.0060
Afang leaf (Gnetum africanum)	5.402	128.3	139.7	11.4	0.019	0.0034

Table 2 Experimental results for samples obtained from Uyo LGA

Samples	Mass (g)	Background count for 10 min	Background + sample count for 10 min	Sample count for 10 min	Corrected count for 1 s	Counts per gram (Bq/g)
Water leaf	6.368	121.0	124.0	3.0	0.005	0.0079
Bitter leaf	9.462	107.7	120.0	9.0	0.015	0.0016
Pumpkin leaf	9.504	127.3	134.0	6.7	0.011	0.0012
Editan leaf (Lasienther africanan)	5.721	118.0	123.7	5.7	0.010	0.0019
Afang leaf (Gnetum africanum)	5.3922	117.3	121.3	4.0	0.019	0.0012

Drilling fluids used for onshore wells are primarily disposed of in reserve pit, while in many areas drilling fluids from offshore wells are primarily disposed of in reserve pit, while in many areas drilling fluids from offshore platforms have been dumped overboard (Reis 1996).

The offshore case is of importance to individuals in Akwa Ibom State because when oil is spilled on water, it spreads out over the water surface and moves with the wind and water current, (Reis 1996). This is the water that plants and animals in the area makes use of. Even though these levels of radioactivity are below the international permissible limit of 0.18 Bq/g, continuous accumulation over periods of time can be dangerous and the effects on humans have not been studied in depth.

3 Conclusion

The results show that the radioactivity levels in the vegetable in Ibeno (an oil producing area) is higher than at Uyo (a non-oil producing area).

Acknowledgements The authors of this article wish to acknowledge the contributions of Effiong, IkpidigheAbasi Paul in the data acquisition.

References

Akpabio, G. T. and Ituen, E. E. (2006) Comparative effect of radioactive radiation on roots in coastal and hinterland locations in Akwa Ibom State Nigeria. Nig. J. Phys. 18(1), 117–120.
Cutnell, T. D. and Johnson, K. W. (1998) Physics, 4th edition. Wiley, New York, pp. 976–977.
Eno, E. E. (1998) Electricity and Modern Physics. Footsteps, Port Harcourt, pp. 204–211.
Holwill, M. E. and Silvester N. R. (1976) Introduction to Biological Physics. Wiley, London, pp. 323–325.
Hornby A. S. (2001) Advanced Learner's Dictionary Oxford University Press. New York, p. 1325.
Mgbenu, A. E, Inyang, A. E., Agu, M. N., Osuwa, J. C. and Ebong, I. D. U. (1995) Modern Physics. Nigerian University Physics Series, Spectrum Books, Ibadan, pp. 188–207.
Nelkon, M. and Parker P. (1977) Advanced Level Physics, 4th edition. Heinemann, London, pp. 946–954.
Reis, J. C. (1996) Environmental Control in Petroleum Engineering. Gulf Publishing, Houston, TX, pp. 261–264.
Sigalo, F. B. and Briggs-kamara, M. A. (2004) Estimate of lonishing radiation levels within selected riverine communities of the Niger Delta. J. Nig. Environ. Soc. 2(2), 159–162.
Starr, C. and Targgart, R. (1998) Biology, 7th edition. Wardsworth Publishing, New York, p. 490.
Tipler, P. A. (1991) Physics for Scientist and Engineers, Vol. 2, 3rd edition, Worth Publishers. New York, pp. 1338–1361.
Wilson, C. L., Loomis, W. E., Steeves, T. A. and Holt, R. (1971) Botany, 5th edition. Winston, New York, p. 99.
Wilson, J. D. and Buffer, A. J. (2000) College Physics, 4th edition. Prentice Hall, Englewood Cliffs, NJ, p. 916.

Influence of Human Activities and Land Use on Heavy Metal Concentrations in Irrigated Vegetables in Ghana and Their Health Implications

E. Mensah, N. Kyei-Baffour, E. Ofori, and G. Obeng

Abstract Anthropogenic activities are major sources of heavy metal pollution which serve as major pathways for plant uptake of heavy metals like cadmium (Cd) and lead (Pb) to enter the human food chain from the soil and irrigation water. This study was conducted to investigate the levels of Cd, Pb, Zn, Fe, Ni and Cu concentrations in sampled vegetables (cabbage and carrots) from two major markets in Kumasi, a metropolis and two producing rural towns along the Accra – Kumasi road. Apart from Ni all other heavy metals in cabbage were far higher than the FAO/WHO permissible values of samples from both urban/peri-urban and rural communities. Cadmium content of the vegetables from the peri-urban communities were extremes (0.5–4.2 mg/kg) and were generally higher than produce from the rural communities with values between 1.6 and 1.9 mg/kg. However, cabbage from Asikam, a rural and mining community contained 2.9 mg/kg of Cd. Lead concentration levels in the sampled vegetables from the peri-urban communities ranged between 6–45 mg/kg whilst values from the rural communities were between 12 and 13 mg/kg. Cadmium and lead concentration levels in the sampled vegetables far exceeded FAO/WHO recommended maximum values of 0.3 and 0.2 mg/kg respectively with samples from urban/peri-urban communities registering higher values than those from the rural towns.

1 Introduction

Anthropogenic activities are major sources of heavy metal pollution which serve as major pathways for plant uptake of heavy metals like cadmium (Cd) and lead (Pb) to enter the human food chain from the soil and irrigation water. Lead leaf contents,

---

E. Mensah (✉), N. Kyei-Baffour, and E. Ofori
Department of Agricultural Engineering, Kwame Nkrumah University of Science and Technology (KNUST), Kumasi, Ghana

G. Obeng
Technology Consultancy Centre, Kwame Nkrumah University of Science and Technology (KNUST), Kumasi, Ghana

for example, are very high in plants growing in urban and industrial areas (Maisto et al. 2004). Heavy metals found in agricultural soils originates from many sources including paints, gasoline additives, smelting and refining of Pb, pesticide production and Pb acid battery disposal (Eick et al. 1999; Paff and Bosilovich 1995), phosphate fertilizers, sewage sludge, wastewater for irrigation, waste from smelting sites and others (Ingwersen and Streck 2005). Cadmium is one of the most mobile and bioavailable heavy metals in soil and may cause human and ecotoxicological impacts even at low concentrations. There is considerable published information about the action of Cd^{2+} on plant growth and on physiological and biochemical processes. Harmful effects produced by Cd might be explained by its ability to inactivate enzymes possibly through reaction with the SH-groups of proteins (Fuhrer 1982). Detrimental effects are manifested in inhibition of photosynthesis and in oxidative stress leading to membrane damage (Prasad 1995).

Application of sewage sludge to agricultural land is a feasible alternative for reutilization residual resource of high nutrient and organic matter contents which represent a good fertilizer and/or soil conditioner for plant and soil (Logan et al. 1997; Wong 1996). Besides, sludge amendment could improve soil physical properties such as soil aeration, water holding capacity and aggregation (Logan and Harrison 1995) while the slightly alkaline property of the sludge buffers against the acidity of acidic soils.

Wastewater and sludge for irrigation or amendment of the agricultural land may lead to build higher concentrations in soils as a result of accumulation. The higher metal levels in soil may cause negative impact on crops, inhibiting the growth in one or other way. One of the most important factors is the pH of the soils. Alkaline pH of the soil would restrict the mobilization of the metal in soil matrix and consequently the metal uptake by crop plant would be controlled, reducing the risk of metal toxicity.

Heavy metals may enter the human body through inhalation of dust, direct ingestion of soil, and consumption of food plants grown in metal-contaminated soil (Sterrett et al. 1996) and/or irrigated with wastewater (Ingwersen and Streck 2005). Prolonged exposure to heavy metals (e.g. Cd, Cu, Pb, Ni and Zn) can cause deleterious health effects in humans (Reilly 1991). In children, Pb has been known to cause decreases in IQ scores, retardation of physical growth, hearing problems, impaired learning, as well as decreased attention and classroom performance. In individuals of all ages, Pb may cause anemia, kidney disease, brain damage, impaired function of the peripheral nervous system, high blood pressure, reproductive abnormalities, developmental defects, abnormal vitamin D metabolism, and in some situations death (Hrudey et al. 1995).

Heavy metal contamination can affect plant health and nutritional value of crops. High Cd concentrations can lead to toxicity symptoms like chlorosis and reduced growth of the leaves of crops. The severity of Cd phytotoxicity is found most evident from dry matter yield in both leaves and roots of crops (Michalska and Asp 2001). The extent of contamination to food crops is likely to increase with intensification of production systems, urbanisation and industrialisation but levels of food contamination are not regularly monitored or controlled. In Ghana vegetable consumption has been on the increase as a result of changes in the eating habits of urban dwellers due to socio-economic changes with time. However, high percentages of vegetables consumed by urban dwellers are produced under wastewater irrigation. These exotic vegetables (cabbage, carrots, lettuce, spring onions, etc.) are produced mainly in the urban/peri-urban communities of Ghana. However, cabbage and carrots may be produced in some locations in a few rural communities. Location influences the type of inputs for vegetable production. Urban/peri-urban produced vegetables are irrigated with wastewater or urban streams that are receptacle of urban effluent, hand-dug wells and in very few situations treated water from the mains while the soils are conditioned and the nutrient levels are improved with sewage sludge. For vegetables produced in the rural communities source of irrigation water is mainly a stream or river and soil nutrient improvement and conditioning is by the application of inorganic fertiliser. The objective of this study was to determine the extent of human activities through the type of input used and land use on heavy metal concentration in irrigated vegetables in Ghana and probable health implications.

2 Materials and Methods

Vegetable samples were collected from two main vegetable markets in Kumasi (Asafo and "European") and two rural towns (Nsutam and Kibi) along the Accra–Kumasi highway between August and September, 2005, for the study. The collected samples were washed and rinsed with distilled water and chopped on a distilled water rinsed kitchen chop board using a washed and rinsed kitchen knife into pieces (of about 25 × 25 mm size for cabbage and 1 mm thickness for carrots). The chopped samples were sun-dried for about 8 h before subjecting them to oven drying for about 24 h at 70 °C. The oven dried samples were milled to pass a 2 mm sieve and were packaged in transparent plastic bags and sealed to prevent moisture ingression. 0.2 g of each plant sample was placed into a teflon beaker and 6 ml of concentrated nitric acid (HNO_3) was added and weighed. The beaker was assembled, placed in a rotor and tightened with torque wrench before placing the rotor in the chamber of an already programmed microwave digester for digestion. A 1-ml aliquot of digested sample was placed in a 15-ml centrifuge tube, diluted with 5 ml of deionized water, and analysed for Cd, Pb, Ni, Zn, Fe and Cu and lead with Agilent 7500c ICP-MS.

3 Results

Figure 1 shows the concentrations of Cd, Pb, Ni, Zn, Fe, and Cu in vegetables from the various sites. Apart from Ni all the other heavy metals concentrations in cabbage were far higher than the FAO/WHO permissible values of samples from both urban/peri-urban and rural communities.

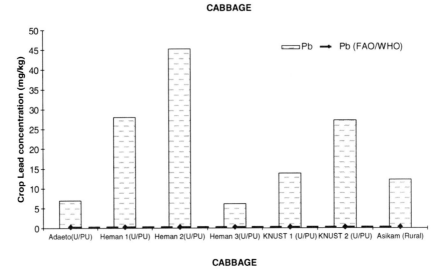

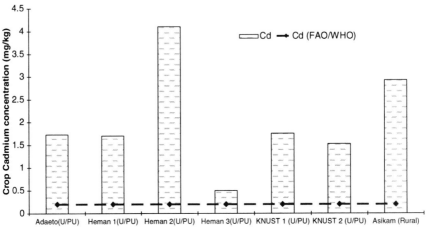

Fig. 1 Relationship between analysed heavy metals concentrations in cabbage samples from both rural and urban/peri-urban communities and permissible FAO/WHO permissible values/guidelines

4 Discussion

The recommended concentration levels of the analysed heavy metals by FAO/WHO in vegetables is shown in Table 1.

Plant samples differed in levels of concentrations of heavy metals based on location. Most of the cabbage sampled had high levels of heavy metal concentration at Heman site 2, a rural community north of Kumasi. The high values of heavy metals concentrations of lettuce and cabbage may be due to the sources of irrigation water (streams) that receive

Table 1 FAO/WHO maximum permissible values of heavy metals in vegetables

Element	FAO/WHO maximum permissible values (mg/kg)
Cd	0.2
Pb	0.3
Ni	67.9
Fe	425.5
Cu	73.3
Zn	99.4

effluents and storm water from Kumasi "Magazine", the largest public sector garage in Africa. The values of Cd and Pb were similar to concentrations of cabbage grown and irrigated with water from streams close to a smelting factory in Addis Ababa, Ethiopia from a study carried out there (Itanna 2002). For the least plant heavy metal concentrations, there were variations for both plants and locations. Cabbage showed the highest concentrations of Cd and Pb which confirms results from a similar study by Petterson (1977). Most of the heavy metals concentration levels recorded were above maximum permitted concentrations by FAO/WHO. Human health implications of heavy metals are determined by accumulated concentration levels from ingestion through food. Food intake is related to body weight and age (Oliver et al. 1995). Also the quantity of vegetable intake is influenced by the level of development of a community in which the people live. In a developing country, it has been recommended that an average intake of carrots and cabbage per person in a day by a vegetarian is 113 g each and 21 g of lettuce (USEPA 2002). The WHO (1989) recommended daily intake of Cd per unit (kg) body weight of an adult is 1 µg. The maximum daily Cd intake, therefore, by a vegetarian of 50 kg weight will be 50 µg. A composite meal of carrots, cabbage and lettuce, for example, weighing 247 g for a 50 kg vegetarian in Ghana will be ingesting between 123.5 and 1,037 µg Cd which is about 2 to 21 fold of the recommended value. Sources of the heavy metals in the vegetables may vary and the sources may be fertilisers, poultry manure and water bodies used for production. The farmers use these inputs based on their availability and affordability. Because the Kumasi metropolis is choked with poultry farms majority of the farmers apply poultry manure to improve the soil. However, due to lack of extension services, the farmers tend to use knowledge from their counterparts in solving problems relating to soil fertility and plant diseases. This has been a major factor in the vegetable production sector in terms of environmental pollution.

5 Conclusion

The study showed that metal concentrations of the vegetables sampled were higher for those produced in the urban/peri-urban communities than those from the rural areas. It could be concluded that the vegetables on the Ghanaian market have high heavy metals concentration levels above the maximum recommended by health authorities and this calls for a thorough investigation to track the source(s) of the metals and measures put in place to ameliorate or eliminate the menace. Soils being used for vegetable production locally are mainly sandy with low pH and organic matter content, favouring high uptake of heavy metals.

References

Eick, M.J., Peak, J.D., Brady, P.V., and Pesek, J.D., 1999. Kinetics of lead adsorption/desorption on goethite: Residence time effect. Soil Sci. 164:28–39.
Fuhrer, J., 1982. Ethylene biosynthesis and cadmium toxicity in leaf tissue of beans *Phaseolus vulgaris* L. Plant Physiol. 70:162–167.
Hrudey, S.E., Chen, W., and Rousseaux, C.G., 1995. Bioavailability in Environmental Risk Assessment. Boca Raton, FL: Lewis.
Ingwersen, J. and Streck, T., 2005. A regional-scale study on the crop uptake of cadmium from sandy soils: Measurement and modeling. J. Environ. Qual. 34:1026–1035.
Itanna, F., 2002. Metals in leafy vegetables grown in Addis Ababa and toxicological implications. Ethiop. J. Health Dev. 16(3):295–302.
Logan, T.J. and Harrison, B.J., 1995. Availability of heavy metals for Brassica Chinensis grown in an acidic loamy soil amended with domestic and industrial sewage sludge. J. Environ. Qual. 24:153–165.
Logan, T.J., Lindsay, B.J., Goins, L.E., and Ryan, J.A., 1997. Plant availability of heavy metals in a soil amended with a high dose of sewage sludge under drought conditions. J. Environ. Qual. 26:534–550.
Maisto, G., Alfani, A., Baldantoni, D., Marco, A.D., and Santo, A.V.D., 2004. Trace metals in the soil and in Quercus ilex L. leaves at anthropic and remote sites of the Campania Region of Italy. Geoderma 122:269–279.
Michalska, M. and Asp, H., 2001. Influence of lead and cadmium on growth, heavy metal uptake, and nutrient concentration of three lettuce cultivars grown in hydroponic culture. Commun. Soil Sci. Plant Anal. 32(3&4):571–583.
Oliver, D.P., Gartell, J.W., Tiller, K.G., Correll, R., Cozens, G.D., and Youngberg, B.L., 1995. Differential response of Australian wheat cultivars to cadmium concentration in wheat grain. Aust. J. Agric. Res. 46:873–886.
Paff, S.W. and Bosilovich, B.E., 1995. Use of Pb reclamation in secondary lead smelters for the remediation of lead contaminated sites. J. Hazard. Mater. 40:39–64.
Prasad, M., 1995. Cadmium toxicity and tolerance in vascular plants. Environ. Exp. Bot. 35:525–545.
Reilly, C., 1991. Metal Contamination of Food. 2nd ed. London: Elsevier.
Sterrett, S.B., Chaney, R.L., Gifford, C.H., and Meilke, H.W., 1996. Influence of fertilizer and sewage sludge compost on yield of heavy metal accumulation by lettuce grown in urban soils. Environ. Geochem. Health 18:135–142.
USEPA, 2002. Children-specific exposure factors handbook. EPA/600/P-00/002B. National Center for Environmental Assessment, Washington, DC.
Wong, J.W.C., 1996. Heavy metal contents in vegetables and market garden soils in Hong Kong. Environ. Technol. 17:407–414.
World Health Organisation (WHO), 1989. Report of 33rd meeting, Joint FAO/WHO Joint Expert Committee on Food Additives, Toxicological evaluation of certain food additives and contaminants No. 24, International Programme on Chemical Safety, WHO, Geneva.

An Economical Solution for the Environmental Problem Resulting from the Disposal of Rice Straw

A.A. El Damatty and I. Hussain

Abstract The disposal of rice straw as a by-product resulting from the cultivation of rice is causing worldwide environmental and health problems. Farmers tend to randomly burn rice straw as the most economical method of disposal. This practice does not only generate smoke, but also breathable dust that contains crystalline silica and other health hazard substances. An environmentally friendly process that produces three valuable products is developed in this research. The process is based on a combustion technology using a special reactor manufactured in Canada. Significant amount of energy is released in the form of steam as a result from this process. The rice straw ash (RSA) resulting from this technology is rich in silica and can act as a mineral admixture that enhances the strength and durability of concrete. Liquid rich in potassium results from the hydrolysis of straw and can be used as a fertilizer. A simulation of the entire process, including hydrolysis, drying, chopping and combustion of the straw followed by grinding of the produced ash was conducted on a reduced industrial scale. A study to assess the enhancement in strength and durability of concrete and its resistance to chemical and corrosion attacks was also conducted. The paper also discusses briefly the main findings of a study funded by the Canadian International Development Agency "CIDA" that was carried out to assess the feasibility of such a project in Egypt.

Keywords Rice straw · silica · mineral admixture · fertilizer · energy

1 Introduction

Rice is one of the most important food items of the world. Global production of rice is estimated to be 618 million tonne during the year 2005 (International Rice Research Institute, IRRI). Each tonne of rice produces an equal amount of straw as

---

A.A.E. Damatty (✉) and I. Hussain
Department of Civil and Environmental Engineering, The University of Western Ontario, London, Ontario, Canada
emails: damatty@uwo.ca; ihussai4@uwo.ca

a by-product. In many countries, farmers dispose of rice straw by random burning in the open fields, which tends to be the most economical method of disposal (Mehta 1977). This practice not only generates smoke but also breathable dust that contains crystalline silica and other hazardous substances. This leads to many environmental problems related to human health and safety and diseases related to lungs and eyes are very common in these areas.

Mehta (1979) suggested that combustion of rice husk, another by-product of rice crop, under controlled conditions of temperature, produce ash (rice husk ash – RHA) which is rich in amorphous silica. This ash can be used as an additive for concrete to improve its strength and durability. Nehdi et al. (2003) used a special reactor, manufactured in Canada, to produce RHA. The ash produced by this reactor is of high quality and is rich in amorphous silica and has been proven to be a good additive for concrete. In the current study, the same reactor has been used to produce ash from rice straw (rice straw ash – RSA).

Egypt is one of the countries that face the challenge of rice straw disposal. An extensive research program was conducted as collaboration between the University of Western Ontario (UWO) in Canada and various Egyptian governmental agencies to develop an economical and environmentally friendly solution for this problem. An industrial process leading to three valuable products was developed in this program. The research was conducted in three phases. The first phase involved laboratory testing. The second phase involved a reduced – scale simulation for various components of the process and the third phase involved conducting a viability study for a full-scale industrial process.

2 Description of the Process

A chemical analysis for a sample of rice straw from Egypt was obtained by conducting a complete oxide analysis and the results are presented in Table 1. The analysis revealed that the ash represents about 15% of the total mass of the straw while the rest of the mass is in form of organic materials. The results show that rice straw is rich in silica. If the combustion of the straw is conducted under controlled conditions, this silica can be in the amorphous state, and in this case, it can have a positive effect in enhancing the mechanical properties of concrete.

The core of the developed industrial process is a special reactor called "Torbed" which uses patented technology to a produce a uniform quality of ash. A preliminary combustion test for rice straw was conducted and a phenomenon described as "centering" was observed during this test. During combustion, the tested straw accumulated together and blocked the reactor preventing from completing the test. This phenomenon is attributed to the presence of a high percentage of alkali in the straw composition in the form of Potassium and Sodium. The equivalent alkalis of the straw defined as $Na_2O + 0.658 K_2O$ is provided in Table 1 showing a value of 10.14% for the tested sample.

Table 1 Inorganic compositions of rice straw (percentage)

SiO_2	TiO_2	Al_2O_3	Fe_2O_3	MnO	MgO	CaO	K_2O	Na_2O	P_2O_5	Equiv. alkali
78.00	0.05	0.51	0.41	0.29	3.38	2.93	10.58	3.18	0.66	10.14

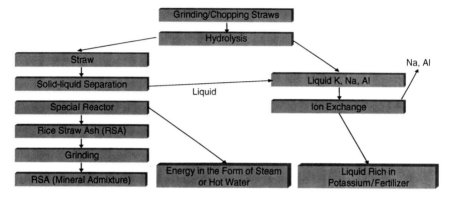

Fig. 1 Flow chart of the entire process

The process starts by chopping the straw to small sizes, typically around 1 in. A hydrolysis process is applied to the straw to reduce the equivalent alkali content to about 2.5%. Saturated straw is then dried to reduce the moisture content to about 35%. The straw is then fed into the Torbed reactor for incineration and the resulting ash is then ground to reach an average particle size of about 7 μm. One tonne of rice straw produces about 150 kg of ash. The calorific value of rice straw is high. The combustion is conducted at an average temperature of 800 °C. During the process, the ash is cooled down to 150 °C and the process of energy exchange leads to the production of hot steam. The combustion of 1 t of straw produces about 5 t of steam. This is equivalent to about 50% of the energy produced by the similar amount of coal.

The liquid resulting from the hydrolysis of straw is rich in potassium, which can be used as an agricultural fertilizer. However, the presence of sodium in the liquid can be harmful to the soil. Either a special ion exchange process can be applied to remove sodium or the hydrolysis can be conducted using a high water to straw ratio such that the amount of sodium can be tolerated. A flow chart of the entire process is provided in Fig. 1.

3 Reduced Scale Simulation of the Entire Process

A reduced scale test that simulates various stages of the proposed process was applied. The following steps were conducted:

(a) Rice straw was first chopped to a fine size.
(b) A large steel tank was constructed to conduct hydrolysis of the chopped straw. A steel screen, with very fine openings size, was constructed and installed inside the tank.
(c) Hydrolysis was conducted to the chopped straw by applying two leaching stages.
(d) It was found that after hydrolysis, the straw had a moisture content of 80%. As such, the straw was dried using a heated oven.
(e) Combustion of the straw was then conducted in the Torbed reactor.
(f) The produced RSA was then ground such that its average particle size was about 7 μm.

Photos illustrating various stages of the process are shown in Figs. 2–5.

Fig. 2 Chopping of rice straw

Fig. 3 Hydrolysis of rice straw

Fig. 4 Torbed reactor

Fig. 5 Liquid from hydrolysis

Table 2 Results of oxide analysis of ash resulting from combustion process

Sample	SiO_2	TiO_2	Al_2O_3	Fe_2O_3	MnO	MgO	CaO	K_2O	Na_2O	P_2O_5	L.O.I	Total	Equiv. alkalis
RSA	84.65	0.12	0.00	3.99	0.19	1.97	3.49	2.44	0.54	0.26	2.0	99.65	2.14

Oxide analysis was conducted for the ash resulting from this process and the results are given in Table 2. RSA can provide a similar function as Silica Fume (SF) and, therefore, its specifications can be compared to SF, provided in the ASTM C 1240-01 standard. In this standard, a minimum ratio of SiO2 of 85.0% and maximum loss on ignition (LOI) of 6.0% has been specified for SF. The results presented in Table 2 show that the LOI for RSA is less than 6% and the silica content was very close to the value specified in the standards.

4 Behaviour of Concrete Incorporating Rice Straw Ash

A total number of five concrete mixtures including a control mixture were prepared. Four concrete mixtures incorporated RSA and SF as partial replacement of mass of cement. A constant water-to-cementations materials ratio (w/c) of 0.4 was used. ASTM type I cement, natural washed gravel with a maximum particle size of 10 mm and silica sand conforming to ASTM C33 was used in the concrete mixtures. The Superplasticizer dosage was tailored in each mixture to achieve a slump of 90 ± 10 mm. Concrete cylinders were tested for compressive strength at ages of 1, 7 and 28 days whereas to measure the durability of concrete, rapid chloride penetrability test was performed at 7 days, following the procedure mentioned in ASTM C 1202.

5 Compressive Strength Tests

Compressive strength results are shown in Fig. 6. An increase in the 1-day compressive strength was achieved by all RSA concretes as compared to the reference concrete, with concrete having 10% replacement of RSA showing the highest 1-day strength. At 7 and 28 days, the strength of all concretes incorporating RSA and SF outperformed that of the reference concrete. At 28 days, concretes with 7.5%, 10% and 12.5% RSA contents showed increase in compressive strength of 12.7%, 18.18% and 23.2%, respectively, as compared to the reference concrete.

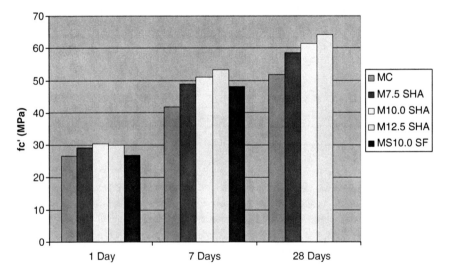

Fig. 6 Compressive strength test results

6 Rapid Chloride Penetrability Test for Durability Assessment

Rapid chloride penetrability tests at 7 days were carried out for all five concretes and the results are presented in Fig. 7 along with the ASTM C1202 classification ranges. All RSA concretes have shown excellent resistance to the chloride ion penetrability and lie either in the very low or low ranges of ASTM classification. Resistance to chloride ion penetrability of RSA concrete is more than that of SF concrete with equal percentage additions.

7 Electrical Conductivity Liquid Resulting from Hydrolysis

Liquid resulting from the hydrolysis of rice straw was tested for electrical conductivity (E.C) in order to measure its salinity and the results are shown is Table 3. This liquid, rich in potassium, can be used as an agricultural fertilizer. E.C is expressed in milli-siemens per centimeter (ms/cm). According to the Department of Agriculture,

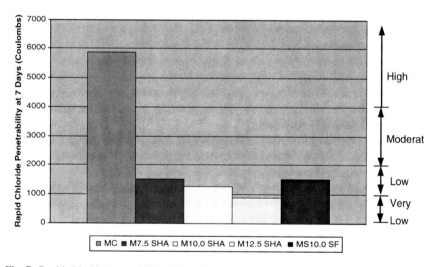

Fig. 7 Rapid chloride penetrability test results

Table 3 Results of electrical conductivity

Stage of hydrolysis	E.C[a] (ms/cm)
I	1.73
II	0.49

[a]E.C value for ordinary tap water is 0.22.

Food and Rural Development of Alberta, Canada (Government of Alberta 2001), the allowed E.C value for a non saline solution should be less than 2 ds/m for a soil depth of 0–60 cm and less than 4 ds/m for a soil depth of 60–120 cm. The results show that the electrical conductivity of the liquid is less than 2 ds/m and therefore, it can be used as agricultural fertilizer.

8   Results of Viability Study

A detailed study was conducted to assess the viability of a full-scale industrial project based on the process described in the study. Although, this project can assist in solving a National problem, its success and sustainability depends to a large extent on the economics of the project. The viability study covered many aspects including the availability and collection of the raw materials, the global demand for RSA as a concrete additive, the potential use of energy, the environmental impact of the project both on the local and global levels and the cost-benefit analysis of the project. Main findings of this study are summarized below:

1. The global annual demand on mineral concrete admixtures will reach a value 1.12 million tonnes by 2013 and the predicted supply of currently available materials can not by far satisfy this demand. As such, a strong market exists for RSA as a concrete additive. The increase in demand on this type of concrete additives is associated with three factors: (a) the natural expansion of the cement and construction industry, (b) the need to optimize the cost effectiveness of concrete by improving its durability and long term performance and (c) the requirement to minimize the environmental impact of cement and concrete production – especially the CO_2 emission.
2. It is more economical to consume the energy resulting from the process either in the form of hot water or steam rather transforming it to electricity, which requires an extra investment to cover the costs of a turbine and a generator. A conducted survey indicated that industries in Egypt are open to the idea of getting their needed supply of steam from a separate rice straw reactor, given that the cost of steam will not exceed its cost based on conventional sources of energy.
3. An environmental impact assessment study revealed that the project has positive impacts on the local level by eliminating many environmental and health problems associated with the current method employed for the disposal of rice straw.
4. On the global environmental level, the process results in emission of carbon dioxide which can be considered as a negative effect in view of the issue of green house emission. However, this hypothesis might be reversed and the suggested process can be considered to have a positive effect on the global environment. The effect of RSA in increasing the strength of concrete leads to a reduction in the structural cross sections and, consequently, a reduction in the consumption of cement. One tonne of RSA can lead to a saving of 3 t of cement. Since the production of cement results in a large emission of carbon dioxide, the net

results associated with the use of RSA in concrete can be a significant reduction in the global emission of carbon dioxide.
5. The value of RSA, energy and liquid fertilizer resulting from processing 1 t of straw is about $84. Given that the cost of 1 t of straw including transportation is about $20; such a project can be economically viable.

9 Conclusions

An industrial process has been developed to produce three valuable products from a waste material, namely from rice straw. A reduced-scale simulation for various components of this process has been conducted. The core of the process is an advanced combustion technology produced in Canada. The three products resulting from this process are: rice straw ash (RSA), significant amount of energy and a liquid fertilizer.

Mechanical properties of concrete specimens reveal the efficiency of the RSA produced using the proposed technology in enhancing the strength and durability of concrete. The processing of 1 t of straw leads to about 5 t of hot steam. This amount of energy is equivalent to 50% of that produced by a similar amount of coal. The third product resulting from the process is a liquid rich in potassium is suitable for use as an agricultural fertilizer. A conducted viability study revealed the strong market demand for RSA as a concrete additive. The combination of RSA, energy and fertilizer makes this project quite viable economically for application in developing countries where rice is cultivated.

References

Government of Alberta, "Salt tolerance of plants", *Agriculture, Food and Rural Development*, 2001.
P. K. Mehta, "Properties of blended cements made from rice husk ash", *Journal of the American Concrete Institute*, V.74, No.9, September 1977, pp. 440–442.
P. K. Mehta, "The chemistry and technology of cements made from rice-husk ash", *Proceedings of UNIDO/ESCAP/RCTT Workshop on Rice-Husk Ash Cement, Peshawar, Pakistan, Regional Centre for Technology Transfer, Bangalore, India*, 1979, pp. 113–122.
M. Nehdi et al., "Performance of rice husk ash produced using a new technology as a mineral admixture in concrete", *Cement and Concrete Research*, V.33, 2003, pp. 1203–1210.

Reversing Africa's Deforestation for Sustainable Development

K. Tutu and C. Akol

Abstract Forests provide goods and services which are not only vital for the survival of the poorest in many African countries, but are also essential for the basic functioning of a wide range of development sectors in the continent. The success of many poverty reduction and economic growth strategies of these countries is thus inextricably linked to the sustainable provision of forest goods and services. This is being severely undermined by chronic and widespread deforestation in the continent. This continues to compromise efforts towards achieving the targets set in millennium development goals, as well as attaining sustainable development in the continent and at the global level. There is therefore an acute need to continue exploring and promoting policy options, measures and models to enhance success in combating deforestation in Africa. This paper explores the significance of deforestation in the continent and helps to advance knowledge and to offer policy and other recommendations to combat this menace in Africa. The paper seeks to contribute to and stimulate cross-sectoral dialogue and actions to combat deforestation. In so doing it will serve to promote and mainstream sustainable forest management principles and measures within the broad agenda of poverty reduction and sustainable development in Africa.

Keywords Africa · sustainable · development · reversing · deforestation · forests · strategies

1 Introduction

The fight against poverty is one of the top most priorities of both the African countries as well as international development policy. So is sustainable development, which is about improving the quality of life for all of earth's citizens without com-

---

K. Tutu (✉) and C. Akol
Food Security and Sustainable Development Division (FSSDD), United Nations Economic Commission for Africa (UNECA) P.O. Box 3001, Addis Ababa, Ethiopia
e-mail: ktutu@uneca.org e-mail: cakol@uneca.org

promising the ability of future generations to meet their needs. Forests and woodland resources are important elements in both poverty reduction and sustainable development strategies of many Africa. In this respect forests provide goods and services that are not only vital for the survival of the poorest of the continent, but which are also essential for the basic functioning of a wide range of development sectors. Forests therefore underpin sustainable development in Africa and at global level. The challenge facing the African continent in particular, is the overwhelming deforestation and degradation of its forests amidst the increasing number of people who are becoming dependent on forest and woodland resources. If immediate solutions to deforestation are not found, efforts to achieve targets set in the millennium development goals as well as attaining sustainable development in Africa will be seriously compromised.

This paper illustrates the significance of forests and explores the extent, causes and consequences of deforestation in Africa. A brief review of strategies undertaken or underway to combat deforestation is made. The paper offers policy and other recommendations to combat the menace of deforestation and foster sustainable development in the continent. It seeks to contribute to and stimulate cross-sectoral dialogue and actions to combat deforestation. In so doing it will serve to promote and mainstream sustainable forest management principles and measures within the broad agenda of poverty reduction and sustainable development in Africa.

2 Extent and Importance of Forests in Africa

There are several definitions of forests according to different organizations. The FAO definition, which is generous, classifies forests as land with a tree canopy cover of more than 10% and an area of more than half a hectare. FAO's definition of forests include natural and plantation forests but excludes stands of trees established primarily for agricultural plantations such as fruit tree and oil palm plantations and trees planted in agroforestry systems.

Forests are estimated to cover an area of 635 million representing 16% of the global forest area and 21% of Africa's land area. One percent of the forest area is classified as forest plantations and protected forests cover an estimated area of 75,885,000 ha (12% of total forest area) (Kidd 2005). The vast majority of Africa's forests are found in West and Central Africa (FAO 2006). However more than 70% of Africa's remaining rainforests are located in the Central African subregion, covering about 720,000 square miles (1.875 million square kilometres). The bulk of forests in this subregion are found in the Congo Basin in the countries of the Democratic Republic of Congo, Congo and Gabon (Butler 2006b). The Congo Basin also contains the second largest continuous tropical rainforest in the world. In Africa, most forest plantations are found in Northern Africa, which is dependent on plantations in the absence of natural forests (FAO 2006).

Forests provide multiple goods and services that are vital for poverty reduction and sustainable development in Africa and at global level. Over two-thirds of

Africa's 600 million people rely directly or indirectly on forests for their livelihoods including food security (CIFOR 2005). It is estimated that forests account for an average of 6% of Gross Domestic Product (GDP) in Africa, which is the highest in the world (NEPAD 2003). About 80% of the energy used in Africa is wood biomass based. Fuelwood is predominantly used in the rural areas, while in urban areas, charcoal is used.

Forests are also important in the recharge of water bodies as well as controlling water flow. In this respect they are vital in the health of communities, agricultural production and electricity power generation, which depend on this water.

Forests and woodlands are critical in biodiversity conservation. Africa is home to 25% of the world's remaining tropical rainforests and contains 20% of the world's biodiversity hotspots (CIFOR 2005). The Congo Basin forest alone harbors the most diverse assemblage of plants and animals in Africa including over 400 mammal species, more than 1,000 bird species, and over 10,000 plant species of which about 3,000 are endemic. Only in Central Africa do forest elephant, gorilla, forest buffalo, bongo, and okapi occur in large numbers across large areas of forest (CBFP 2005).

Forests are vital indeed in combating land degradation and desertification and mitigating climate change, which pose serious threats to agricultural production among other economic sectors.

3 The Threat of Deforestation

According to FAO, deforestation is the conversion of forests to another land use (such as agriculture, pasture, water reservoirs and urban areas) or the long-term reduction of the tree canopy cover below the minimum 10% threshold.

There are many root causes of deforestation. No one factor takes place independently, and most studies point to a combination of factors as the key driving forces behind deforestation (Clausen et al. 2003). The most cited ones in Africa are the high levels of poverty, hunger, high population growth rates, conflicts, market forces, poor resource and economic development policies, and insufficient long-term funding. The more visible direct causes of deforestation are conversion of forests into agricultural land, large-scale shifting cultivation (slash and burn), logging, fuelwood collection including clear-cutting for charcoal production (Allen and Barnes 1985). The others include forest conversion for permanent pasture, open pit mining and large-scale mining operations, large roads and infrastructure projects, wildfires that destroy forest canopy, dam construction, volcanic eruptions, chemical defoliants and urban expansion (Roper 1999). Both the underlying and direct causes of deforestation operate in an interlinked manner.

Although the direct and indirect environmental impacts of commercial logging may be significant, it is generally not thought to be the predominant cause of tropical deforestation (Burgess 1993). However, it should be noted that in many African countries, logging does not take place in the prescribed way, neither is there replanting by loggers with the result that most logging results in deforestation. In a paper done for the world Bank on the consequences of log ban in Ghana, it was found that lack of in-depth processing of timber logs led to more timber being cut and increase in deforestation (Tutu et al. 1992).

In Africa, deforestation constitutes a crisis that is extensive. Between 2000 and 2005, Africa posted the second highest rate (4 million hectares per year) of loss of its forests (FAO 2006). In terms of loss of its rainforest, Africa lost the highest percentage of rainforests during the 1980s, 1990s, and early 2000s of any biogeographical realm (Butler 2006b). Africa's forests are the most depleted of all tropical regions with only about 30% historical stands still remaining. http://www.unep.org/geo/geo1/ch/ch2_3.htm. FAO data shows that tropical deforestation rates increased by 8.5% from the 1990s to 2000–2005 while loss of primary forests may have increased by 25% over the same period (Butler 2006a).

West African rainforests are highly fragmented. The only large forest blocks remaining are in the border zone between Liberia and Ivory Coast. In Ghana, small remnants rainforest patches are restricted to protected areas (Butler 2006b).

There are varying degrees of rates of deforestation among countries. For example the deforestation rates range from 0.4% in the Democratic republic Congo (DRC), 0.5% in Liberia, 0.6% in Sierra Leone, 1% in Ivory Coast, to 1.3% in Ghana http://www.worldwildlife.org/bsp/publication/africa/127/congo_06.html.

In many African countries, between 60–85% of the total land area, which was originally covered with humid tropical forests, has been deforested. In Ghana, for instance, the total forest cover that stood at 8.2 million hectares at the turn of last century (1900s) decreased to 1.6 million hectares by 2000. According to Butler (2005), in the period between 2000 and 2005, Nigeria lost 55.7% of its primary forests, topping the world in deforestation of this type of forests).

With these alarming rates of deforestation, many countries in Africa therefore face a danger of incurring high social, economic and environmental costs from deforestation. For instance, it is estimated that about 20% of annual green house gases (GHG) emissions result from deforestation and forest degradation (Clausen et al. 2003). In a recent global warming conference held at UNEP in Nairobi, losses from extreme weather were estimated to reach $1 trillion in a single year by 2040 (Wallis, 2006). This is attributable to climate change and global warming (Reuters 2006). In Zambia for example, ECA (2002) quotes an estimated cost US$15 million as requirement over a 5-year project cycle for mitigating climate change brought about by deforestation in the country. Wong et al. 2005 report that deforestation has increased the frequency of floods and droughts, and unsustainable clearing of land for agriculture has contributed to drops in soil fertility in Mozambique.

4 Strategies and Measures Taken Combat Deforestation

Over the years many initiatives and strategies with a bearing on sustainable forest development in general and combating deforestation in particular have been undertaken in Africa and at the global level. These have had varying degrees of success in combating deforestation and fostering sustainable forest management as a whole. Never the less they have provided useful lessons learnt that serve to inform the future direction in combating deforestation. Some of the main initiatives include the following:

(a) The FAO, World Bank and other donors initiated the Tropical Forestry Action Plan (TFAP), to combat deforestation through the development and implementation of forest action plans in individual tropical countries. Many African countries were supported through this large initiative that begun in 1985. According to Winterbottom (1990), 'the needs to control deforestation and to reclaim lost forestlands' were greater than ever before, 5 years into the implementation of TFAP. Rice et al. (2001) states that, 'despite the high volume of spending, the TFAP eventually came to be seen as a failure because it was not curbing deforestation; increasingly, the plan was viewed in many countries simply as a way of promoting more forest development.' Other flaws identified by the assessment of the TFAP by the World Resources Institute (WRI) in 1990 were that: the needs and rights of forest dwelling people were not stressed in the original plan and it was wrongly assumed that increased funding with the forest sector would solve problems whose roots reach deep into economic and social policies made and observed out side the forest sector (Winterbottom 1990). The WRI assessment concluded then, that increased attention to policy reform both within and outside the sector, and improved land-use planning and coordination with agricultural and other development programs could help turn the tide against uncontrolled deforestation and wasteful depletion of tropical forest resources (Winterbottom 1990). To a great extent these concerns and challenges remain relevant today in terms of enhancing the success of other forestry programs in addressing deforestation.

(b) Community forestry has been a driving force in the management of both natural and man-made forests since the 1970s (Clausen et al. 2003). Tree planting on farmlands, agroforestry, community-based natural forest management (CBNFM) within existing natural forests and collaborative forest management (CFM) have been promoted. These strategies are aimed at not only restoring and increasing forest cover but also to enhance community benefits from the management of trees and forests.

Through these strategies, communities have been given access to forest reserves land to plant trees. Controlled access to natural forest resources has also been granted. Sharing of revenue and other forest co-management schemes have also evolved. In order to enhance success and community benefit from forest management and development programs, some of the key lessons that have emerged from these initiatives are that: clear community land tenure and user rights for resources are extremely important; attention should be paid to increasing access to markets; community organizational and enterprise capacity development is necessary; and ethnic and community diversity issues especially in relation to forest groups need to be addressed (Clausen et al. 2003; Arnold 2001; Sayer et al. 1997).

(c) National forest programmes (NFP) have been promoted internationally as fundamental tools that provide a policy and planning framework for translating the principles of sustainable forest management into domestic action. About two thirds of African countries have NFPs at different levels of implementation (FAO 2006). The NFPs have resulted in the adoption of more robust forest policies and legislation and establishment of more accountable forest administrations. The policies and legislations embrace sustainable development principles including multi-stakeholder participation in the management and development of forest. The case in point is the new Uganda forestry law resulting from the NFP. This law requires the designation of national tree planting days. There is however a slow pace in implementing the priority interventions identified in the NFP processes. This is partly due to limited funding resulting from the low priority accorded to the sector and the scanty integration of NFP priorities into the poverty reduction strategies at national level.

(d) Market based instruments such as payment for forest ecosystem services and forest certification have recently emerged as important tools in promoting sustainable forest management. In this respect, the importance of forests as a vehicle to offset climate change has become widely recognized. However a lot remains to be done to foster a broader understanding of the role and to operationalize schemes on payment for forest services and forest certification in relation to combating deforestation in Africa. With respect to adopting forest certification in developing tropical countries, inadequate land ownership and/or tenure rights, poor governance, high cost of forest certification and limited access to markets for certified forest products are listed among the main constraints (Becker).

While it has not been possible to quantify the contribution of each of these various strategies to tackling deforestation, FAO (2006) indicates that during the period 2000–2005, there was a slight decline in the net loss of forests in Africa owing to forest planting, landscape restoration and natural expansion of forests in the region.

5 Moving Forward in Reversing Deforestation

Reversing deforestation entails the halting of further net loss of forest cover and ensuring forest restoration and the development of forest plantations. In these endeavours, causes of deforestation must be addressed taking into account lessons learnt to date. In order to foster sustainable development social, economic and environment aspects need to be kept in sharp focus in forestry programs. The required strategies to reverse deforestation should therefore encompass the development of a clear understanding and appreciation, and the quantification of the multiple benefits of forests and trees to the poor and to a range of development sectors. Information quantifying the benefits of forest goods and services in Africa is scanty and limited to selected goods and services, and more over in a few countries. The required strategies should also secure and guarantee equitable and increased flow of benefits from forest management and development to key stakeholders. Further more these strategies should seek to engage the local community, the private sector and government sectoral agencies to participate and invest in forest management and development.

Against this background, the following mutually interactive strategies and actions are recommended to improve the success of efforts to reverse deforestation in Africa.

1. A conscious and scaled-up effort to transform agriculture and rural livelihood will be one of the most important strategies to address poverty and deforestation in Africa. Agriculture should be transformed by using relevant technologies at the production level and ensure value addition in all the chains of the industry. Farmers should be resourced to undertake intensive cultivation by using appropriate and high productivity technologies. This will ensure the reduction in extensive cultivation, increased agricultural productivity and output, increased incomes, poverty eradication and above all reduced deforestation.
2. Sustainable livelihood in the rural and urban areas must mean a change in the energy use. The use of fuelwood and charcoal must be reduced to the lowest minimum through energy substitution. In many African countries, there are several rural areas that have been electrified. Unfortunately, to a large extent, it is being used for lighting not as energy for household consumption and production purposes. Some rough calculations have shown that it costs slightly less per household to use gas rather than charcoal that the urban households prefer to use. They are unable to make the substitution because of initial high fixed cost of buying gas stoves. Through community partnerships, poor people must be assisted to form associations for rural credit mobilization (such as the Susu system in Ghana) to enable them purchase gas and electric stoves. Innovative hire-purchase (credit) system should be introduced to enable urban and rural households purchase these stoves.

In cases where the cost of electricity or gas is the constraint, subsidies that are equal to the cost of deforestation should be put on them to enable as many people do the energy switch. In many cases however, it is impossible to use gas or electricity for household energy purposes. In this case, improved cooking stoves, fast growing trees for fuelwood must be encouraged.

3. The gap between preaching and practice should indeed be reduced. More assistance provided to the forest sector to combat deforestation should be targeted at integrated interventions at local governments, the private sector and the local community levels. This is urgently required to translate the new policies into action and to test the efficacy of these policies in combating deforestation. And more over this has the potential to directly tackle deforestation on site and meet livelihood needs of the forest dependent communities.
4. Local communities and the private sector should be attracted to participate in sound forest management and development activities through policies that encourage greater community and private sector tenure and access to forest land and resources for use and tree planting activities. In implementing these policies the practice of restricting participatory forestry to degraded or poorer areas of

the forests should be avoided. Leasing of forests and offering of concessions and contracts to the local community should be promoted. In this regard the development of local community organizational and forest enterprise management capacity is essential. The success of these policies can be enhanced if they are implemented in tandem with those that fight corruption and promote openness and accountability in the forest sector.
5. More holistic solutions involving other sectors in combating deforestation are needed. Linkages between forestry and other sectors and in particular with poverty reduction need to be further determined and elaborated through targeted research. The political leadership in particular and other key stakeholders need to be to be familiarized with these linkages using high impact communication strategies and tools. This is key to the development of political leadership, which will enhance high level and cross-sectoral dialogue and appreciation of measures to combat deforestation. It will also facilitate the integration forest programs in other sectoral programs and therefore the mobilization of multi-sectoral and multi-tier support and action to combat deforestation.
6. In particular the understanding of and schemes for payment for forest ecosystem services need to be developed. Economic valuation of forest ecosystem services should therefore be promoted and undertaken. In combating deforestation through forest ecosystem services based schemes such as carbon sequestration that have global level benefits, local community needs that ought to be met from other services and goods provided by the forest, should remain in focus and planned for.
7. Unlike many social and economic development interventions, initiatives to combat deforestation like other environment related programs need long-term investments. Forestry programs in general and especially those targeted at combating deforestation should therefore be designed with long-term and adequate resource commitments.
8. The emergence of Regional and subregional initiatives such as the Environmental Initiative of the New Partnership for Africa's Development (NEPAD), the Central African Forest Commission (COMIFAC), and the Congo Basin Forest Partnership (CBFP) present opportunities to galvanize action to combat deforestation. They have a great potential in addressing cross-boundary issues related to deforestation. These initiatives however need to be linked and coordinated with national development process to accelerate their implementation and achieve greater impact on deforestation.

6 Conclusion

Deforestation threatens the very basis of economic growth and poverty reduction in the Africa and undermines prospects of sustainable development. The chronic and widespread deforestation in Africa is a result of diverse and intertwined economic, social, environmental and political pressures. The task of reversing deforestation is complex one. Despite the numerous forestry initiatives that have been undertaken to among others things combat deforestation, little progress has been made to date.

In order to make a significant dent on deforestation, sustained long term multi-sectoral and multi-layered strategies and approaches are need. Policy frameworks that encourage community participation through securing local community benefits from forest management and development should be promoted to underpin these measures. In this regard, empowerment of the community and the private sector, through organisational capacity building should not be relegated in forestry programs. Many interventions should be targeted at field level. It is crucial to integrate and address the needs of the local community in forestry programs of global dimension such as mitigating climate change.

Investment in measures to tackle deforestation and ensure sustainable forest management can and should be enhanced by improving access to micro credits, and increase access to markets and opportunities for value addition to forest products. Opportunities for increased resources for tackling deforestation are presented by the emerging schemes for payments for forest ecosystem services, forest certification and regional and sub regional initiatives.

References

Allen, J.C. and Barnes, D.F. (1985) The causes of deforestation in developing countries. Annals of the Association of American Geographers 75:163–184.
Arnold, J.E.M. (2001) Forestry, poverty and aid. CIFOR Occasional Paper No. 33.
Barnes, R.F.W. (1990) Deforestation trends in Tropical Africa. African Journal of Ecology 28:161–173.
Becker, M. (2004) Barriers to forest certification in developing tropical countries. Paper for Master's in Forest Conservation (M.F.C.), University of Toronto, Faculty of Forestry. Accessed from http://www.forestry.utoronto.ca/pdfs/becker.pdf
Burgess, J.C. (1993) Timber production, timber trade and tropical deforestation. Ambio, Stockholm [AMBIO] 22(2–3):136–143.
Butler, R.A. (2005) Nigeria has worst deforestation rate, FAO revises figures. Accessed on December 20, 2006 from *Mongabay.com*. Website: http://news.mongabay.com/2005/1117-forests.html
Butler, R.A. (2006a) A world imperiled: Forces behind forest loss. Accessed on June 22, 2006 from *Mongabay.com*. Website: http://rainforests.mongabay.com/0801.htm#tables
Butler, R.A. (2006b) Afrotropical realm: Environment profile. Accessed on December 20, 2006 from *Mongabay.com*. Website: http://rainforests.mongabay.com/20afrotropical.htm
CBFP (the Congo Basin Forest Partnership) (2005) The forests of the Congo Basin: A preliminary assessment. Website: http://www.cbfp.org/docs_gb/forest_state.pdf
CIFOR (2005) Contributing to African development through forests. Strategy for engagement in sub-Saharan Africa. Website: http://www.cifor.cgiar.org/publications/pdf_files/Books/PCIFOR0501.pdf
Clausen, R., Gibson, D. et al. (2003) USAID's Enduring Legacy in Natural Forests: Livelihoods, Landscapes and Governance Volume Two: Study Report. USAID.
ECA (United Nations Economic Commission for Africa) (2002) Economic impact of environmental degradation in Southern Africa. ECA/SRDC/SA/PUB/03.

FAO (2006) Global Forests Resources Assessment: Progress Towards Sustainable Forest Management. Rome.

Kidd, M. (2005) Forest issues in Africa. Website: http://www.joensuu.fi/unep/envlaw/materi2005/kidd.doc

NEPAD (2003) Action plan for the environment initiative. New Partnership for Africa's Development, Midrand. Website: http://nepad.org/2005/files/reports/action_plan/action_plan_english2.pdf

Reuters (2006) Disasters losses may top one trillion dollars per year by 2040. In Environmental News network (ENN). Website: http://www.enn.com/today.html?id=11659

Rice, R.E., Sugal, C.A., Ratay, S.M., and Fonseca, G.A. (2001) Sustainable forest management: A review of conventional wisdom. Advances in Applied Biodiversity Science, Washington, DC: CABS/Conservation International, Vol. 3, pp.1–29.

Roper, J. (1999). Deforestation: Tropical Forests in Decline, Canadian International Development Agency, Hull, Quebec, Canada.

Sayer, J.A., Vanclay, J.K., and Byron, N. (1997) Technologies for sustainable forest management: Challenges for the 21st century. CIFOR Occasional Paper No. 12.

Wallis, D. (2006) Reporting for Reuters on UNEP Conference on Climate Change in Nairobi, November 15, 2006.

Winterbottom, R. (1990) Taking Stock: The Tropical Forestry Action Plan After Five Years. WRI.

Wong, C., Roy, M, Duraiappah, A.K. (2005) Linkages between poverty and ecosystem services in Mozambique. *Connecting Poverty and Ecosystem Services: Focus on Mozambique*. IISD. Website: http://72.14.203.104/search?q=cache:wtnFmBdVVMsJ:www.eldis.org/static/DOC20108.htm±IISD±deforestation±poverty&hl=en&gl=us&ct=clnk&cd=10

Tutu, K. Asante, E., and Barfour, O. (1992) Consequences of Log Ban in Ghana. Technical paper for the World Bank, Accra.

Websites:

http://www.mongabay.com/rates_africa.htm. African Deforestation & Forest Data

http://www.unep.org/geo/geo1/ch/ch2_3.htm. Major Environmental Concerns

http://www.worldwildlife.org/bsp/publication/africa/127/congo_06.html. Deforestation in Central Africa: Significance and Scale of the Deforestation.

Life Cycle Assessment of Chocolate Produced in Ghana

A. Ntiamoah and G. Afrane

Abstract Life Cycle Assessment (LCA) has lately emerged as a comprehensive tool for environmental management, and is also becoming increasingly important in the development of cleaner production schemes. Conducting an LCA involves collecting data on raw material and energy consumption and on waste emissions to air, water and land. Data is collected for every stage of the life cycle of the product, from mining or cultivation of the raw materials through to processing, transporting, consumption and disposal. Based on a relevant functional unit for the system under study, the collected data is aggregated and modeled into a life cycle inventory, which in turn is classified and characterized to determine the environmental impacts of the entire system. The International Organization for Standardization (ISO) has standardized the process for conducting LCA in their ISO 14040 series. Many companies have turned to a life-cycle approach in an attempt to properly assess the full environmental impact of their products.

This paper presents an environmental life cycle analysis of chocolate produced in Ghana. The study was conducted in accordance with the international ISO procedural framework for performing and presenting LCA results. The product's life cycle stages, involving the cocoa supply chain, i.e. cocoa beans production, beans transportation and storage and the industrial processing of beans and chocolate manufacturing stages, were studied. The total environmental impacts associated with chocolate production and the relative contribution of each life cycle stage to the impacts are presented and discussed. The functional unit on which the analysis was based is the production of 1 kg chocolate.

A. Ntiamoah
Chemical Engineering Department, Kwame Nkrumah University of Science
& Technology, Kumasi, Ghana
e-mail: ntigh@yahoo.com

G. Afrane (✉)
Koforidua Polytechnic, Koforidua, Ghana
e-mail: gafrane@yahoo.com

Keywords LCA · chocolate production · cocoa supply chain · environmental impacts

1 Introduction

Cocoa, *Theobroma cacao*, ranks third in global agricultural export commodities following coffee and sugar. About 70% of the world's cocoa is grown in West Africa, with Ghana being the second largest producer, coming after her western neighbour Côte d'Ivoire. Chocolate manufacturing consumes about 90% of the world's cocoa production. Worldwide demand for chocolate is increasing due to its perceived nutritional and health benefits (*www.cabi-commodities.org*; ICCO 2004). Currently Ghana processes about 35% of its cocoa beans into butter, liquor, cake, powder and chocolate for both the local and international markets. Chocolate made from Ghana's premium quality cocoa beans, is acclaimed to be one of the finest in the world (Awua 2002; Appiah 2004).

With the growing awareness and concerns about natural resource depletion and environmental degradation, the impact of production processes on the environment and lifestyles has become a key modern issue. The impact of food production has attracted increasing interest because of the scale and relevance of food production. Not only the content and quality of foods are important to consumers these days, but also the environmental impacts of farming, processing and transporting to the market are also becoming important issues (Ellingsen and Aanondsen 2006). There is thus a growing need for the development of methods to better understand and address the environmental impacts associated with products and processes. One of the techniques gaining popular acceptance as an environmental management tool is life cycle assessment (LCA). LCA addresses the environmental aspects and potential environmental impacts throughout a product's life cycle from raw material acquisition to production, use, end-of-life treatment, recycling and final disposal (i.e. cradle-to-grave) (ISO 14040). As an internationally accepted tool, guided by international standards in ISO 14040 (1997), ISO 14041 (1998), ISO 14042 (2000) and ISO 14043 (2000). LCA is being used by many organizations to help in effective environmental decision-making. This study aimed to introduce LCA to the cocoa industry in Ghana and specifically conduct LCA on chocolate, which is the main product derived from cocoa.

1.1 Relevance of the Study

The study will provide LCA data to the Ghana Cocoa Board (COCOBOD) and other stakeholders of the Ghanaian cocoa industry. This information can be used to prioritize measures that can be taken to improve the environmental performance of the industry. It will also improve on the competitiveness of locally produced cocoa products on the increasingly environmentally sensitive global market.

2 Goal and Scope of the Study

The goal of the study was to identify and quantify the potential environmental impacts associated with chocolate production, focusing attention on the cocoa supply chain. The following objectives were identified:

(i) To contribute to LCA awareness creation and application in developing countries
(ii) To identify and quantify key environmental impacts along the chocolate production chain, and to assess the relative contribution of each stage of production to the identified environmental impact categories
(iii) To suggest a number of improvement options or impact reduction measures based on the results obtained

The main ingredients in chocolate are cocoa butter and cocoa liquor, which are obtained from the cocoa beans. In addition, sugar, milk, vanillin and emulsifiers may be added. However, only the environmental impacts resulting from the acquisition of cocoa butter and cocoa liquor (which constitutes about 47.85% of the chocolate studied) was included in this work. Hence the system was sub-divided into four main life cycle stages as follows: cocoa production on the farms, transportation of cocoa beans to processing factory, industrial processing of cocoa into cocoa liquor and butter, and mixing of cocoa liquor, cocoa butter and other ingredients to produce the chocolate. The following aspects were excluded from the study due to lack of relevant data: distribution and consumption phase of the product, including disposal of packaging and expired food. The functional unit chosen is the production of 1 kg chocolate bar. All the inputs and outputs in the Life Cycle Inventory (LCI) and impact scores produced in the Impact Assessment phase were expressed with reference to the functional unit.

3 Life Cycle Inventory (LCI) Analysis

This phase involves identifying and quantifying the inputs and outputs associated with each of the life cycle stages considered. The inventory analysis consists of two major steps; data collection and data processing.

3.1 Data Collection

Data for cocoa production were collected by paying site visits to farms and through interviews with cocoa farmers and researchers at the Cocoa Research Institute of Ghana (CRIG). Background data on production of fertilizers and pesticides, transportation and electricity generation were included using the eco-invent database and the GaBi 4 LCA database. Emissions due to fertilizer and pesticide usage were quantified using estimation methods (Hauschild 2000; Heathwaite 2000). A state owned cocoa processing company, which is considered state of the art, was selected for detailed study and data gathering on cocoa processing and chocolate manufacturing.

3.2 Data Processing

The data collected were adjusted to values that relate to the functional unit. The data were then modeled into environmental inputs and outputs; and aggregated to result in an inventory table, which is a collection of all normalized (scaled) values for all inputs and outputs for all the different life cycle stages considered in the study.

4 Life Cycle Impact Assessment

Classification and Characterization, which are the mandatory steps in the life cycle impact assessment phase, according to ISO 14042 (2000), were applied to the inventory data to assess their impacts on human health and the environment. The following impact categories of concern were assessed: global warming, acidification, eutrophication, photochemical ozone creation, freshwater aquatic ecotoxicity, terrestrial ecotoxicity, human toxicity and ozone layer depletion. The life cycle impact assessment method developed by the Centre for Environmental Science, University of Leiden, called the CML 2001 method and the GaBi 4 LCA software were used for this phase.

5 Results and Discussion

The characterization results (the overall impact scores) for the production of 1 kg chocolate in Ghana based on the CML 2001 method, is presented in Table 1 and illustrated graphically in Fig. 1. As shown in Fig. 1, the most significant impacts associated with the chocolate production chain in Ghana are freshwater aquatic eco-toxicity, human toxicity and global warming potentials. The relative contributions of the various life cycle stages to the overall impact scores are also presented in Fig. 2.

The cocoa production stage was found to make almost 100% contributions to the impact categories of freshwater aquatic eco-toxicity and human toxicity. In 2001, the government of Ghana initiated moves to increase cocoa production. These moves include mass spraying of cocoa farms with pesticides and distribution of specially formulated fertilizers across the country. These toxicity impacts were found to be mainly caused by the pesticides and the heavy metals present in the

Life Cycle Assessment of Chocolate Produced in Ghana

Table 1 The overall environmental impact score for the production of 1 kg chocolate in Ghana, in absolute values, based on the CML 2001 method

Environmental impact category	Overall impact score	Unit
Acidification potential (AP)	9.7351E-03	kg SO_2-equiv
Eutrophication potential (EP)	9.1568E-04	kg PO_4^{3}-equiv
Freshwater aquatic ecotoxicity potential (FAETP)	5.0797E + 00	kg [a]DCB-equiv
Global warming potential (GWP)	3.5602E-01	kg CO_2-equiv
Human toxicity potential (HTP)	4.4426E + 00	kg [a]DCB-equiv
Ozone layer depletion potential (ODP)	4.9805E-09	kg [b]R11-equiv
Photochem. ozone creation potential (POCP)	9.3002E-04	kg Ethene-equiv
Terrestric ecotoxicity potential (TETP)	6.3796E-03	kg [a]DCB-equiv

[a]DCB is 1, 4 dichlorobenzene, [b]R11 is trichlorofluoromethane.

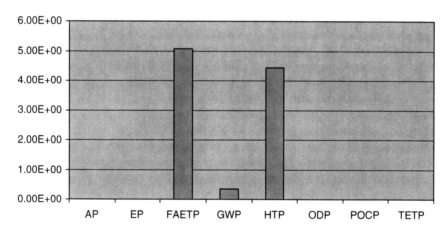

Fig. 1 Characterization results for the production of 1 kg chocolate in Ghana

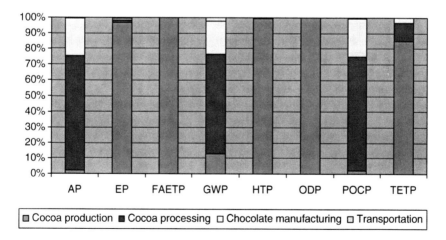

Fig. 2 Relative contributions by life cycle stages to overall impact scores

fertilizers used in cocoa production. Industrial processing of cocoa beans however made the largest contribution to global warming potential (63.70%). This impact was found to be mainly caused by the use of diesel for steam generation and roasting of cocoa beans. Chocolate manufacturing and transportation are relatively the most environmentally friendly of the life cycle stages, since their contributions to the impact categories considered were not very significant. The transportation stage is the least important in terms of environmental impacts, whilst chocolate manufacturing is intermediate between cocoa processing and transportation.

6 Conclusions and Recommendations

This has been an initial investigation to identify and quantify the key environmental impacts associated with chocolate production in Ghana. Even though the actual chocolate manufacturing stage can be said to generate less environmental impacts, some significant impacts namely, freshwater aquatic eco-toxicity, human toxicity and global warming occur during the cultivation and processing of the main raw materials. The study therefore recommends that environmental improvement actions should be focused on the cocoa production and processing stages of the product's life cycle, particularly on agrochemicals and fossil fuels inputs. In this respect, the effort being made by government through the Cocoa Research Institute of Ghana (CRIG) in the development of cost-effective biological control and an integrated pest management system, adequate soil fertility management, continued farmers education and extension services, and introduction of early-bearing, high-yielding and disease-tolerant cocoa varieties, in a programme dubbed Cocoa High Technology Project (COCOA HI TECH) which is now on a pilot basis in the Eastern Region of the country (Appiah 2004), is highly welcome. Early implementation of these strategies across the country would further enhance the environmental friendliness of cocoa production in Ghana.

Future studies could expand the system boundary to include the product distribution and consumption phase of the life cycle and also include the environmental impacts coming from the production of packaging, and other ingredients in chocolate such as sugar and milk. This will provide a comprehensive picture of the overall environmental burdens associated with the life cycle of the product.

In carrying out this study some European LCA databases namely, the Swiss Ecoinvent LCA database and the GaBi 4 LCA database, were consulted for certain background data that were not locally available. These included data on production of pesticides, fertilizers, diesel and electricity, as well as emissions resulting from transportation. According to Mungkung et al. (2006), this can limit the representativeness and hence the reliability of LCA results.

The results of LCA are being used for product improvement, product comparison, eco-labelling and public policy formulation among others. Clearly developing countries cannot be left behind in the adoption of LCA as an analytical tool if we want to produce environmentally and socially acceptable goods for the current international market. The major challenge now is to develop ready-to-use LCA databases (providing inventory data on the various energy generation systems, and some widely used intermediate products, etc.) that are relevant to the conditions prevailing in these countries, as has been done in several developed countries. These data can then be adapted by local LCA practitioners to help reduce the time and cost involved in collecting data along a product's entire life cycle. The UNEP/SETACT life cycle initiative is helping in this direction. The aim of the initiative is to facilitate a global application of the tool, both in rich and poor countries, and in big companies and SMEs.

Acknowledgments The UNEP/SETAC Life Cycle Initiative awarded the GaBi 4 LCA software used for this project. The authors would like to thank the following people for their contributions: Frank Asante (Cocoa Processing Company, Tema), Gyima Gyamfi (Cocoa Swollen Shoot Virus Disease Control Division of COCOBOD, Kumasi), Kofi Acheampong and James Buabeng (CRIG) and Dr. Lawrence Darkwah, chemical engineering department, Kwame Nkrumah University of Science & Technology, Ghana.

References

Appiah, M.R. (2004): Impact of cocoa research innovations on poverty alleviation in Ghana. Ghana Academy of Arts and Sciences Publication, Accra.

Awua, P.K. (2002): The success story of cocoa processing and chocolate manufacturing in Ghana – The success story that demolished a myth. David Jamieson, Saffron Walden, Essex.

Ellingsen, H. and Aanondsen, S.A. (2006): Environmental impacts of wild caught cod and farmed Salmon – A comparison with chicken. Int J LCA 11(2): 122–128.

Hauschild, M. (2000): Estimating pesticide emissions for LCA of agricultural products. In: Weidema, B.P. and Meeusen, M.J.G. (eds) Agricultural Data for Life Cycle Assessment, Vol. II. The Hague: Agricultural Economics Research Institute (LEI).

Heathwaite, L. (2000): Flows of phosphorous in the environment: Identifying pathways of loss from agricultural land. In: Weidema, B.P. and Meeusen, M.J.G. (eds) Agricultural Data for Life Cycle Assessment, Vol. II. The Hague: Agricultural Economics Research Institute (LEI).

ICCO (2004): International Cocoa Organisation. 2003/2004 annual report. Available from http://www.icco.org.

ISO 14040 – Environmental management – Life cycle assessment – Principles and framework (1997).

ISO 14041 – Environmental management – Life cycle assessment – Goal and scope definition and inventory analysis (1998).

ISO 14042 – Environmental management – Life cycle assessment – Life cycle impact assessment (2000).

ISO 14043 – Environmental management – Life cycle assessment – Life cycle interpretation (2000).

Mungkung, R., Udo de Haes, H. A., and Clift, R. (2006): Potentials and limitations of life cycle assessment in setting ecolabelling criteria: A case study of Thai shrimp aquaculture product. Int J LCA 11(1): 55–59.

Sustainable Production of Traditional Medicines in Africa

O.A. Osunderu

Abstract Since the dawn of history, mankind has been actively experimenting with a variety of available plants as means for food, alleviation of pain and to safeguard its health and promote improved quality of life. Over the years, traditional knowledge, science and technologies have accumulated to form a rich background of cultural heritage.

In Africa, over 70% of the population depends on traditional medicine (Federal Ministry of Science and Technology, 1985) because the rich resources of traditional remedies and practitioners are available and accessible.

However most Government/policy makers and financial investors do not pay much attention to herbal medicine as a means of creating wealth and transforming lives, especially for the poor people living in these countries. As the medicinal plants in use are neither classified as food or cash crops of health concern. Also no policy is in place to ensure the sustainability of traditional medicine and the protection of the environment.

This essay is focused on the role of medicinal plants in creating markets and transforming lives. In Nigeria so much emphasis has been placed on revenues generated from crude oil to the detriment of other revenue sources like herbal medicine.

Keywords Medicinal plants · traditional · medicine · resources · market · health and prosperity

1 Introduction

The WHO in its Alma-Ata declaration of 1978, gave due recognition to the role of traditional medicine(TM) and its practitioners (TMPs) in achieving comprehensive and affordable health care delivery, previously tagged "Health for all in the 21st century" although the situation remains "Health for some".

O.A. Osunderu (✉)
Hermon Development Foundation, P.O. Box 17331, Ikeja, Lagos, Nigeria
e-mail: kosunderu@yahoo.co.uk; *hermonest@gmail.com*

Traditional medicine refers to health practices approaches, knowledge and beliefs incorporating plant, animal and mineral based medicines, spiritual therapies, manual techniques and exercises, applied singularly are in combination to treat, diagnose and prevent illness or maintain well-being. Traditional medicine which includes herbal medicine is also known as Complementary/Alternative medicine.

Traditional medicine plays a primary role in people's health, as they have for thousands of years there is a wide range of therapies and practices, varying greatly from country to country and from region to region. The most well known are the Ayurveda of India and traditional Chinese medicine and these systems of medicines have now spread to other countries.

Traditional medicine has always maintained its' popularity worldwide. For more than a decade, there has been an increasing use of traditional medicine (inform of complementary and alternative medicine) in most developed and developing countries.

2 Objectives

The objectives of this essay include the following:

- Provide information to economic and financial policymakers, the international financial community and/or international domestic investors on the huge economic opportunities that herbal medicine provides in Africa, particularly Nigeria, the most populous black nation in the world
- Provide insight into current Herbal Medicine Development (HMD) research in Nigeria which includes innovative and data-supported research resulting from the author's own professional and academic work
- Develop and support the implementation of leading-edge HMD initiatives that will ensure sustainable development
- Design a balance between conceptual and practical considerations for private-sector involvement in developing countries, and the effect of that involvement on development and the attainment of the Millennium Development Goals (MDGs) by the Nigerian government

The federal republic of Nigeria is situated along the golf of guinea, in the eastern part of the West African subcontinent. It extends over an area of 923,768 km^2 making it the tenth largest country in the world. Nigeria has a wide diversity of habitat, ranging from arid areas, through many types of forest, to swamps. Associated with the varied zones is an array of plant and animal species. The major vegetation formation are the mangrove forest and coastal swamps, fresh water swamps, low land rain forest, derived savanna, northern guinea savanna, Sudan savanna, Sahel, mountain, sub-montane forest and grassland.

In Nigeria, traditional medicine plays a significant role in meeting the health care needs of the majority of Nigerians. It also provides a livelihood for a significant number of people who depend on it as their main source of income, 75% to 80% of the Nigerian population uses the services of traditional healers (Federal

Government of Nigeria 2002). It can be extrapolated that 75% to 80% of Nigerians use herbal medicine because medicinal plants are the primary sources used by traditional healers in Nigeria (WHO 2006). The percentage could even be higher because almost all Nigerians eat herbal vegetables.

Several medicinal plants of global importance originate in the country. For example Calabar bean (Physostigma venenosum) was traditionally used in Nigeria as an "ordeal poison" in trials of wrong doers, from it the major component physiostigmine (eserine) and its derivatives, very important drugs have been discovered, and are now used against intra ocular pressure (glaucoma) (Landis 1996).

Nigeria has been ranked 11th in Africa for plant diversity. Out of the 5,000-plant species that exists in the country, 205 are considered endemic making the country the ninth highest among the 12 African countries in the level of endemic species (Ref FGN, 2002). With an estimated population of over 120 million people, distributed among over 250 distinct ethic groups or tribes, the country is unique in having high cultural diversity and a significant share of the global biological diversity.

3 Distribution of TMPs in Nigeria

In a study carried out by the author and her team in South West Nigeria 2005, the following results were obtained:

3.1 Methodology

Interviews and short questionnaires were used to elicit information from the practitioners. 126 practitioners were interviewed in three states Lagos, Ondo and Oyo states.

3.1.1 Results

One hundred and twenty-six questionnaires were distributed to TMPs in the South Western geographical zones of Nigeria comprising Oyo, Ondo and Lagos States. Sixty-eight percent of the respondents were men while 32% were women (Fig. 1).

4 Traditional Medicine Practices in Nigeria Include

4.1 Traditional Birth Attendants (TBAs)

Traditional delivery, obstetrics and gynecology are services rendered by traditional birth attendants in Nigeria. They are widely practiced in the rural and urban areas of

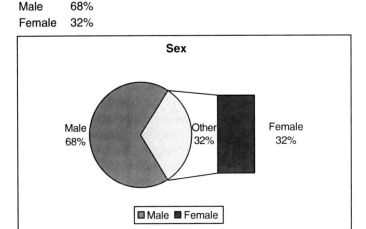

Fig. 1 Gender distribution of TMPs in Nigeria

the country and are highly patronized by our people due to increasing costs of health care in our conventional health institutions. Their positive contributions to health care delivery system can not be disputed. The services of traditional birth attendants are relatively cheap, readily accessible, personalized and does not carry out unnecessary surgery as a line of treatment, as practiced by conventional medical practitioners. However, this practice has to be improved upon.

The number of TBAs far out number that of their allopathic counter parts. The National Demographic and Health survey report (NDS 1999) indicate that only 37% of birth take place in conventional Health centers or hospitals. By inference 63% of Births are handled by Traditional medicine practitioner.

WHO seminars have recently highlighted the roles of TBAS in African society and many countries like Nigeria have arranged formal training in hygiene and obstetrics for these practitioners to improve their effectiveness. Some of the herbs used by TBAs include the root of *Carica papaya L.* chewed with seven seeds of *meleguata pepper* during labour, which is suppose to cause an immediate delivery of the baby. The bark of *Blighia sapida koeng* ground and mixed with locally made black soap is used for bathing through out the period of pregnancy. This ensures an easy delivery process for the pregnant woman (Sofowa 1982).Thus prevents surgeries.

4.2 Traditional Bone Setters and Massage Therapists

Traditional bone setting and Massage therapy are services rendered by traditional medicine practitioners and they are part of our culture and accumulation of our peoples' experience in the management of fractures, pains and other related problems through the ages while making use of herbs. They need to be financially empowered in view of the immense contributions they make to the health care delivery system. There is no ailment known to man which nature has not provided remedy or cure for (WHO 2006).

Traditional bone setting is a specialized section of traditional medicine which uses;

- Fingers in the assessment of the extent of damage to a broken bone often without the use of x-ray.
- Application of herbal formulations to relieve pains.
- Application of local splint/caste to immobilize the bone in the management of fracture (WHO 2002, 2006), while traditional massage therapy is the application of gentle but firm pressure with the tip of the fingers and palms in order to relieve pain, stress, aching muscles, nerves, ligaments and tendons. Traditional bone setting and massage therapy are widely practiced in the rural areas due to their cheapness, availability and accessibility. They provide services that require no amputation as a line of treatment.

5 Global Market for Traditional Medicines

It is obvious that Traditional Medicine (TM) has maintained its' popularity in all regions of the developing world and its use is rapidly spreading in industrialized countries. Countries in Africa, Asia, and Latin America use TM to help meet most of their primary health care needs. In industrialized countries, adaptations of traditional medicines are termed "complementary" or "alternative" (CAM).

The global market for herbal medicines currently stands at over 60 billion US Dollars annually and is growing steadily (Federal Ministry of Science and Technology, 1985; Brevoort 1998). Nigeria like some other African countries with their vast bioresources and biodiversity are not making any contribution to the global market because the relevant stake holders are not showing enough interest to harness these vast resources to improve the lives of their citizens while creating markets. More than half of the populations live in poverty while these God-given resources waste away (Brinker 1998).

6 Sources of Medicinal Plants in Africa Include

- Home gardens and agricultural lands
- Botanic gardens
- Few proven cultivation practices for medicinal plants e.g. silviculture although this is minimal in developing countries although the potentials for this form of supply is enormous
- Traditional Medicine Practitioners also harvest from the wild e.g. natural forest remote and marginal lands

The result below was obtained from a survey conducted in South-south Nigeria and South-west Nigeria respectively in 2006 by the author on TMP's sources of medicinal plants:

South-south Nigeria		
Sources of herbs plant parts	Forest	45%
	Garden	24%
	Market	31%
Frequency of harvest	Daily	37%
	Weekly	48%
	Monthly	15%
South-west Nigeria		
Sources of herbs/plant parts	Forest	48%
	Home garden	15%
	Market	37%
Frequency of harvest of herbs	Daily	42%
	Weekly	46%
	Bi-monthly	7%
	Monthly	5%
Method of harvesting applied	Uprooting	30%
	Plucking/pruning	46%
	Debarking	24%

6.1 Current Herbal Medicine Development and Research

The government of Nigeria desires to maintain and encourage the growth and development of traditional medical practice through coordination and control.

These Government agencies like NAFDAC, NIPRID etc., NGOs and tertiary institutions work together, to facilitate national and international cooperative and collaborative research development and promotion effort in traditional medicines. They research, collate and document all traditional medicine practices and products in order to preserve the nation's positive indigenous knowledge on traditional medical science and technology development and promote safe, efficient traditional therapies and facilitate their integration into the national health care delivery system (Holmstedt 1972).

7 Accessing the Natural Medicine Market

The World Health Organization (WHO) estimates that one-third of the world's population has no regular access to essential modern medicines, in some parts of Africa, Asia, and Latin America, as much as half of the population faces these persistent shortages. However, in these same situations the rich resources of traditional remedies and practitioners are available and accessible.

7.1 Ways of Generating Income from Natural Medicine

7.1.1 Medicinal Plant Trade

An economic venture. The demand for products from the medicinal plants trade has been on the increase worldwide (Quansah 1998; Sergi and Kadriye 1995; Sofowa 2002). The trade so far relies heavily on wild collections to meet the ever-increasing demands. Nat Quansah reported that in Madagascar during 1996, one primary supplier earned 67.5 US Dollars for providing 225 kg of the raw material (raw plant) of *Sigesbeckia orientailis* to two sellers. These sellers sold the plant material to a total of 1,500 people and made 450 US Dollars. Thus, 225 kg of *Sigesbeckia orientialis* yielded 517.5 Dollars in a market. Also the market value of 399,447 kg of 11 species of medicinal plants and/or their extracts in 1995 was estimated at 1,382,000 US Dollars. Nigerian plants such as the *Ocimum spp., Vermonia amygdalina, cympopogon citratus* have great marketing potentials because of their multipurpose uses as spices, ornaments, medicines etc.

7.1.2 Horticulture and Perfumery

African plants are valued ingredients but have been traded only at the lower end of business, for example, many Africa pelargonium species are offered for sale in other countries. Revenue generated from the sales of pelargonium species in Belgium, Netherlands and Germany approximately 6 billion US Dollars annually (Rosonaivo 1996).

7.1.3 Pharmaceuticals

The array of medicines derived from plants is impressive – analgesics, diuretics, laxative, anti-cancer and anti-parasitic components, cardiovascular, anti-inflammatory and anti-viral drugs, just to mention a few. It has been estimated that as many as 75% to 90% of the worlds rural people rely on herbal traditional medicine as their primary health care (WHO 2006) and this is a source of incomer for the growers such plants and practitioners. African flora is potential for new compounds with pharmacological activities. Such efforts have led to the isolation of several biologically active molecules that are in various stages of development as pharmaceuticals.

7.1.4 Cultivation

The time has come for economic & financial policymakers, the international financial community and/or international domestic investors to start large scale cultivation of medicinal plants especially those with proven efficacy and safety. Most of these plants grow well even on deteriorating soil.

To ensure the production of standard plant drugs with acceptable quality, uniformity and purity, factors such as, sources and identity of the plant, its physical characteristics, its chemical constituents, the pharmacological and biological activities of the crude drug and method of preparation, uses and storage, amongst others, need to be identified and documented.

8 Drawbacks of Herbal Medicine Markets in Africa (Nigeria)

Drawbacks to our traditional medicine heritage and medicinal plants that can serve as raw materials for traditional and modern drug production:

- Rapid destruction and degradation of our environment
- Loss of useful medicinal plant species
- The passing away of the custodians of knowledge on our medicinal plants
- The lack of documentation of our medicinal, aromatic and pesticidal plants and traditional medical knowledge

9 The Steps Forward

Careful identification and documentation of an inventory of Nigeria's Medicinal Plants, their usefulness and the knowledge base for their utilization amongst other factors is expedient. An inventory/database on available plant raw materials, among others, should be available. These will stimulate the processes for the development of legislative and administrative framework for their conservation and as raw materials for exploitation.

10 Conclusion

In order to effectively protect the environment, there is need for government to recognize and formulate policies ensure the uniformity of herbal medicine practices. Factors such as, sources and identity of the plant, physical characteristics, chemical constituents, the pharmacological and biological activities of the crude drug and method of preparation, uses and storage, amongst others, need to be identified and documented.

The development of herbal medicine will create markets and ultimately improve the lives of the people who depend on it as a source of livelihood; the Traditional Practitioners, the cultivators/gatherers, the market women (hawkers), the manufacturers of packaging materials, increase the income generated by the government and attract foreign investment. It will also contribute positively to the attainment of a befitting primary health care for the citizenry.

References

Brevoort, P. 1998. The booming US botanical market: A new overview. Herbal Gram 44:33–48.
Brinker, F. 1998. Herb contraindications and drugs interactions. Sandy, OR: Eclectic Medical Publications.
Federal Government of Nigeria. 2002. National biodiversity strategy and action plan Abuja. Government Press 2–7
Federal Ministry of Science and Technology, Federal Republic of Nigeria. 1985. Report of the national investigative committee on traditional and alternative medicine, Abuja. Government Press.
Federal Ministry of Health, Nigeria Demographic and Health Surveys, 1999.
Holmstedt, B. 1972. The ordeal bean of old Calabar; the pageant of physostigma venonosum in medicine. Swain Ted. Plants in the development of modern medicine. Cambridge, MA: Harvard University Press, pp. 303–360.
Landis, P. 1996. Market report. Herbalgram 36:69. American Botanical Council and the Herb Research foundation USA.
Quansah, N. 1998. Medicinal Plants: A Global Heritage 16: 160. Proceedings of the International Conference on Medicinal Plants for Survival.
Rasoanaivo, P. 1996. Plantes Medicinales et Aromatiques a Valeur economiques a Madagascar.
Sergi, H. and Kadriye, T. 1995. The strength and weakness of Turkish bone setters. World Health Forum 16(2) 203–205.
Sofowa, A. 1982. Medical plants and traditional medicine in Africa. Wiley.
World Health Organization. 2002. Traditional medicines – Growing Needs and Potential. WHO policy respective on Medicine, Geneva (WHO/EDM/2002. 4).
World Health Organization. 2006. WHO report on traditional medicine, my documents/WHO traditional medicine.htm 15/06/2006.

Microbial Risk Assessment: Application and Stages to Evaluate Water Quality

M.T.P. Razzolini, W.M.R. Günther, and A.C. Nardocci

Abstract This works presents the stages of microbial risk assessment and its application in the evaluation of drinking and recreational water quality and further the risk of diseases attributable to pathogens present in these waters. In Brazil, infectious diseases were responsible for 5.1% of deaths, the fifth place of mortality cause. Most affected people are children, elderly and immunocompromised population. According to Brazilian Health Ministry, from 1995 to 1999, environmentally-caused diseases occupied 3.4 million beds in hospitals. All of these reports showed a very public health concerning scenario.

This tool can provide bases to establish tolerable (acceptable) risk and thus defining the protection level to human health for each hazard, in this case, pathogens microorganisms. The stages are hazard identification, exposure assessment, dose-response relationship and risk characterization. The hazard identification is related to the presence of microorganism and toxins and their association with specific diseases. The exposure assessment includes the intensity, frequency and duration of human exposure to a specific agent. The aim of dose-response relationship is to acquire a mathematical relation between amount of microorganism (concentration) and adverse effect on human health. Risk characterization represents the integration of the previous stages. Risk assessment is a tool used for decisions making and providing information to take measures of control and interventions, as well as to evaluate the impacts of these actions. It provides support to a decision-making process based on scientific results in several areas of the knowledge. In Brazil, risk assessment is a recent area of research, but promising for the management of water's quality such as catchments points and recreational waters, in special in periurbans areas of metropolitan regions, which show precarious sanitary conditions.

M.T.P. Razzolini (✉), W.M.R. Günther, and A.C. Nardocci
Environmental Health Department, School of Public Health of University of São Paulo,
Av. Dr. Arnaldo 715 1o andar 01246-904 São Paulo – SP – Brazil
e-mail: razzolini@usp.br

Keywords Microbial risk assessment · water quality · risk management

1 Introduction

In recent years it has been observed that an increasing number of waterborne outbreaks have occurred due to water resource contamination by sewage and solid waste that carry pathogenic microorganisms. These microorganisms do represent a public health concerning which of them can survive long enough to cause human health disturbs. Out of them, emerging and reemerging pathogens deserve special attention. According to World Health Organization, in developing countries it was estimated that 94% of the diarrhea burden of disease is attributable to environmental and associated with risk factors such as unsafe drinking-water and poor sanitation and hygiene, on average, children in developing countries lose eight-times more healthy life, per capita, than their peers in developed countries from environmentally-caused diseases. Infectious diseases which contributing to the environmental burden of disease among children from 0 to 14 years old are intestinal nematode infections (1.5%) and diarrhea disease (29%) (WHO 2006). Bartram (2006) reported that 1.8 million people, mostly children, die of diarrhea every year. In Brazil, infectious diseases were responsible for 5.1% of deaths, the fifth place of mortality cause. Most affected people are children, elderly and immunocompromised population. According to Brazilian Health Ministry (Brasil 2004), from 1995 to 1999, environmentally-caused diseases occupied 3.4 million beds in hospitals. All of these reports showed a very public health concerning scenario.

According to Hass et al. (1999), the preventing of infectious diseases transmission from human exposure to contaminated food, water, soil and air remains a major role of environmental and health professionals. In this way, microbial risk assessment is a very useful tool to evaluate drinking and recreational water quality, applying it and further to determine the risk of diseases attributable to pathogens present in waters sources. Besides, this tool can provide bases to establish tolerable risk and thus defining the protection level to human health for each hazard, in this case, pathogens microorganisms. Establishing tolerable risk is an important issue to be considered because of the risk perception occurs in different ways dependable on affected people, magnitude of the adverse effects, how people are habituated to face the adverse effects and amount of people affected (Peña et al. 2001), moreover this perception can be diverse by the general population, politics, researches and managers. Undoubtedly, microbial risk assessment allows an improvement in the capacity to evaluate and control the risks as well as in the making-decision process to minimize risks as choosing the best sanitary barriers option. Acquiring all this knowledge would be an advance to protect human health, mainly in developing countries where there are many areas with evident environmental vulnerability.

The goal of this work is present the stages of microbial risk assessment and its application in the evaluation of drinking and recreational waters quality.

2 Microbial Risk Assessment Application

Risk assessment process consists of estimating the probability that an event can occur and the probable magnitude of its adverse effects (Gerba 2000). In relation to microbial risk assessment, it can be said that it is a process which evaluate the probability of an adverse effect on human health after exposure to pathogenic microorganisms or contact with a source (water, soil, air, food) with pathogens presence as well as their toxins. The advantages of using this kind of analysis is estimate the results of exposure to infectious microorganisms as well as express it quantitatively in terms of probability of infection, morbidity (disease) and/or death. With these results is possible to do prevision about the expected number of infectious diseases, diseases and fatalities due of a determined exposure. In addition, microbial risk assessment can provide bases to establish tolerable risk and thus defining the protection level to human health for each relevant pathogen.

United States Environment Protection Agency (US EPA), due to occurrence of waterborne infectious agents in surface waters catchments points such as *Giardia* cysts and enteric viruses developed the Surface Treatment Rule (STR). The STR, based on microbial risk assessment, required that all drinking water plants be capable of removing 99.0% of *Giardia* cysts and 99.9% of enteric viruses, to get to the infection tolerable risk because of these etiologic agents was not superior to 1 per 10,000 exposed persons annually (10^{-4} per year) (Gerba 2000). The same approach can be transferred to other sort of waters as recreational or else wastewater with potential reuse.

World Health Organization (WHO) in the latest publication of Guidelines for drinking water quality and Guidelines for recreational waters environments – Coastal and Freshwaters, consider microbial risk assessment as a way to estimate risks to human health related to microbiological quality of drinking and recreational waters as well as permit to translate risk of developing a specific illness to disease burden per case expressed in DALY (Disability-adjusted life-year) (WHO 2004). It is interesting for the reason that it allows to compare severity among diseases and microbial agents.

3 Microbial Risk Approach

Microbial risk assessment is characterized by a framework which consists of problem formulation and hazard identification, exposure assessment, dose-response assessment and risk characterization. Each component aims to characterize the entire scenario and its impact of human health as shown in Fig. 1.

Accordingly Soller problem formulation involves all stakeholders and its point is to identify the purpose of the risk assessment, critical issues and the treatment of the obtained results to protect public health (Soller 2006).

Identifying hazard is a step that consist of acquiring information from clinical studies and epidemiological and surveillance studies. All aspects to lead to potential hazards reach drinking and recreational waters and furthermore human being affecting their health condition should be considered. This information

MICROBIAL RISK ASSESSMENT: APPLICATION AND STAGES TO EVALUATE WATER QUALITY

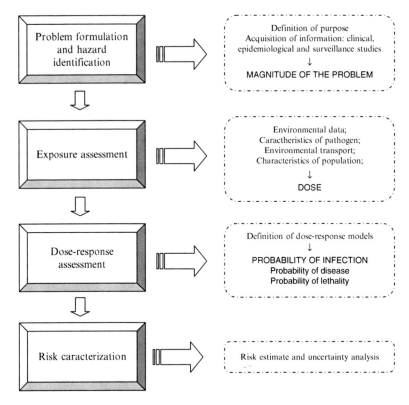

Fig. 1 Microbial risk assessment approach

gives the magnitude of the problem and how it impacts affected population health. It is not easy to do because a part of affected population does not look for medical attention for several reasons (mild symptoms or difficulty medical accesses, for instance), asymptomatic persons occurrence, unsatisfactory register data as source of infection and infectious agent, inappropriate pathogen method detection, medical attendance quality discrepancies among different areas around the country and some times microorganism infectivity and lethality are not clear.

The exposure assessment refers to the measure process or intensity, frequency and duration of human exposure to a specific infectious agent. It is used to estimate the microorganism concentration which is corresponding to a single exposure or a total of organisms in a set of exposures (Haas et al. 1999; Gerba 2000). The exposure can happen via inhalation, ingestion of water or food or dermal. In this step is important to consider the population affected, real or potential, and the contamination sources as well as transport mechanisms and biotransformation and moreover all pathogenic microorganism pathways until reach its host including ingress via. In this case, the amount of drinking water is daily ingested or volumes ingested of water during recreation in swimming pools or beaches for that reason regional ingestion water data is important to conduct the exposure assessment.

Dose-response assessment is a result from experimental studies to develop a relationship between the level of microbial exposure and the likelihood of occurrence of an adverse consequence (Haas et al. 1999). These data are currently available for waterborne pathogen from studies using healthy adults volunteers (Gale 2001) to identify the infectivity to specific pathogens, on the other hand there are not studies with more susceptible population such as children, elderly, immunocompromised. The consequences from an infection process are highlighted: (a) asymptomatic illness, (b) acute symptomatic illness – mild to moderate symptoms, days loss from work and heath care cost and (c) chronic infections as, for example, hemolytic uremic syndrome or stomach ulcers. Dose-response models should approach the probability of infection as well mortality probability. Even though, in some cases data about pathogens infectivity are not available so mathematical models are useful to estimate dose-response relationship such as Poisson distribution and Exponential distribution as shown in Table 1.

Risk characterization integrates all information acquired in previously steps – problem formulation and hazard identification, exposure assessment and dose-response relationship –to calculate the probability of infection of the exposure by drinking and or recreational waters contaminated with pathogens. Besides that, this stage consists of uncertainty evaluation and risk communication.

In addition, the quantitative risk estimate is useful to obtain a burden disease expressed by DALYs (Disability-adjusted life-year). In countries such as Brazil, with so difference in socio-economical and sanitary conditions among its regions the DALYs calculating is interesting to give the adequate significance for each burden disease in specific areas avoiding distortions. It has a special meaning to risk management phase when financial support should be addressed and public policies elaborated to improve waters sanitary quality.

Table 1 Dose-response parameters for pathogen ingestion mathematical models studies (adapted from Gerba (2000))

Microorganisms	Best mathematical model
Echovirus 12	Beta-Poisson
Rotavirus	Beta-Poisson
Poliovirus 1	Exponential/Beta-Poisson
Poliovirus 3	Beta-Poisson
Cryptosporidium	Exponential
Giardia lamblia	Exponential
Salmonella	Exponential
Escherichia coli	Beta-Poisson

4 Final Considerations

In Brazil, microbial risk assessment is a recent area of research, but promising for the management of water's quality such as catchments points and recreational waters, in special in periurbans areas of metropolitan regions, which show precarious sanitary conditions that lead a high waterborne diseases incidence. In Brazil and others developing countries microbial risk assessment is useful to allow to maker decision to define more realistic goals to these poor areas considering not only minimizing risk but to evaluate benefit-cost of intervention actions.

On the other hand, developing countries present limiting structure to implement microbial risk assessment such as lack of truthful data and specific data for each studied area. It should be highlighted in view of applying imported information or models from others countries can increase the uncertainties and impair the use of this tool.

To sum up, it should be emphasized how important is the criteria flexibility in decisions demand institutions with political powerful and technically well prepared. In developing countries these aspects do not happen very often.

References

Bartram, J. (2006) Water, health and the millennium development goal. In: IWA World Water Congress. 10–14 September 2006, Beijing, China.

Brasil (2004) Ministério das Cidades. Saneamento Ambiental 5. Cadernos MCidades Saneamento Ambiental. Brasília.

Gale, P. (2001) A review – developments in microbiological risk assessment for drinking water. J Appl Microbiol, 91, 191–205.

Gerba, C.P. (2000) Risk assessment. In: Maier, R.M., Pepper, I.L., and Gerba, C.P. (Eds) Environmental Microbiology. San Diego, CA: Academic Press, pp. 557–570.

Haas, C.N., Rose, J.B., and Gerba, C.P. (1999) Quantitative Microbial Risk Assessment. 1st ed. New York: Wiley.

Peña, C.E., Carter, D.E., and Ayala-Fierro, F. (2001) Evaluación de Riesgos y Restauración Ambiental. Southwest Hazardous Waste Program – a Superfund Basic Research and Training Program at the College of Pharmacy, The University of Arizona.

Soller, J.A. (2006) Use of microbial risk assessment to inform the national estimate of acute gastrointestinal illness attributable to microbes in drinking water. J Water Health, 4(2), 165–186.

WHO (World Health Organization) (2004) Guidelines for drinking-water quality. Volume 1 – Recommendations. 3rd ed. Geneva.

WHO (World Health Organization) (2006) Preventing disease through healthy environments – Towards an estimate of the environmental burden of disease. Executive summary.

Benefits and Dangers of Nanotechnology: Health and Terrorism

Y.A. Owusu, H. Chapman, T.N. Dargan, and C. Mundoma

Abstract The first goal in this paper is to discuss the fundamental principles, applications, advantages, and disadvantages of nanotechnology with a view of promoting the importance and validity of nanotechnology in the developed countries as well as the emerging developing countries in Africa and elsewhere. The second goal is intended to provoke critical thinking, analysis, medical applications, environmental and economic issues or implications involving nanotechnology. The third goal is to discuss the potential security threat that would pose world peace, should nanotechnology, nanodevices, nanomaterials fall into the wrong hands.

Keywords Nanotechnology · quantum dots · nanites · United Nations · cancer

1 Introduction

The discoveries in the field of nanotechnology of its existence and applications over the past four decades are staggering. If the dream of a nanoage is within two decades according to nanotechnology prognosticators, then life as its known today will be changed forever. The science of manipulating matter at its most fundamental level, atoms and molecules, would be the discovery of the millennium. The National Science Foundation (NSF) of the United States of America is massively funding research in nanotechnology. Neal Lane, Director of NSF, said: 'If I were asked for an area of science and engineering that will most likely produce the breakthroughs of tomorrow, I would point to nanoscale science and engineering, often called simply *nanotechnology*.' Nano is a one billionth unit of measurement and derives from the Greek word for dwarf. The

---

Y.A. Owusu, H. Chapman, T.N. Dargan (✉)
FAMU-FSU College of Engineering, Florida A & M University, Tallahassee, Florida (USA)
e-mails: owusu@eng.fsu.edu; hanschap@eng.fsu.edu; urccet@eng.fsu.edu

C. Mundoma
Institute of Molecular Bio-Physics, Florida State University, Florida (USA)
e-mail: cmundoma@sb.fsu.edu

focus of nanotechnology ranges from developing nano-sized chips for supercomputers to engineering molecular robots (nanorobots) that perform specific functions.

More specifically, nanotechnology deals with the creation of functional materials, devices and systems through the control of matter and exploitation of novel phenomena and properties on nanometer scale (Merkle 1992). For an example, if atoms are re-arranged in coal, that structure converts itself into a diamond. Also, if the atoms that are found in sand are re-arranged and added a few other elements, that structure becomes computer chips or solar panels. Of particular significance is the emerging field of nanotechnology-based biomedical techniques.

2 Medical Applications of Nanotechnology

For many different reasons, there has been great effort to improve and discover new manufacturing processes, products, and systems, which will aid physicians to treat and prevent asthma, cancer and other chronic diseases. As far fetched, as this may seem, this is what has been discussed in this paper: the application of nanotechnology for the treatment and prevention of chronic diseases, such as asthma and cancer. Two main applications are presented.

2.1 Application of Gold Nanoparticles (GNPs) Used as Biomarkers for Detecting Metastasis of Cancer Tumor

Due to their biocompatibility and excellent optical properties, gold nanoparticles are finding increasing application in a variety of areas including DNA detection and cancer diagnostics. The gold nanoparticles can be conjugated to an antibody for epidermal growth factor receptor (EGFR) proteins. These EGFR cells which are found on the surface of most cancer cells respond to the presence of gold due to the ability of the gold nanoparticles to absorb light. Figure 1 shows the optical effect of gold nanoparticles on cancer cells due their high affinity for cancer cells. In addition to these developments, colloidal gold nanoparticle platforms have been designed and used as a tumor targeting ligand and a cancer therapy (Paciotti et al. 2005). A new synthesis has been developed for the tumor-associated, cell-surface carbohydrate moiety, known as the Thomsen-Friedenrich antigen (T-antigen) (Svarovsky et al. 2005).

2.2 Use of Nanoparticles (I.E.: Quantum Dots) to Aid in Imaging of Cancer Cells for Site-Specific Drug Delivery

A variety of nanomaterials are finding viable applications including drug delivery as well as targeting and imaging of the onset of malignant tumor formation in tissues (Owusu and Owusu 2000). It is estimated that the global market for drug

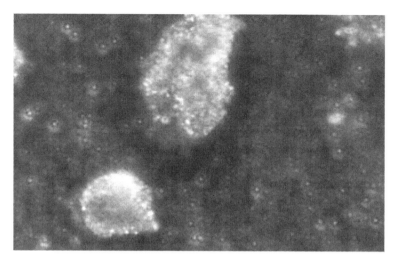

Fig. 1 SEM image showing the high affinity of gold nanoparticles for cancer cells. The gold nanoparticles stick to cancer cells and make them shine (El Sayed 2005)

delivery products is $33 billion per year and growing at a rate of 15% per year (Hill 2005). Recent studies have suggested that nanoparticle drug delivery might improve the therapeutic efficacy of anticancer drugs and allow the simultaneous monitoring of drug uptake by tumors (Kukowska-Latallo et al. 2005). Research in the design and applications of semiconductor nanocrystals, known as Quantum Dots (QDs) is gaining prominence. Quantum dots are the nanoparticles that are recently emerging as an alternative to organic fluorescence probes in cell biology and biomedicine. QDs are monodispersed semiconductor nanocrystals (size range: 2–10 nm) covered with a stabilizing monolayer. QDs have several predictive advantages such as ability to absorb light within a broad band of wavelength but emit at much narrower bands. QDs have high stability, and possess superior imaging capabilities. These unique properties allow simultaneous excitation of different sizes of quantum dots with a single excitation light source with simultaneous resolution and visualization as different colors (Fig. 2). In the area of biomedical applications, water-soluble and biodegradable QDs that have been encapsulated with glycopeptides in the form of receptors and ligands have been shown to bind to living cells (Loo et al. 2004). Recently, highly luminescent encapsulated quantum dots with cadmium tellurium (CdTe) core structure (Fig. 3) have been synthesized in a one-pot aqueous synthetic method (Svarovsky et al. 2005).

The goal of this current research work is to exploit the luminescence properties of QDs to aid in imaging for identifying cancer cells for site specific drug delivery. T-antigen bearing quantum dots are used to locate small metastasis in the midst of normal tissue. In addition, QDs can significantly improve clinical diagnostic tests for the early detection of malignant tumors. Encapsulated QDs

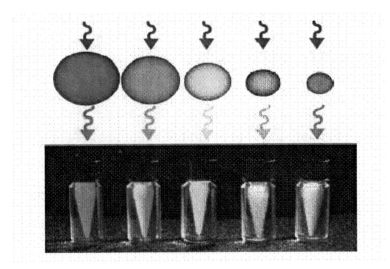

Fig. 2 Schematic representation of the effect of size on the color of QD particles (Five different QD solutions can be excited with a single wavelength)

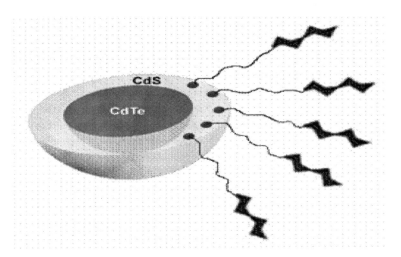

Fig. 3 Quantum Dot structure showing CdTe core and cds shell formation. The linkers are the ligand attachments (Svarovsky et al. 2005)

can be used to identify disease markers similar to currently widely used Magnetic Resonance Imaging (MRI).

Of particular significance is the emerging field of nanotechnology-based biomedical techniques. Nanoparticle-based techniques are promising novel tools that present the ability for earlier, faster and more selective diagnosis of cancer at much lower costs and reduced side-effects than traditional methods.

2.3 Nanoparticles for Malignant Tumor Diagnostics and Drug Delivery

A variety of nanomaterials are finding viable applications in the area of targeting and imaging of the onset of malignant tumor formation in tissues. The targeting capability of nanoparticles is influenced by particle size, surface charge, surface modification, and hydrophobicity whilst their general performance in vivo is influenced by morphological characteristics, surface chemistry, and molecular weight. Nanoparticles can be designed for site-specific delivery of drugs since they act as potential carriers for several classes of drugs such as anticancer agents and antihypertensive agents.

2.4 Design and Synthesis of Gold Nanoparticles

Due to their biocompatibility and excellent optical properties, gold nanoparticles are finding increasing application in a variety of areas including DNA detection and cancer diagnostics. They are also useful for identification of protein-protein interactions (Salata 2004). Gold nanoparticles functionalized with oligonucleotides have been used as probes in DNA detection (Daniel and Astruc 2004). The gold nanoparticles can be conjugated to an antibody for epidermal growth factor receptor (EGFR) proteins. These EGFR cells which are found on the surface of most cancer cells respond to the presence of gold due to the ability of the gold nanoparticles to absorb light. The gold nanoparticles display a much higher efficacy. QDs acting as multivalent fluorescent tags (Fig. 4) can be used to seek out specific lectins and antibodies in a multiplex fashion to detect various disease states or harmful pathogens simultaneously in high throughput for medical diagnostics. Small spherical particles of average diameters between 5 and 20 nm have been prepared by UV radiation of gold ions. Larger particles (20–110 nm) have been formed by irradiation coupled by reduction with absorbic acid (Sau et al. 2001). Figure 5 shows schematic representation of protein complexes formation in relation to cell genomics. Figure 6 shows the effect of gold nanoparticles (GNPs) on lung specimens compared with treatment with phosphate buffered saline (PBS) solution. Those treated with GDPs look better than those treated with PBS (as shown in Fig. 6).

2.5 Expected Outcome, Intellectual Merit, and Broader Impact of Current Research

It is envisioned that quantum dots and gold nanoparticles sensors outlined in this paper would:

- Improve sensitivity and selectivity of nanosensors in targeting of Epidermal Growth Factor Receptor (EFGR) proteins commonly present at the surface of most cancer cells (since EFGR proteins are at the same molecular level as the cancerous cells)

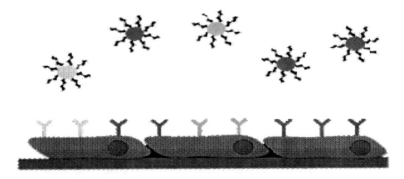

Fig. 4 The multivalent fluorescent tagging property of quantum dots (QDs)

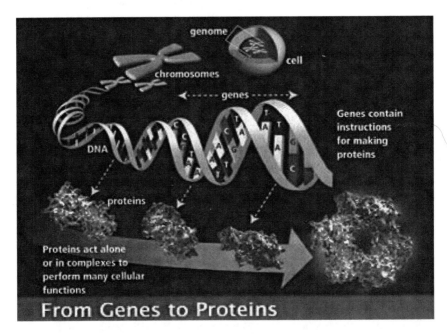

Fig. 5 Schematic representation of protein complex formation in relation to cell genomics (U.S. Department of Energy Human Genome Program, Available at: http://www.ornl.gov/hgnis, 2008)

- Lead to a more effective screening and detection of cancerous tumors
- Improve selectivity of drug injection into individual cells

It is likely that this advanced technology will soon result in rapid and widespread introduction of manufactured nanomaterials and devices into commerce.

The progress in this research endeavor is expected to uniquely and substantially enhance our capability in our efforts toward the development of nanorobotic devices (Owusu et al. 2006) that will be programmed to mitigate against the growth

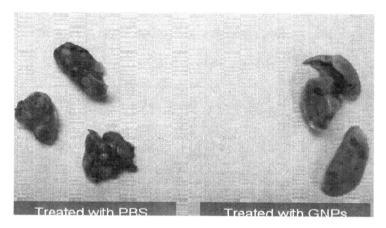

Fig. 6 Comparison of lungs treated with phosphate buffered saline (PBS) solution (left) and gold nanoparticles, GNPs (right)

of malignant cells in the human body that are responsible for diseases like cancer, asthma, etc. The expected outcomes outlined above coincide with the research goals of Research Center for Cutting Edge Technologies (RECCET) aimed at providing positive implications to the human interface between incurable diseases and nanotechnology (Owusu and Owusu 2000). The set-up in Fig. 7 is to aid in generating ultra-violet radiation for fabrication of nanosensor particles by the authors of this paper at the Research Center for Cutting Edge Technologies (RECCET) in Tallahassee, Florida, USA. This is a small portion of an elaborate experimental set up and sophisticated equipment servicing the Research Center.

2.6 Real World Applications of Nanotechnology in Medical Field

This is a proposed structure of a simple nano robot and diamondiod sphere for selective transport method as envisioned for controlled release of oxygen by Drexler (1992) in Figs. 8 and 9, respectively (1992). The concept in Figs. 10 and 11 will help to mitigate heart attacks and strokes by pumping oxygen to where it is needed during an attack or stroke.

2.7 Simulation of Selective Oxygen Nanotransport Device (SOND)

Nanodevices can be of great use in the medical field just for the purpose of preserving human life. Nanodevices are on the scale in which all things are created, there by, have the ability of accomplishing feats that no drug itself can do. This is due to

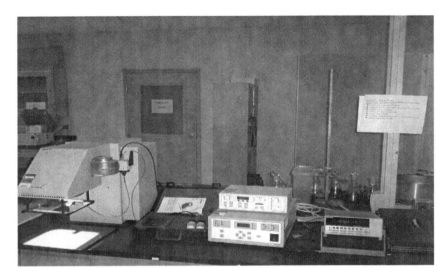

Fig. 7 Solar simulator and its controls along with the electrometer at Research Center for Cutting Edge Technologies (RECCET)

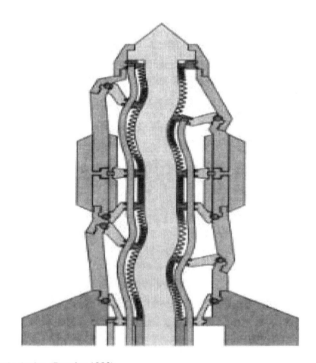

Fig. 8 Robotic device (Drexler 1992)

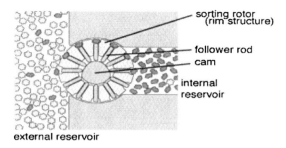

Fig. 9 Selective transport method (Drexler 1992)

the fact that nanodevices are extremely precise. In fact, nanodevices have the ability to form a structure whether living or nonliving, atom by atom. With this level of precision, nanodevices will not only be able to cure, but will also be able to prevent diseases. Specifically, a nanodevice such as Oxygen Nanotransport Device being developed at the Research Center (RECCET) by the authors of this paper can be used to prevent horrible medical ailments such as heart attacks, aneurism, and strokes. Heart attacks, aneurism and strokes are caused by the lack of oxygen going to the heart and/or the brain. Schematic views of the Selective Oxygen Nanotransport Device (SOND) combined with the artificial holding reservoir are shown in Figs. 10 and 11. The device can collect and redistribute oxygen in the blood stream in order to prevent abnormally low levels of oxygen or shortage of oxygen in the blood stream to the brain and/or the heart.

3 Terrorism Issues of Nanotechnology

The purpose of the following sections is to discuss the potential dangers and security threats that may be posed to world peace should nanotechnology materials and devices fall into the wrong hands for the purpose of terrorism and domination of world power. The role of the United Nations Regulatory Committee, as a watchdog for monitoring the use of nanotechnology, is also discussed.

3.1 The Right Tools in the Wrong Hands

Like computers, nanotechnology and programmable assemblers could become ordinary household objects. It is not too likely that the average person will get hold of and launch a nuclear weapon, but imagine a deranged person or a terrorist launching an army of nanorobots programmed to kill anyone with brown eyes and curly hair or people with blue eyes and straight hair. Even if nanotechnology remains in the hands of governments, think what a Stalin or a Hitler could do or few

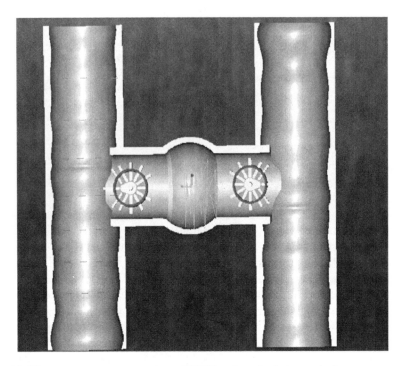

Fig. 10 The oxygen nanotransport system (SOND) and arteries in human body

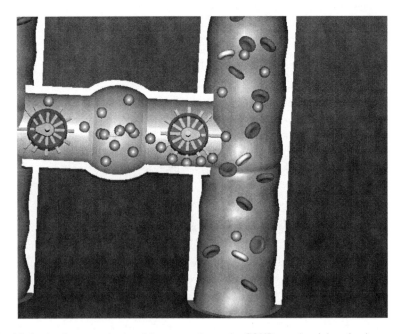

Fig. 11 Lack of oxygen in the right artery shows the SOND on the right releasing oxygen molecules (ball shaped particles)

countries will lord over other countries (as it is currently with nuclear weapons). Vast armies of tiny specialized nano killing machines could be built and dispatched in a day. Nano-sized surveillance devices or probes could be implanted in the brains of people without their knowledge.

3.2 Potential Threat of Bio-terrorism

The most serious aspect of the down side of nanotechnology, yet, is the potential threat of bio-terrorism. Matter is being manipulated at its most fundamental level, and this is the level at which diseases could be manufactured. Trillions of deadly, poisonous, and infectious nanites (nanorobots) could be manufactured in a nano-laboratory and be released on a sector of humanity wiping out whole populations, should this technology fall into the hands of terrorist organizations or some power seeking countries.

3.3 Role of United Nations for Nanotechnology Security Around the World

It is with this in mind that the authors of this paper feel strongly that the need for a world governing body or regulatory organization on nanotechnology is imperative. This proposed body, formed, constituted and run by the United Nations will focus its energies on the down side of nanotechnology. Figure 12 shows a proposed organizational chart of what such a body will look like. It is in this vain that the proposal to establish this International Steering Committee on Nanotechnology (ISCON) as a governing body to regulate the development, deployment, and the use of nanotechnology is of utmost importance. The governing body should take the form of a Committee of the United Nations and will comprise members from developed as well as developing countries. Lawyers, Doctors, Scientists, the Clergy as well as Government appointed officials will form the core of such a committee with 12 members; and each being also an active member of a sub-committee responsible for different aspects of the regulatory function. The chairperson of this committee will be appointed every 4 years by the United Nations Security Council and will report directly to the office of the United Nations (UN) Secretary General.

It is not the purpose of the authors of this paper to dampen the natural initiatives, innovations, and discoveries in this field; but one cannot over emphasize the need to keep this potentially deadly form of technology from falling into the wrong hands. Nanotechnology must be seen as one of the tools of science, engineering, and technology to foster a sense of brotherhood, goodwill, and peace among nations while forging greater ties among peoples, rather than for destructive means.

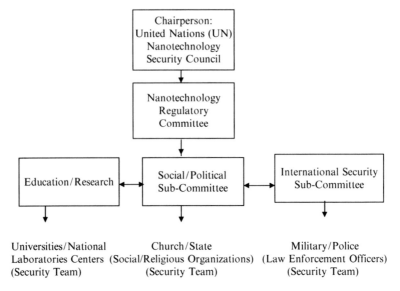

Fig. 12 Proposed organizational chart of United Nations Regulatory Committee on nanotechnology and its applications

4 Concluding Remarks

It is not a matter of if, but when nanotechnology becomes a household application. The nanoage is here and it will revolutionize the status quo on planet earth, including its International Security Profile. Had Albert Einstein foreseen the use to which his relativity theory would have been put in 1945, in Hiroshima, Japan, he probably would have had second thoughts about the way he disseminated his discoveries.

As General Douglas Macarthur said in his farewell speech before a joint sitting of the US Congress after the end of World War II on April 19, 1951:

> Men since the beginning of time have sought peace. Various methods through the ages have been attempted to devise an international process to prevent or settle disputes between nations. .... We have had our last chance. If we will not devise some greater and more equitable system, our Armageddon will be at our door. The problem basically is theological and involves a spiritual recrudescence, an improvement of human character that will synchronize with our almost matchless advances in science, art, literature, and all material and cultural developments of the past two thousand years. It must be of the spirit if we are to save the flesh.

We need to heed General Macarthur's warning and learn from the lessons of history and use nanotechnology solely for peaceful means.

References

Daniel, M.-C. and Astruc, D., Gold Nanoparticles: Assembly, Supramolecular Chemistry, Quantum-Size-Related Properties, and Applications Toward Biology, Catalysis, and Nanotechnology, *Chem. Rev.* (Review) 2004, 104(1): 293–346.

Drexler, K., Nanosystems: Molecular Machinery, Manufacturing, and Computation, Wiley, 1992.

El-Sayed M. (Georgia Institute of Technology), Science Daily, Gold Nanoparticles May Simplify Cancer Detection: Available at: http://www.gatech.edu/news- room/release.php?id = 561, May 10, 2005.

Hill, D.M., Applications of Glyconanoparticles Biotechnology, Innovations in Pharmaceutical Technology, 2005 [Available at: http://www.inpharm.com/static/intelligence/pdf/MAG_365480.pdf].

Kukowska-Latallo, J.F., Candido, K.A., Cao, Z., Nigavekar, S.S., Majoros, I.J., Thomas, T.P., Balogh, L.P., and Baker, J.R., Nanoparticle Targeting of Anticancer Drug Improves Therapeutic Response in Animal Model of Human Epithelial Cancer, *Cancer Res.* 2005, 65(12): 5317–5324.

Loo, C. et al., Nanoshell-Enabled Photonics-Based Imaging and Therapy of Cancer, *Technol. Cancer Res. Treat.* February 2004, 3(1): 33–40.

Merkle, R., Self Replicating Systems and Molecular Manufacturing, *J. Brit. Inter. Soc.* 1992, 45: 407–413.

Owusu, Y.A., Chapman, H., Mundoma, C., and Williams, N., Novel Nanoparticles for Cancer Detection and Prevention, NSTI Nanotech Conference, Boston, MA, May 2006.

Owusu, Y.A., Advanced Manufacturing Processes and Materials, Textbook in Progress, 2005.

Owusu, Y.A. and Owusu, N.A., Fabrication of Nanorobot for Medical Applications in Asthma Treatment and Prevention, 32nd Southeastern Symposium on System Theory, FAMU-FSU College of Engineering, March 5–7, 2000, pp. 46–52.

Paciotti, G.F., et al., Colloidal Gold Nanoparticles: A Versatile Platform for Developing Tumor Targeted Cancer Therapies, Nano Science and Technology Institute, May 05, 2005.

Salata, O.V., Application of Nanoparticles in Biology and Medicine, *J. Nanobiotechnol.* 2004, 2: 3.

Svarovsky, S.A. and Barchi, J.J., Journal of American Chemistry, 2005.Author Queries

Part II
Sustainable Energy and Fuel

Environment Friendly Biodiesel from *Jatropha curcas*: Possibilities and Challenges

C. Baroi, E.K. Yanful, M.F. Rahman, and M.A. Bergougnou

Abstract Bio-diesel development from *Jatropha curcas* (JTC), a tropical plant, is currently being carried out in various parts of the world. High oil content of the JTC seed, high cetane number of the JTC biodiesel, its drought resistant characteristics, its toxicity, which makes it unwanted by both humans and animals, and its various other uses render this plant an extremely promising source for bio-fuel development in arid areas and rural communities in the developing world, in particular. Catalytic hydrocracking of various other vegetable oils, such as sunflower oil and soybean oil, has been reported. Chemical composition of JTC seed oil shows similarities in fatty acid composition to these oils. A two-stage catalytic hydrocracking process is proposed to convert JTC oil to high-cetane number biodiesel. If successful selectivities are obtained at the bench scale, this technology could be implemented in existing refineries in developing countries without major modification.

Keywords *Jatropha curcas* · biodiesel · catalytic hydrocracking

---

C. Baroi (✉) and M.F. Rahman
MESc. Candidate, Department of Civil and Environmental Engineering, University of Western Ontario, London, Ontario, N6A-5B9, Canada
e-mails: cbaroi@uwo.ca; mrahma49@uwo.ca

E.K. Yanful
Professor and Chair, Department of Civil and Environmental Engineering, University of Western Ontario, London, Ontario, N6A-5B9, Canada
e-mail: eyanful@eng.uwo.ca

M.A. Bergougnou
Professor Emeritus, Department of Chemical and Bio-chemical Engineering, University of Western Ontario, London, Ontario, N6A-5B9, Canada

1 Introduction

Environmental pollution and non-renewability of fossil fuels are some reasons why alternative fuels are gaining wide attention. Biodiesel is carbon-neutral, non-toxic and has a low emission profile compared to conventional petroleum (Meher et al. 2006). *Jatropha Curcas* (JTC) is a multipurpose plant with many attributes and considerable potential as an energy crop (Openshaw 2000). High viscosity and lower volatility are some of the problems with plant oils that have to be overcome before using them as fuel for conventional diesel engines (Meher et al. 2006). To date, transesterification is the most widely reported method of converting vegetable oils to biodiesel (Shah 2004). Some researchers have noted that catalytic hydrocracking of vegetables oils, such as soybean oil and sunflower oil, can produce successful synthetic diesel (Filho et al. 1992). Since JTC oil and the above mentioned oils have similar fatty acid compositions, it can be assumed that catalytic hydrocracking of JTC oil might produce high quality biodiesel. Researchers at the University of Western Ontario are planning to transform JTC oil to premium biodiesel using a two-stage catalytic hydrocracking process. This paper attempts to summarize the envisioned program. It also looks into the possibilities and challenges of JTC cultivation, the proposed technology and its application in the developing regions of the world.

2 *Jatropha curcas* – Origin and Description

JTC is a drought resistant tropical plant that belongs to the genus *Euphorbiaceae* (Gubitz et al. 1999). The JTC plant is a small tree or a large shrub that can reach a height of up to 5 m. The life- span of the JTC plant is more than 50 years (Henning 2000). This plant can be grown in low to high rainfall areas (Openshaw 2000) and is mainly used as living fence as it is toxic to both humans and animals (Gubitz et al. 1999). JTC has significant potential as an energy crop because its seed contains about 55% oil (Forson et al. 2004). With mechanical oil expellers, up to 75–80% of the oil can be extracted. With a hand press only 60–65% of the oil can be obtained (5 kg of the seeds give about 1 l oil) (Henning 2000). The properties of JTC oil are given below in Tables 1 and 2.

3 Possibility of Producing Biofuel from *Jatropha curcas*

Attempts have been made to produce biodiesel by transesterification of JTC oil and successful results were obtained. Conversion of JTC oil to useful fuels may also involve short contact time pyrolysis, catalytic cracking and catalytic hydrocracking. Table 2 shows that the calorific value of JTC oil is close to that of diesel oil and that the cetane value of JTC oil is slightly higher than that of diesel oil. The sulphur content of JTC oil is also much smaller than that of diesel oil in developing countries.

Table 1 Physico-chemical characteristics of Jatropha oil (JTC) (Akintayo 2004)

Parameter	JTC
Colour	Light yellow
Free fatty acid (mg/g)[a]	1.76 ± 0.10
Acid value (mg KOH/g)[a]	3.5 ± 0.1
Saponification value (mg KOH/g)[a]	198.85 ± 1.40
Iodine value (mg iodine/g)[a]	105.2 ± 0.7
Mean molecular mass	281.62
Unsaponifiable matter (%)	0.8 ± 0.1
Refractive index (25 °C)	1.468
Specific gravity (25 °C)	0.919
Hydroxyl value	2.15 ± 0.10
Acetyl value	2.16 ± 0.10
Viscosity (30 °C) cSt	17.1

[a]Values are mean ± standard deviation of triplicate determination.

Table 2 Comparison between Jatropha oil and diesel oil (adapted from Kandpal and Madan 1995)

Parameter	Diesel oil	Jatropha oil
Specific gravity	0.8410	0.9180
Sulphur (% wt)	1.2	0.13
Calorific value (cal/g)	10,170	9,470
Flash point (°C)	50	240
Cetane value	50	51

Table 3 shows that JTC oil contains mainly unsaturated and long chain (C_{18}) fatty acids. Linoleic acid is the dominant fatty acid in the oil. Soybean oil has a similar type of fatty acid composition and linoleic acid (60%) is also the dominant fatty acid in it (Filho et al. 1992). Sunflower oil and canola oil also contain unsaturated and long chain (C_{18}) fatty acids. Linoleic acid and oleic acid are the dominant fatty acids in sunflower oil and canola oil respectively. Sadrameli and Green (2007) reported that thermal cracking of canola oil in the temperature range 300–500 °C resulted in 38.1% to 14.8% of liquid products in which aromatics were 7.7–9.0%, whereas the gaseous product was 15–75% of the total product. Another study used $ZnCl_2$ catalyst to crack sunflower oil within the temperature range 610–690 K (Demirbas and Kara 2006). Soybean oil was thermally decomposed (pyrolysis) in the presence of N_2 and air using ASTM standard method D86-82 and the products contained nearly equal amounts of alkanes and alkenes and approximately 2% of aromatics. Soybean oil was also catalytically hydrocracked in the presence of $NiMo/\gamma Al_2O_3$, giving 8.7% of gaseous products, 66.6% alkanes, 11.9% cycloalkanes, 4.3% alkylbenzenes (aromatics), 0.5% acids, 5.4% water and 2.4% losses (Filho et al. 1992). Several other vegetable oils were equally hydrocracked in the presence of $NiMo/\gamma Al_2O_3$ catalyst and it was observed that the yield of gases in the converted material increased with the

Table 3 Fatty acid composition of *Jatropha curcas* (JTC) oil (Adapted from Adebowale and Adedire 2006; Filho et al. 1992; http://www.scientificpsychic.com/fitness/fattyacids1.html)

Composition	Weight percentage		
	Sunflower oil	Soybean oil	JTC oil
Palmitic acid (C16:0)	7.0	5.0	11.3
Stearic acid (C18:0)	5.0	1.6	17.0
Oleic acid (C18:1)	19.0	29.8	12.8
Linoleic acid (C18:2)	68.0	60.0	47.3
Arachidic acid (C20:0)	–	–	4.7
Arachidoleic acid (C20:1)	–	–	1.8
Behenic acid (C22:0)	–	–	0.6
C24:0	–	–	4.4

decrease in the fatty acid chain length and the degree of unsaturation of the vegetable oils (Filho et al. 1992). The degree of unsaturation in sunflower oil is greater than that of canola oil, which is why gaseous products are higher in canola oil. The oil composition is an important factor for the alkane content in the converted material. The higher the unsaturation, the higher is the formation of cycloalkanes and alkylbenzenes and the less will be the alkanes (Filho et al. 1992). Catalytic acidity and pore size affect the formation of aromatic and aliphatic hydrocarbons (Demirbas and Kara 2006). Unsaturated oil yields less gaseous products and hydrocarbons are produced by the reduction reaction of carboxyl groups rather than by decarboxylation of carboxyl group (Filho et al. 1992).

$$\text{Reduction: } C_{17}H_{35}COOH + 3H_2 \rightarrow C_{18}H_{38} + 2H_2O$$

$$\text{Decarboxylation: } C_{17}H_{35}COOH \rightarrow C_{17}H_{36} + CO_2$$

Decarboxylation can be either thermal or catalytic but reduction is only a catalytic process. In catalytic hydrocracking, a dual function catalyst is used. Its acid sites catalyze the cracking reaction while the metallic sites catalyze hydrogenation. Thus, when large unsaturated molecules are cracked into small molecules at the same time they are converted to saturated small molecules by hydrogenation. In catalytic hydrocracking, higher temperature favors chain cleavage but the formation of alkylbenzene probably originating from cycloalkanes is enhanced. Increase in hydrogen pressure favors the reduction of the carboxyl groups of fatty acids, with formation of water (Filho et al. 1992). The presence of carboxyl groups is undesirable, because they contain oxygen, which is corrosive to metals. When soybean oil was catalytically hydrocracked, the alkane content of the liquid product increased and the acid content was decreased substantially. On the basis of these results, it can be inferred that catalytic hydrocracking may yield better results for converting JTC oil into premium quality synthetic diesel with, hopefully, a higher cetane number.

4 Proposed Approach

The proposed plant would consist of two stages. The first stage would be a hydrotreating stage used for deep desulphurization, denitrogenation and possible deoxidation of raw JTC oil. The main products would be 'clean JTC oil', hydrogen sulfide, ammonia, water and smaller hydrocarbon fragments generated by a limited amount of JTC oil hydrocracking. The reactions are exothermic. Catalysts would use oxides of cobalt, nickel, vanadium and tungsten on alumina.

Hydrocracking would take place in the second stage on the 'clean JTC oil' coming out of the first stage distillation system. The catalyst would probably be a noble metal on a molecular sieve base. Sulfur and nitrogen are strong poisons for such a catalyst. The latter would be highly selective for compounds giving a high cetane number for JTC bio-diesel and would also minimize consumption of expensive hydrogen.

Figure 1 shows a simplified flow-plan for the first stage (hydrogenation). Raw JTC, fresh hydrogen and recycle hydrogen would be preheated against hot reactor products in a heat exchanger train and brought up to reactor temperature in a furnace. The hot combined liquid and gas phases would trickle down through one or several beds of catalyst particles of a size in the millimeter range. Cold quenching recycle hydrogen coming out of the high-pressure separator would cool the reactor. The hot reactor product would finally be cooled in the cooling heat exchanger train and in the water cooler train and go to a high-pressure separator. The gas phase would give the hydrogen recycle. The liquid phase would be depressurized in the low-pressure separator and go to the distillation train. The second stage (hydrocracking stage) would be similar to the first stage, and would have the clean JTC oil as the feedstock.

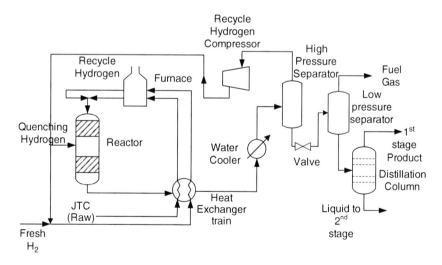

Fig. 1 Simplified flow-plan of the first stage for the proposed hydrocracking of JTC oil

Depending on the required bio-diesel specifications and with appropriate catalysts, only the first stage might be needed.

5 Challenges, Possibilities and Conclusion

The proposed program would start at the bench scale to test various hydrogenation and hydrocracking catalysts. Temperatures would be in the range 300–500 °C and pressure would be around 10 MPa (100 atm). Hydrogen diffuses through metals giving possibly explosive mixtures in air. Therefore, the laboratory should be well ventilated and have explosion proof electrical equipment. Material balances should be high. The life of a catalyst should be in the order of 1 year. Many of the refineries in the developing countries already have hydrocracking units installed in their refineries. If successful results are obtained at the laboratory scale, the process could be implemented in those countries without major modification of the existing refineries and could be operated by local manpower. However, the cost of the equipment, catalysts and analytical devices could be significant. In existing hydrogenation units (one stage) or hydrocracking units, co-processing of JTC oil with petroleum cuts might be cheapest.

Biomass supplies the bulk of the total energy demand in many of the world's poor nations (Ackom and Ertel 2005). Approximately 60% of the total energy consumption in Africa in 1995 came from biomass and it is estimated that, by 2030, approximately 823 million people will be relying on biomass in Africa (Ackom and Ertel 2005). This over-reliance on biomass could lead to serious environmental degradation and desertification in the developing world. Promotion of the multipurpose JTC would help remote rural communities address their energy needs and minimize environmental degradation. In many African communities JTC seeds are harvested by women and used for making soap, candle and medications. By selling these products, women in those rural areas could generate extra-income, which inherently would promote women empowerment and, also, high-value added unique compounds could be extracted from hydrogenated or hydrocracked JTC oil.

One important aspect of JTC diesel production is that it would not compete with food production, as JTC is toxic to both humans and animals and also because JTC plant can grow in dry wastelands with minimum input. The set up of JTC oil converting industries would create employment opportunities, provide a source of income for farmers and suppliers of feedstock and would, eventually, be a great source of revenue for governments. Also, it will reduce the dependency on fossil fuels and minimize crude petroleum import costs. Currently JTC cultivation is being promoted in a number of countries, including India, Ghana, Mali, Malawai, Zimbabwe, Egypt and Burkina Faso (Ackom and Ertel 2005), and some of those countries are good candidates for large scale industrial production of JTC biodiesel and other products. Comparatively cheaper JTC biodiesel would encourage farmers to use biodiesel for irrigation purposes and also could be used by the transportation sector and industry. Government and donor agencies would have to come up with necessary planning and appropriate financial support to subsidize establishment costs for industry to develop JTC biodiesel resource. This would eventually stimulate economic empowerment through employment and poverty reduction (Ackom and Ertel 2005), and promote sustainable and appropriate development in the developing and drought prone regions of the world.

References

Ackom, E.K.; Ertel, J., 2005. An alternative energy approach to combating desertification and promotion of sustainable development in drought regions. Forum der Forschung, 18, 74–78.

Adebowale, K.O.; Adedire, C.O., 2006. Chemical composition and insecticidal properties of the underutilized *Jatropha curcas* seed oil. African Journal of Biotechnology, 5(10), 901–906.

Akintayo, E.T., 2004.Characteristics and composition of *Parkia biglobbossa* and *Jatropha curcas* oils and cakes. Bioresource Technology, 92, 307–310.

Demirbas, A.; Kara, H., 2006. New options for conversion of vegetable oils to alternative fuels. Energy Sources, Part A, 28, 626–629.

Filho, G.; Rocha, N.; Brodzki, D.; Djega-Mariadassou, G., 1992. Formation of alkanes, alkylcycloalkanes and alkylbenzenes during the catalytic hydrocracking of vegetable oils. Fuel, 72(4), 543–549.

Forson, F.K.; Oduro, E.K.; Donkoh, E.H., 2004. Performance of Jatropha oil blends in a diesel engine. Renewable Energy, 29, 1135–1145.

Gubitz, G.M.; Mittelbach, M.; Tranbi, M., 1999. Exploitation of the tropical oil seed plant *Jatropha curcas* L. Bioresource Technology, 67, 73–82.

Henning, R., 2000. The Jatropha booklet – a guide to the Jatropha system and its dissemination. GTZ-ASIP-Support-Project Southern Province, Choma. http://www.jatropha.de/documents/jcl-booklet-o-b.pdf. Visited: 1600 hours, 21 June, 2006.

Kandpal, J.B.; Madan, M., 1995. Jatropha curcus: A renewable source of energy for meeting future energy needs. Renewable Energy, 6(2), 159–160.

Meher, L.C.; Vidya, S.D.; Naik, S.N., 2006. Technical aspects of biodiesel production by transesterification – a review. Renewable and Sustainable Energy Reviews, 10, 248–268.

Openshaw, K., 2000. A review of Jatropha curcas: An oil plant of unfulfilled promise. Biomass and Bioenergy, 19, 1–15.

Sadrameli, S.M.; Green, A.E.S., 2007.Systematics of renewable olefins from thermal cracking of canola oil. Journal of Analytical and Applied Pyrolysis, 78(2), 445–451.

Shah, S., 2004. Biodiesel preparation by lipase-catalysed transesterification of Jatropha oil. Energy & Fuels, 18, 154–159. http://www.scientificpsychic.com/fitness/fattyacids1.html. Visited: 1244 hours, 20 June, 2007.

Thermal Utilization of Solid Recovered Fuels in Pulverized Coal Power Plants and Industrial Furnaces as Part of an Integrated Waste Management Concept

G. Dunnu, J. Maier, and A. Gerhardt

Abstract Solid Recovered Fuels (SRF) are highly heterogeneous mixtures generated from high calorific fractions of non-hazardous waste materials intended to be fired in existing coal power plants and industrial furnaces (CEN/TC 343 2003). They are composed of a variety of materials of which some although recyclable in theory, may have become in forms that made their recycling an unsound option. Their use is regulated under EU regulations and requires specification for commercial and regulatory purposes. The use of waste as a source of energy in itself is an integral part of waste management, as such within the framework of the European Community's policy-objectives related to renewable energy, an approach to the effective use of wastes as fuel sources and energy recovery from wastes is outlined in documents such as the European Waste Strategy.

This work involves a characterization step for SRF especially for co-firing in pulverized coal power plants for the purpose of generating heat and electricity. The nature of SRF requires a thorough understanding of their combustion properties before optimal energy recovery can be realized. The characterization process involves among other things lab-scale experiments that critically examine the fuel concerning their physical and chemical properties. The de-volatilization, ignition and combustion processes associated with different types of SRF are also studied using a thermo-gravimetric analyzer (TGA). Based on these experiments, suggestions are made for a successful application of SRF in power plants and industrial furnaces. Finally, an overview of the potentials of waste materials as fuel and a source of energy is discussed.

Keywords Solid recovered fuels (SRF) · municipal solid waste (MSW) · integrated waste management concept · co-combustion/co-firing · thermo-gravimetric analysis

---

G. Dunnu(✉), J. Maier, and A. Gerhardt
Institute of Process Engineering and Power Plant Technology (IVD), University of Stuttgart Pfaffenwaldring 23, 70569 Stuttgart, Germany
e-mail: dunnu@ivd.uni-stuttgart.de

1 Introduction

The production and thermal utilization (energy recovery) of Solid Recovered Fuels (SRF) from bio-waste, residues, mixed- and mono waste streams is fast becoming a key element in an integrated waste management concept, as the deposition of untreated waste stream in landfills is not an option in the future as outlined in the European Landfill directive (1999/31/EC). The key elements of such waste management concept would be recycling/reuses, mechanical- and biological processes (M/B), solid recovered fuel production and municipal/hazardous waste incineration (MSWI). The targeted hierarchy (Fig. 1) within this concept would be material recovery, energy recovery and final disposal according to the directive (1999/31/EC).

The implementation of solid recovered fuel production in an integrated waste management concept demands a potential market for the products, which can be found in the energy sector, and in product-oriented sectors such as cement or lime industries by substituting fossil fuels. Figure 2 illustrate the process scheme for SRF utilisation in industrial processes and power plants (Gawlik and Ciceri 2005).

By definition, Solid Recovered Fuels are highly heterogeneous mixtures generated from high calorific fractions of non-hazardous waste materials intended to be

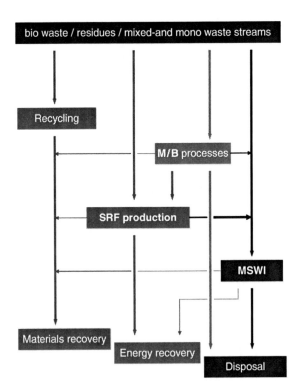

Fig. 1 The targeted hierarchy within the waste management concept

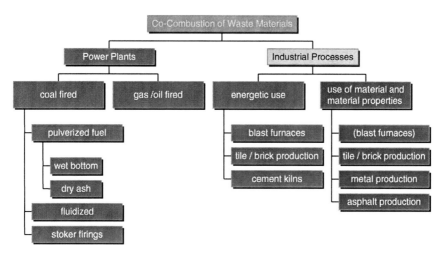

Fig. 2 Process scheme for SRF utilisation

fired in existing coal power plants and industrial furnaces (CEN/TC 343 2003). They are composed of a variety of materials of which some although recyclable in theory, may have become in forms that made their recycling an unsound option. Their use is regulated under EU regulations and requires specification for commercial and regulatory purposes. Figures 3a, b below shows a type of SRF called *Substitutbrennstoff aus Siedlungsabfaellen* (SBS®) –high calorific fraction generated from municipal solid waste (MSW)– and its percentage composition (Dunnu et al. 2006).

Co-combustion of SRF such as SBS® in existing coal fired utilities can have a huge impact on a power plant performance. Due to different physical and chemical properties of these fuels in comparison with coal, a number of processes including milling, feeding, combustion, heat transfer, and steam production can be affected. With the current knowledge of co-combustion, it is not possible to predict the impact of a single SRF or their mixtures on a power plant performance. With the knowledge of the physical, chemical and biogenic characteristics of the SRF, some of the negative impacts of co-combustion on the performance of power plants can be avoided and subsequently, carrying out expensive large-scale tests and trials in order to determine the suitability of a solid recovered fuel at certain plants can be minimized. Characterization of SRF is therefore essential for the purpose of quality assurance and quality management system for the smooth operation of plants and other industrial furnaces.

2 Waste-to-Energy a Better Option

Waste disposal and the availability of cleaner energy sources are two major issues facing Europe and the rest of the world. Both landfills and the emission of greenhouse gases present serious health and environmental threats. Finding solutions to

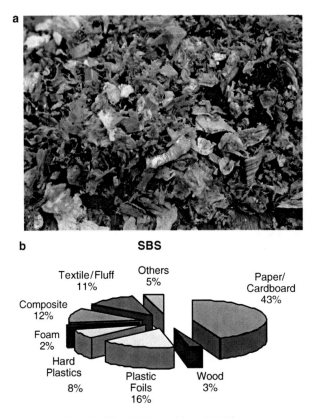

Fig. 3 (a) Solid recovered fuel (SBS®); (b) Composition of SBS®

these threats therefore advance waste-to-energy (WTE) concept as a potential option that should not be overlooked. The production of SRF from non-hazardous bio-waste, residues and other mixed- and mono waste streams represents a potential solution to both cleaner burning fuels compared to fossil fuels and waste materials that are currently disposed of in landfills.

2.1 Environmental Benefits

Using waste as an energy source would reduce the amount of fossil fuels used. Under normal conditions, 1 t of waste can generate 3.5 MW of energy, as much energy as contained in 300 kg of fuel oil (Fitze 2002). If this kind of energy is recognized as partially renewable resources, then fossil fuels can to some extent be substituted. The environmental benefits would be enormous if municipal solid wastes (MSW) now going to landfills are turned into SRF and subsequently combusted in thermally efficient Waste-to energy facilities. Approximately 1.4 billion tonnes of MSW worldwide are disposed in landfills, deep enough to generate biogas of approximately 50% methane and 50% CO_2. The annual generation of methane is estimated at 62 million tonnes, of which less than 10% is recovered in controlled landfill equipped to capture biogas (Themelis 2006). As a matter of fact, methane is 20 times more potent as a greenhouse gas than carbon dioxide (Veltzé 2005). As a result, if the contribution of landfill non-captured methane to greenhouse gas emission is considered, it will be clear that a gradual move from landfill to WTE is one of the low-hanging fruits in reducing global emission of greenhouse gases.

Figure 4 shows the projected situation of the German waste market 2006 with an expected capacity deficit of approximately 4.5 Mt. Co-firing SRF in energy production throughout Europe, even in small thermal shares, offers enormous potential as a sustainable, efficient and environmentally friendly waste-to-energy technology. The specific CO_2 emission based on initial SRF quality (BGS® quality 2, LHV 13.5MJ, 0.4wt% Cl_{th}) with a biogenic fraction of 50wt% is 0.04 kg CO_2/MJ, compared to 0.12 kg CO_2/MJ for brown coal (Kronberger 2001).

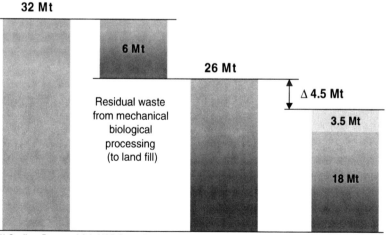

Fig. 4 Projected situation ion of Germany waste market 2006 (PROGNOS 2004)

3 Experiments

3.1 Fuel Analysis

Particle size and reactive surface area of fuel have the largest influence on a combustion process, as all matter and energy transport take place (Kock 2002). Both renewable and fossil fuels consist of a combustible organic fraction and an inert inorganic part (mineral matter). The main elements in most solid fuels are carbon (C), hydrogen (H), oxygen (O), sulphur (S) and nitrogen (N). Besides, there exist minor constituents like chlorine (Cl), potassium (K), and sodium (Na). The share of these elements in solid fuels varies, influencing the pyrolysis and combustion characteristics. Particularly the share and nature of the inorganic compounds and trace metals affect the fuel reactivity (Jenkins et al. 1998). Fuels suitable for co-firing with coal include not only typical biomass fuels like wood and straw, but also waste streams from agriculture, industrial and municipal sources.

The characterization of SRF involves the vdetermination of the physical/mechanical, chemical and biological components of a given fuel. While it is important to determine the biogenic characteristics of SRF because it enable the calculation of the CO_2 emission savings, this work is limited only to characterizing SRF based on their physical/mechanical properties, and major chemical elements.

The SRF used for the characterization were milled down to particle size 1 mm with a rotor-sieve cutter. Because of the heterogeneous nature of this type of fuel, a representative sample can easily be achieved with a finer particle size distribution and furthermore minimise the heat and mass transport resistances in the boundary layer and within the particles. The proximate and ultimate analyses of different fuels were conducted. The fuels included SBS, shredded rubber tyre, paper/plastic mixture, and demolition wood. In addition, base line analyses of lignite and hard coal as referenced fuels were also included.

3.2 Thermo-Gravimetric Investigation

The behaviour and release of volatile components of SRF and the referenced fuels were investigated by thermo-gravimetric analysis (TGA). With the standardised methods according to DIN51718-DIN51720, the mass fractions of water, volatiles, fixed-carbon and ash were determined. For the combustion behaviour the temperature range in which the volatiles are released are of interest. The TGA analyzer was charged with SRF test portions each of about 1 ± 0.1 g. With a heating rate of 5 °C/min, the test portions were heated up to a temperature of 900 °C under a reducing atmosphere (N_2). At 106 °C the temperature was kept constant to evaporate the moisture content of the fuel, the heating continues when the mass deviation of the samples was less than 0.1%. The weight loss curve of the heat-up ramp from 106 °C to 900 °C with Nitrogen atmosphere gives information about the release of the volatile matter which is composed mainly of hydrocarbon groups.

4 Results

The results from the proximate and ultimate analysis presented in Table 1 shows that demolition wood has the largest percent of volatile matter followed by SBS, shredded tyre, lignite, and hard coal in the respective order. The de-volatilization behaviour of the respective fuels investigated with a thermo-gravimetric analyzer is also presented in Figure 5.

Table 1 Fuel analysis

	SBS	Wood	Shredded tyre	Lignite (DE)	Hard coal (SA)
Proximate analysis, % (as received)					
Moisture, 106°C, %	1.7	5.7	0.62	10.2	3.6
Volatile matter, %	77.8	77.9	52.8	48.6	21.6
Fixed carbon, %	13.1	0.022	22.7	36.6	56.2
Ash, 550°C (SRF); 815°C (coal), %	7.3	16.1	23.8	4.6	18.6
Ultimate analysis, % (as received)					
C, %	52.7	48.2	83.3*	57.5	65.3
H, %	8.1	6.5	6.9*	5	3.9
N, %	0.6	<0.3	<0.3*	0.52	3.5
S, %	<0.3	<0.3	2.7*	<0.3	1.3
Cl, % LHV, J/g	25,053	18,780	29,095	22,350	24,931

*Without metal constituent.

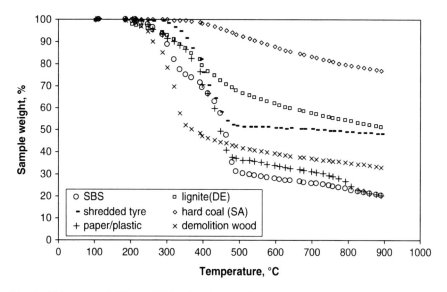

Fig. 5 TGA curves of different SRF and coal comparing weight loss due to devolatilization

The TGA curves shows that the temperature at the onset of primary de-volatilization for all the SRF is around 220 °C and ends at about 490 °C. Large percentage of the volatile matter is released at much lower temperature as compared to the primary fuels, lignite and hard coal, and it occurs within a narrow temperature range. Demolition wood exhibited maximum primary weight loss at 350 °C whereas that of paper/plastic, SBS, and shredded tyre occurred at 490 °C. Volatiles released at low temperatures favours particle ignition and improve stable flame conditions, therefore the release of large amounts of combustible matter at low temperatures show that a good ignition can be expected for all the SRF tested.

5 Conclusion

- The concept of using waste as a source of fuel to be co-fired as part of an integrated waste management concept with the aim of energy recovery and not disposal is an effective way to reduce the amount of landfill un-captured methane emission which is more potent a greenhouse gas than CO_2 (combustion product).
- The biogenic fraction of SRF makes it partially renewable and partially CO_2 neutral.
- The thermo-gravimetric analysis performed on the SRF is an essential tool to determine the suitability of fuel to be co-fired in power plants or industrial furnaces, however, it is not a blue print since other factors such as slagging, fouling, sintering, and corrosion potential also contribute significantly as to whether a particular SRF can successfully be used. Not withstanding this, the TGA provided very useful information concerning the de-volatilization of the SRF. All the waste fuels investigated showed an early de-volatilization and this occurs at relatively low temperatures. This phenomenon substantially predicts a good ignition if such SRF are to be co-fired with coal in power plants and industrial furnaces.

In sum, it should be clearly stated that the number one objective remains the removal of waste in an environmentally correct manner, without forgetting that energy recovery can take the place of other fossil energies, and in doing so reduce the impact of vhuman activity on the environment.

References

Draft business plan of CEN/TC 343, 2003, Solid Recovered Fuels.
Dunnu G., Hilber T., and Schnell U., 2006, Advanced size measurements and aerodynamic classification of solid recovered fuel particles. Energy & Fuels, 20(4): 1685–1690.
Fitze U., 2002, Les ordures – source d'énergie. UMWELT Review, no. 3/02.
Gawlik B. M. and Ciceri G., 2005, QUOVADIS waste-to-fuel conversion, DG JRC Workshop, Ispra, ISBN92-894986110.

Jenkins B. M., Baxter L. L., Miles Jr. T. R., and Miles T. R., 1998, Combustion properties of biomass. Fuel Processing Technology, 54: 17–46.
Kock O., 2002, Regenerative Brennstoffe – Charakterisierung des Brennverhaltens. 7. Fachtagung Thermische Abfallbehandlung, Band 20, Technische Universität Dresden, pp. 195–209.
Kronberger R., 2001, Waste to recovered fuel – cost-benefit analysis, GUA Gesellschaft für umfassende Analysen GmbH, Wien.
Prognos AG in VGB-PowerTech (Oct. 2004); LAGA Bericht zum 63. UMK (Aug. 2004).
Themelis N., 2006, Better together: gas turbine cogeneration improves energy recovery from WTE plants. Waste Management World, July–August: 97–105.
Veltzé S. A., 2005, Editorial, proper management of methane at landfills is key to alleviating the greenhouse gas problem. Waste Management World, March/April: 6 of 96.

Biogas Production from Organic Waste in Akwa IBOM State of Nigeria

E.E. Ituen, N.M. John, and B.E. Bassey

Abstract In view of the need for environmental management, waste recycling and alternative energy resources, there has been on-going work on biogas production with a locally fabricated digester in Akwa Ibom State of Nigeria. The first result shows that $0.032\,m^3$ of biogas was produced from 180 l of poultry manure mixed with same volume of moisture in 16 days. Using the same concentration of cow dung in a repeat experiment, $0.015\,m^3$ of biogas was produced in 7 days. There was interruption in the gas production in the second case due to excessive wetness of weather. Further investigation indicated that the volume of gas evolved might also depend on the ambient temperature as shown in the second experiment. A maximum volume of $0.006\,m^3$ was got for the maximum temperature $51\,°C$ and a minimum volume of $0.00\,m^3$ got for the minimum temperature of $22\,°C$. We also observed that the gas yield with poultry manure was higher than that with cow dung.

Keywords Biogas · poultry manure · cow dung · biodiversity · digester

1 Introduction

Biogas (energy from plant/animal origin) is now receiving much attention mainly because of the imminent world energy crisis. Conventional energy from fossil fuels (coal, petroleum and gas) is depleting due to total dependence on it by man and yet it is non-renewable. Biomass has been found as a reliable alternative to conventional energy because it is a renewable energy source from solar energy. However, solar

---

E.E. Ituen (✉), N.M. John, and B.E. Bassey
Department of Physics, University of Uyo, Uyo, Akwa Ibom State, Nigeria

E.E. Ituen
Department of Soil Science, University of Uyo, Uyo, Akwa Ibom State. NigeriaMailing Address: Department of Physics, University of Uyo, PMB 1017 Uyo, Akwa Ibom State, Nigeria,
e-mail: enoituenus@yahoo.com

energy has low efficiency in direct energy conversion to other forms. Through the process of photosynthesis, solar energy is absorbed by green plant tissue to provide energy. This reduces carbon dioxide in the atmosphere and then forms carbohydrates, which are then utilized as energy sources and raw materials for all other synthetic reactions in the plant. Thus, solar energy is captured and stored in the form of chemical energy. Land and marine vegetation absorb an estimated 4.2×10^{14} kW h of solar radiation per year while world's energy consumption in this second millennium, is expected to reach 3×10^{14} kW h per year (Stout 1979).

The energy absorbed in the visible sunlight is in the form of electromagnetic radiation and it presents in packets or quanta; each light quantum $E\lambda$ of wavelength λ given by $\lambda E\lambda = hc$; c being the speed of light. Cellulosic/lignin plant material presents a vast, untapped supply of energy. It is renewable but has a low content of sulphur and other pollutants. The conversion of cellulose or lignin into energy provides about 4,500 kcal per kilogram of dry matter (petroleum oil yields 9,500 kcal per kilogram which is the standard reference). Stored plant energy may be released by drying the material and burning it directly or various processes may be utilized to obtain potential fuels (biogas) such as ethanol, methane, or other gaseous or liquid fuels including biodiesel. The processes used to derive these fuels may require different forms of the plant/organic materials.(Ituen 2005).

There is danger of pollution to the global environment as a result of emissions from internal combustion engines using conventional fuels (fossil fuels). There are also problems such as global warming, acid rain, air and soil pollution as a result of these emissions. Plants add no net carbon dioxide (CO_2) to the environment since they emit that which was used in photosynthesis. They have low sulphur content as indicated earlier. Therefore, bio-fuels (biogas, bio-ethanol for spark-ignition engines, and bio-diesel for diesel engines) have many advantages over conventional fuel. They are oxygenated fuels (10% oxygen content) and so will burn properly and contribute to clean air strategies.

Thus, developed and developing nations are heavily embarking on the development of biofuels. For instance, India and China appear to be leading countries in biogas production and utilization. Other countries are South Africa, Australia, United States and Europe. A lot of money is spent on research, especially on biodiesel. Apart from being a replacement fuel or diesel, biodiesel can be blended with diesel in certain proportions without any adverse effects in engine performance. Rather it cuts down on the poisonous petroleum diesel emissions. In many parts of the United States, it is a law for trucks to blend their diesel with about 20% biodiesel for anti-environmental pollution. Diesel engines are also found to cause high environmental temperature, which contribute to grain failure in the hinterland of many nations (Ituen 2005).

2 Significance of the Work

'Nigeria is yet to venture intensively into this resourceful environmental bioremediation through biodiversity. Thus, inefficient waste management is posing a terrible threat in this society and more so, with the increasing poultry and diary projects all over the country and multiplicity of automobiles producing poisonous emissions due to the use of petroleum fuels. In particular, danger of environmental pollution is very critical in the Niger-Delta region where petroleum oil exploration activities and gas flaring are predominant.

3 Biogas

Biogas is composed almost exclusively of methane and carbon dioxide with traces of H_2S, N_2, H_2 and CO. The energy value of biogas is typically 400–700 BTU/ft^3 as compared to 1,000 BTU/ft^3 for natural gas. The breakdown of cattle manure into biogas is accomplished by three types of bacteria: (1) hydrolytic, (2) transitional, and (3) methanogenic. Hydrolytic bacteria are utilized in the first steps of production by reducing macromolecules like proteins, fats and carbohydrates into much smaller molecules such as amino acids, sugars, acids, and alcohols. Transitional bacteria further reduce these molecules into acetic acids, H_2 and CO_2. The final step of breakdown is accomplished by methanogenic bacteria, which reduce the molecules into methane (CH_4) and carbon dioxide (CO_2) (Engler and McFarland 1997). Hansen et al. (1998) states that acetate-utilizing methanogens are responsible for 70% of methane produced in biogas reactor.

Biogas production is also a temperature-dependent process (Misra et al. 1992). The process takes place in either psychrophilic (<250 °C), mesophilic (25–400 °C) or thermophilic (45–600 °C) temperatures. Though some ammonia is necessary for biogas production, the combination of the two (thermophilic temperature and excessive free ammonia), can inhibit and destroy the bacteria that is necessary for biogas production as reported by (Angelidaki and Ahring 1994).

Hasen et al. (1998) showed that free ammonia is determined by three parameters; total [NH_4], temperature and pH. Angelidaki and Ahring (1994) also found that reducing the temperature to 550 °C when the NH_3 loading rate was high increased stability within the process and also increased biogas yield. This result was measured by a significant reduction in the volatile fatty acids of the effluent (Kottwitz and Schulte 1982).

4 Prospect of Biogas Technology

Not only does biogas technology meet the thermal energy needs of rural communities, it also has other significant benefits, such as:

- Improved air quality.
- Improved health and reduced respiratory infections.
- Better management of animal dung human excrement and other solid wastes.
- Reduced ground water pollution.
- Reduced deforestation and resultant soil erosion.

- Reduced depletion of solar nutrients.
- Reduced green-house gas emissions.
- The slurry provides an organic fertilizer material and for increased crop production or can be sold to generate income.

5 Materials and Method

The practical aspect of this paper was performed at John Ker Nigeria Organo-Mineral Fertilizer Company. This company is located at No.112B Old Itu Road, Ibiakpan Village, in Ikot Ekpene Local Government Area of Akwa Ibom State.

The major materials used for the practical work include water and a freshly obtained poultry manure from a deep littler system. This manure was collected from Gracib Investment Limited at No. 104B Umuahia Road, Ikot Ekpene Local Government Area of Akwa Ibom State.

6 Apparatus

This included: a drum digester, a weighing balance, a water reservoir, rubber tubes, a ½ in. plastic hose, a bucket, a pair of pliers, spanners, polythene bags, a masking tape and a stirrer (Fig. 1).

Fig. 1 A locally fabricated digester

7 Procedure

The poultry manure was weighed in an 18-l bucket and poured ten times into an empty container. Water was also weighed in the same container and was poured into a larger container containing poultry manure. The stirrer was used to stir the mixture of manure and water thoroughly. The content of the container, which is fresh poultry manure and water in the ratio of 1:1 by volume, was scooped into the digester through the inlet channel. The digester was filled to about 360l with a space called scum to facilitate the formation of the biogas. The cover was tightened to prevent air from entering. The valve was removed and the gas vent was connected through the 0.5 in. plastic hose to the rubber tube through which the gas was to store. The set up was allowed to stand for a period of time for thorough observation and the result is as shown in Table 1.

Experiment 1: 1:180l of waste with 180l of moisture.
Experiment 2: A repeat experiment with same concentration of cow dung.

The experiment was repeated all over with same concentration of manure-water ratio of 1:1.There was also an attempt to compare the temperature of the digester contents with the volume of biogas produced each day. The result is as shown on Table 2 above.

Table 1 Biogas production at 2 days interval for poultry manure

Days	Total volume (m^3)	Cumulative volume (m^3)
2	0.0026	0.0026
4	0.0027	0.0053
6	0.0170	0.0223
8	0.0023	0.0246
10	0.0026	0.0272
12	0.0020	0.0292
14	0.0016	0.0308
16	0.0012	0.0320

Table 2 A repeat experiment using cow dung manure

Days	Volume (m^3)	Cumulative volume (m^3)	Temperature (°C)
1	0.000	0.000	30
2	0.001	0.001	43
3	0.002	0.003	46
4	0.006	0.009	51
5	0.004	0.013	48
6	0.001	0.014	44
7	0.001	0.015	39
8	0.000	0.015	27
9	0.000	0.015	22

8 Analysis and Discussion

The first result shows that the total volume of $0.032\,m^3$ of biogas could be obtained from 180 l the poultry manure within 16 days. This result cannot be standardized because the amount of biogas produced depends on numerous factors like nature of waste, pH value, ambient temperature, loading rate, retention time, and toxicity (Angolidaki and Alung 1994; Hansen et al. 1998). It explains why the graph of volume versus number of days does not give a constant straight line (Fig. 2). The above fact is confirmed by the second result which shows a total volume of $0.015\,m^3$ of biogas with same concentration of cow dung. As a further confirmation this second experiment could produce gas for 7 days only and then it stopped thereafter. The reason is that a season of excessive rainfall had set in during the course of the latter experiment. The extreme wetness of the weather and high humidity hindered continuous anaerobic action. This is a serious pointer to the need for proper insulation of the digester as described in literature.

This second result gives evidence of the dependence of volume of gas on temperature. The maximum volume of $0.006\,m^3$ was given by the maximum temperature of 51 °C and the minimum volume of $0.00\,m^3$ occurred at the minimum temperature of 22 °C. The gas yield with poultry manure is much higher than that with cow dung which confirms the literature.

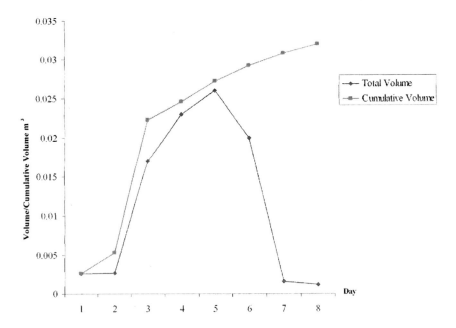

Fig. 2

9 Conclusion and Further Work

With the available equipment in Nigeria, some biogas could be produced with Poultry manure and cow dung. A suitable thermal insulator to shield the digester is a major requirement for consistency in production despite change of climatic conditions. This work is still in progress to discover conditions for maximum yield of the biogas.

References

Engler, R. and McFarland, J. 1997. Diary manure digestion research and demonstration project. In Proceedings of Workshop #1 Livestock Waste Streams: Energy and Environment. Austin, TX: Texas Renewable Energy Industries Association, pp. 64–68.
Hansen, H., Irini A., and Birgitte, K. 1998. Anaerobic digestion of swine manure: Inhibition by ammonia. Water Research 32:5–12.
http://www.afdc.gov/altfuel/nio.geberal.html 2002
http://web1.muse.mus.edu./intext/biogas%20article.htm#Top 2005
http://web.epa.gov/agstar/fraq.html 2005
http://www.habmigern 2003.into/biogas/methane-digester html
Ituen, E., U. U. 2005. Biomass, a potential renewable energy source for Nigeria, Unpublished Public Lecture at Cross River State University of Technology, CRUTECH, Calabar, Nigeria.
Kottwitz, D.A. and Schulte, D.D. 1982. Tumble mix anaerobic digestion of dry beef manure, ASAE paper presented at the 1982 winter meeting, American Society of Agricultural Engineers.
Stout, F. 1979. Energy for World Agriculture F.A.O, Rome.
St. Joseph, M.I., Misra, U., Sonja, S., Amarika, S., and Pandy, G.N. 1992. A new temperature controlled digestion for biogas production. Energy Conservation Management 33:983–986.

West African Gas Pipeline (WAGP) Project: Associated Problems and Possible Remedies

E.O. Obanijesu and S.R.A. Macaulay

Abstract Global focus is gradually turning away from crude oil as a major source of energy to natural gas due to its abundant availability, environmental friendliness and cost effectiveness, this has effectively increased the transboundary pipeline networks with minimal consideration to the impact at which the offshore segment of such projects could have on the environment.

This paper considers Nigeria's present engagement in transboundary transportation of 11.3 billion cubic meters per day (11.3 BCMPD) of natural gas to Benin, Togo and Ghana for thermal and industrial uses through a 1,033 km pipeline network out of which 617 km is a submerged offshore pipeline network. The study is necessitated by the alarming frequency at which hydrocarbon pipeline failure occurs in Nigeria with the resulting economy, environmental and human consequences. It was discovered that any failure along the offshore segment of the pipe-length poses high risk of hydrate formation and dissolution of some constituents which could result to problems ranging from behavioral nature (e.g. fish excitement, increased activities and scattering in the waterbody) to chronic poisoning, fire outbreak, loss of human lives and livestock and climate change.

Development of pragmatic management scheme, robust leak detection model and predictive model on natural gas flow pattern in waterbody are recommended.

Keywords Natural Gas · submerged offshore pipeline · hydrate formation · climate change · ARPU

E.O. Obanijesu (✉)
Chemical Engineering Department, Ladoke Akintola University of Technology,
Ogbomoso, Nigeria.
e-mail: emmanuel257@yahoo.com.

S.R.A. Macaulay
Chemvical Engineering Department, Ladoke Akintola University of Technology,
Ogbomoso, Nigeria.
e-mail: sramac2@yahoo.com

1 Introduction

Global demand for petroleum products is in the increase due to increase in population and industrialization. Oil and natural gas today constitute the most important primary energy sources; their share in total consumption is over 65% (Janovic 2005). Measured in financial indicators, 90% of chemical products in industrially developed countries pertain to organic chemical products, while production based on oil products and natural gas participates in the obtaining of over 98% of basic organic chemical feeds (Janovic 2005). Presently however, global focus is now turning away from crude oil as the major source of energy to natural gas due to its economical and environmental friendliness. This has increased the transboundary pipeline networks with minimal consideration to the impact at which the offshore segment could have on environment.

Natural gas's world proven reserves estimate is put at 207,890 billion cubic meters (bcm) and grows at an estimated 5% annually (Oyekunle 1999). The former Soviet Union (FSU) and the Middle East have 70% of the total gas reserves while the United States of America's natural gas demand for 2003 is put at 786.32 bcm (Hill 2005). Africa's proven gas reserves in 1995 was totaled about 6.3 trillion cubic meters (tcm) with potential reserves estimated at 17.65 tcm in the year 2010.

About 78% of African natural gas resources is deposited at the Niger-Delta region of Nigeria with others at Libya, Algeria, Egypt, Angola, Mozambique, Namibia, Tunisia and Tanzania. This gas which is of low utilization presently in Nigeria is flared off to the environment (Fig. 1), resulting into heating effects, production of noise, soot and particulate at the region, as well as economy wastage

Fig. 1 A gas flaring scene at Niger-delta region of Nigeria

to the country. Current gas flared at the region is estimated at 56.5×10^6 M^3/day (2 billion standard cubic feet per day) which is about a quarter of the gas flared globally and over 50% of the flaring level by OPEC member countries (Obanijesu and Sonibare 2005). To minimize this flaring, the WAGP project was proposed by World Bank to develop the abundantly available natural gas fields in Nigeria.

WAGP project involves the transportation of 11.3 BCMPD (450 MMSCFD) natural gas supplies from Nigeria to power generators and industrial consumers in Ghana, Benin and Togo for thermal and industrial uses through an additional installation of 617 km of offshore pipeline and 57 km of onshore pipeline from an offtake at the existing 359 km Escravos-Lagos natural gas system.

The WAGP project, estimated to cost $550 million includes pipeline installation and metering, pressure regulation, gas scrubbing and compression facilities. The project is undertaken by a joint venture project of NNPC, ChevronTexaco West African Gas Pipeline Co. Ltd, Shell Overseas Holdings Ltd and Takoradi Power Co. Ltd. The transboundary project whose construction work commenced in 2005 is expected to last for minimum of 20 years after takeoff before renewal.

Of the new 674 km pipe-length to be constructed, 57 km of the pipe-length would transport the gas from Alagbado (Lagos) to Badagry beach in Nigeria while 617 km transboundary portion will run offshore from Badagry beach through Cotonou (Benin), Lome (Togo) to Takoradi Power Station (Ghana), which is the final terminal of the pipeline system. Another 80 km offshore pipe-length would be run from Takoradi to Effasu in Ghana.

Specifically, the existing Escravos-Lagos pipeline with 24″ internal diameter has the capacity for about 4.24×10^9 m^3 (2.5 BSCFD) and the tie-in point to the proposed project is at Alagbado West of Lagos with pipeline of internal diameter of 30″. The Alagbado-Badagry would be 30 km onshore and 27 km offshore. This will be an extension of the existing Escravos-Lagos Pipeline System. The onshore portion will commence at Alagbado and terminate at the Atlantic Beach in south of Ajido near Topol Badagry in Lagos State after crossing the Badagry Creek. The Badagry Lagoon Creek and the Atlantic beach crossing will be by horizontal directional drilling or if not feasible, it will be by trenching and burial operations.

The onshore pipeline system will include a mid-line relief system located off the Otta-Idi Iroko highway near Canaan Land in Ogun state. An 18,000-horse power onshore compressor station would be put in place at the Lagos beach from where 20″ by 51 km pipeline is connected and extended 15 km offshore before turning west across the Nigeria-Benin border to Cotonou. While maintaining the 15 km offshore extension, the internal diameter is reduced to 16″ for 276 km to Tema where a further size reduction to 12″ for another 233 km to Takoradi is made (Fig. 2) before another 12″ by 80 km offshore pipeline from Takoradi to Effasu in Ghana is linked. Along base route, additional lateral spurs, averaging 15 km will extend from the base pipeline route to landfall regulating and metering facilities at Cotonou, Lome and Tema. The pipeline system will have laterals of 8″ to Cotonou, 10″ to Lome and 18″ to Tema. At Cotonou, a compression station operating between 40 and 150 MW shall be installed. The Cotonou-Tema pipe-length with the initial cost estimate of $130 million is designed for conveying 2.26×10^9 m^3 (80 MMSCFD)

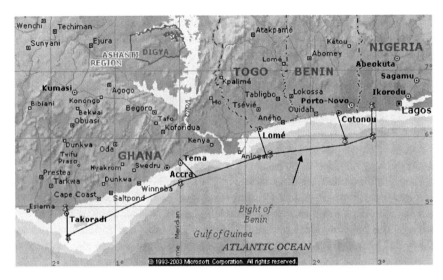

Fig. 2 View of the WAGP project from Nigeria to Ghana with laterals at Cotonou, Lome, Tema and Takoradi

natural gas initially with the ultimate expectation of 3.39×10^9 m^3 (120 MMSCFD).

The ultimate destination in Ghana is Volta River Authority Plant in Takoradi. The pipe-length with the initial cost estimate of $106 million (Tema-Takoradi) is designed to supply 1.70×10^9 m^3 (60 MMSCFD) of natural gas to Takoradi Power Plant with gas demand of 1.22×10^9 m^3 (43 MMSCFD) and the additional Hydro Power Plant in Effasu. In Tema, it is expected to supply VALCO power plant whose gas demand is 0.60×10^9 m^3 (21 MMSCFD). The specific status of each segment is summarized Tables 1–6. The offshore segment of the pipeline is to be submerged inside the water at between 26 and 70 m.

Pipeline transportation of natural gas is globally acceptable. Shell Transmission owns or has an interest in 17 Gulf of Mexico natural gas pipelines in operation or under construction with capacity of almost 9 billion cubic feet per day (Shell 2004). Also, Equitable Gas Company, USA supplies over 250,000 customers in Southwestern Pennsylvania while providing natural gas distribution services to over 260,000 residential, commercial and industrial customers located mainly in the city of Pittsburgh and surrounding municipalities in Southwestern Pennsylvania through pipeline network systems (Equitable 2004). With the successful operation of the existing pipeline networks, more countries are getting involved thereby promoting cross-boarder networking. Saipem s.p.A started building a 370 km, 48 in natural gas export pipeline between Qatar and UAE in 2004. This network was scheduled to be completed in 2006 (Saipem 2004) while Libya equally started constructing pipelines to Egypt and Tunisia for natural gas supply in 2004 (Buddy 2004). Most of these gas projects

Table 1 Benin power plant

Item	Comment
Operator	CEB
Capacity	40–150 MW
Status	Preliminary studies done

Table 2 Segment 2 of the proposed pipeline system (Cotonou to Tema)

Item	Comment
Capacity	80 MMSCFD (2.26×10^{9} ^{3}) (initial)
	120 MMSCFD (3.39×10^9 m^3) (final)
Length	276 km
Diameter	16 in.
Cost	$130 million (initial)

Table 3 Operating capacity of Valco power plant in Tema

Item	Comment
Operator	Valco
Capacity	Up to 100 MW
Gas demand	21 MMSCFD (0.60×10^9 m^3) (max.)
Type:	Combined cycled gas turbine
Status	Under evaluation
Remarks	Valco considering supplemental power to allow them to restart their 25% of capacity that is currently idle

Table 4 Segment 3 of the proposed project (Tema to Takoradi)

Item	Comment
Capacity	60 MMSCFD (1.70×10^9 m^3)
Diameter	12 in.
Length	233 km
Cost	$106 million (initial)

Table 5 Takoradi power plant

Item	Comment
Operator	Volta river authority
Capacity	40 MMSCFD (1.13×10^9 m^3)
Type	Combined cycle gas turbine
Type	Start-up: 2Q 1997
Status	Under construction
Initial fuel	Light crude oil
Financing	World Bank
Remarks	Two-100 MW GE turbine to be installed

Table 6 Tema power plant

Item	Comment
Operator	GNPC
Phase 1	
Capacity	131 MW (simple cycle)
Expanded capacity	200 MW (combined cycle)
Start-up	Funding pending
Phase 2	
Capacity	120 MW

involve offshore transportation. In recent times however, considerable public and regulatory attention is focused on the rate of pipeline failure in Nigeria. Pipeline failure is the failure of the pipe body due to metallurgical or processing abnormalities. Apart from pipeline vandalization, WAGP project is susceptible to failure through corrosion, defect welding, incorrect operation, defective pipe and malfunction of equipment amongst others (Obanijesu et al. 2006). After a failure along any part of the offshore portion of the project, the conveyed fluid escapes into the immediate environment (which is waterbody) to cause hazardous impacts on the ecosystems. Each accident leads to intensive vertical flow of the gas from the leak point to the surface. This is otherwise referred to as vertical dispersal. This vertical dispersal is often affected by the ambient current of the sea.

2 Associated Problems

2.1 Hydrate Formation

Natural gas boils at −162 °C, hence, its release into waterbody results in formation of hydrates. Hydrates are formed during the interaction of many components of the gas with water under certain combination of high pressure and relatively low temperature. They appear in form of crystallized solid (like compressed snow). The hydrate formation sometimes accompanies vertical flow of the fluid resulting to problems ranging from behavioral nature (e.g., fish excitement, increased activity, and scattering in the water) to chronic poisoning depending on the quantity of the gas and the total period of exposure (Patin 2004).

2.2 Dissolution of Component

In its untreated state, natural gas sometimes contains relatively small but undesirable quantities of carbon-dioxide (CO_2) (Table 7), hydrogen sulfide (H_2S), O_2 etc. (Stress 2003), thereby posing high risk to the people around and the living organism in the

Table 7 Molar composition of Nigerian natural gas (Sonibare and Akeredolu 2004)

Chemical compound	Molar composition (%)		
	Kokori field study	Utorogu gas plant	Sapele West field
Methane	68.42	90.19	68.14
Ethane	7.65	6.94	14.22
Propane	11.27	2.09	10.27
N-butane	4.00	0.361	3.23
I-butane	4.42	0.414	2.38
N-pentane	0.94	0.005	0.75
I-pentane	1.55	0.007	1.07
Hexane	0.18	–	–
Nitrogen	0.16	–	–
Carbon-dioxide	1.02	–	–

water. The release of this gas into the waterbody poses high risk of dissolution of acidic constituents which is taken in by the sea animals. Through the bioaccumulation process, man takes in the poisoned fish which could be very detrimental to his health. The acidified water could be taken up by plants within the watercourse through their roots to impair their growth. This dissolution also reduces the reproduction rate in fish which also reduces the availability for human consumption thereby leading to increase in price. Dissolved CO_2 indirectly affects seawater temperature, salinity, ice-cover, turbulence and current while all these abiotic effects have biotic consequences (US DOE 1985).

2.3 Loss of Human Life and Livestock

Due to high pressure of discharge at the point of failure and insolubility of most of the components, the gas is ultimately released to the atmosphere at a particular distance away from the point of discharge either through vertical transport or by water inversion to cause air pollution which could lead to deaths. Typical of this is the Lake Nyos (Cameroon) incidence of August 1986 where an enormous volume of carbon dioxide (CO_2) was released from an underwater pipeline, killing about 1,700 people (Clarke 2001) and livestock up to 25 km away; and Lake Monoun incidence of 1984 where a smaller release of CO_2 killed 37 people (Steven 2000).

2.4 Climate Change

Nigerian natural gas contains between 68% and 90% methane (CH_4) (Table 7) which is insoluble in water. Much of this gas is expected to have been oxidized in the water by bacteria before reaching the water surface (Petit et al., 1999; Dickens

2001) but with the pressure at which the gas is been discharged from the point of failure, this could not be achieved.

With 8.4 years global mean atmospheric lifetime (Prather et al. 1995), 12 years perturbation lifetime (Schimel et al. 1996) and average transport distance by wind in the lower atmosphere of 500–1,000 km per day (NEGTAP 2001), it is easy to see how substantial quantity of CH_4 will be exchanged between countries in Africa.

CH_4 is a greenhouse gas (GHG) which aids climate change. The dangers posed by atmospheric abundance of CH_4 increase from 1,520 ppb in 1978 to 1,745 ppb in 1998 (Prather and Ehbalt 2001) has led to global desired declination which has been observed annually over the last two decades. Global methane production from energy source (in Tg [CH_4/year]) is estimated as 75 (Fung et al. 1991), 97 (Hein et al. 1997), 110 (Lelieveld et al. 1998), 89 (Houweling et al. 1999) and 109 (Olivier et al. 1999). Based on its atmospheric burden and climatic impacts (Alcamo 1994), CH_4 was among the GHGes listed in the Kyoto Protocol. Hence, through an accidental discharge of natural gas through WAGP project, this desire would be jeopardized.

CH_4 oxidizes in stratosphere with OH, Cl and O leading to in-situ source of stratospheric water vapor (Equation (1)).

$$CH_4 + O_3 \rightarrow CH_3^{\cdot} + OH + O_2$$
$$CH_4 + OH \rightarrow CH_3 + H_2O$$
$$CH_3 + Cl \rightarrow CH_3Cl$$
$$CH_4 + Cl \rightarrow CH_3Cl + H^{\cdot} \qquad (1)$$

Oxidation of CH_4 is a source of mid-stratospheric H_2O and currently causes its abundance to increase from about 3 ppm at the tropopause to about 6 ppm in the upper stratosphere. Water vapor in the lower stratosphere is a very effective GHG. Baseline levels of stratospheric H_2O are controlled by temperature of the tropical tropopause, a parameter that changes the climate (Moyer et al. 1996; Rosenlof et al. 1997; Dessler 1998; Mote et al. 1998). Finally, stratospheric ozone depletion increases tropospheric OH (Bekki et al. 1994; Fuglestvedt et al. 1994) which reacts with CH_4 to produce more water vapor. CH_4 also reacts with Cl atoms in the troposphere (Singh et al. 1996).

An assessment undertaken by IPCC (1991) generated some speculations about possible climate change impacts on fish population and aquatic life such as drastic declination in the availability of king crab stock in the eastern North Pacific such as eastern Bering Sea and Gulf of Alaska (Wooster 1982) and the rapid disappearance of California Sardine (Ueber and McCall 1982) amongst others.

2.5 Flammability

Natural gas is 35% lighter than air and therefore any leakage from gas pipeline would pass through the underwater and then dissipate into the air to explode or otherwise burn in the presence of ignition source thus affecting large area due to the resulting fire.

3 Proposed Control Measures

With all these problems resulting from any failure along the pipe-length, the followings are proposed as control measures.

3.1 Development of a Robust Predictive Leak Detection Model

This model should be in place in order to quickly detect and locate any failure along the pipe-length. Various predictive models are available. Watanabe et al. (1993) developed a model capable of detecting any leakage as small as pinhole in a pipeline through sound wave medium while Abhulimen and Susu (2004) developed a predictive model to detect pipeline failure using pressure drop. However, the two models considered liquid hydrocarbon. A related model should be developed with natural gas as the fluid.

NS (2005) developed a natural gas pipeline corrosion predictive model with CO_2 as the corroding agent while Obanijesu et al. (2007) developed a related model with H_2S as the corroding agent. The two models used thermodynamic properties of the fluid to develop a model that could predict the corroding rate of a pipeline per year. As a preventive means, these models are capable of predicting when a pipe should be replaced. Nonetheless, the models made use of some physical data, (e.g. temperature and pressure) while some parameters were taken to be constant over a few kilometers to reduce the iteration of measurement some of which could be costly. Also, the results presented were calculated without full consideration being given to corrosion inhibitors (like the introduction of glycol). A more robust model based on Nigerian gas composition should be developed for this project. Such model should be able to predict when any of the involved pipe-length should be replaced, thus, minimizing such an accident. Where gases such as CO_2, H_2S, O_3, and other organic and inorganic gases are present, these two models could be modified to accommodate the contributions of each agent to corrosion rate of the pipeline. Where inhibitors are present (which is common in pipeline systems) their effects must be evaluated separately.

3.2 Development of Predictive Model on Natural Gas Flow Pattern in Waterbody

Since the failure rate could only be minimized and not fully prevented, there is a dire need for a proactive and a comprehensive study of such scenario to identify the spots of high risk of pollution by preparing spill contingency plans for such accident and an important element in such plan is the use of mathematical models to predict the transport (trajectory and fate) of the gas in waterbody. The model should be developed around the four main mechanism of transport that occur to the leaked gas which are vertical bubble transport (bulk flow), vertical turbulent diffusion, dissolution of components and horizontal advection by ambient currents. Such model should be simulated using an appropriate numerical scheme with full consideration of downstream and upstream boundary conditions. A computer package should be used to solve the resulting algorithm and such model should be made available to the operator's Accident Response Plan Unit (ARPU) in predicting the spread point (location) of the gas with time in the waterbody.

3.3 Development of Pragmatic Management Scheme

A pragmatic management scheme should be developed by the operator. This scheme should include compensation for any affected individual or community in case of accident. Immediate response and instant arrival of ARPU personnel to such point of accident should be considered. For this to work perfectly, the personnel must be well trained and equipped with necessary materials. Inventory of the available materials should be regularly taken (probably quarterly) for immediate replacement. Since offshore problem is being considered, the Unit must also be equipped with helicopters (probably three) in case of multiple accident at the same time.

To ensure effectiveness, various types of insurance schemes be considered for each of the ARPU personnel.

4 Conclusion

During the course of this study, it has been discovered that maximum utilization of the abundantly deposited natural gas reserve in Nigeria through the WAGP project is necessary in order to drastically minimize her global contribution to environmental degradation through flaring. However, in as much as this project is financially and environmentally beneficial to Nigeria as a country, any failure along the offshore segment could be disastrous to Africa as a continent since the resulting accidental discharge could result into loss of lives and livestock and global warming amongst others. It is therefore important to prevent failures along the pipe-length as much as possible. The best possible means of achieving this could be by full commitment of the operators by investing in research and development to develop various relevant scientific models and management schemes.

References

Abhulimen, K.E. and Susu, A.A. (2004), "Liquid Pipeline Leak Detection System: Model Development and Numerical Simulation", Chem. Eng. J., **97**(1), 47–67

Alcamo, J. (ed) (1994), "IMAGE 2.0: Integrated Modelling of Global Climate Change", Spec. Issue Water Air Soil Pollut., **76**(1–2)

Bekki, S., Law, K.S., and Pyle, J.A. (1994), "Effect of Ozone Depletion on Atmospheric CH_4 and CO Concentrations", Nature, **371**, 595–597

Buddy, I. (2004), "BPA Membership", Pipeline and Gas Technology, June 2004

Clarke, T. (2001), "Taming Africa's Killer Lake", Nature, **409**(February), 554–555

Dessler, A.E. (1998), "A Reexamination of the 'Stratospheric Fountain' Hypothesis", Geophys. Res. Lett., **25**, 4165–4168

Dickens, G. (2001), "On the Fate of Past Gas: What Happens to Methane Release from a Bacterially Mediated Gas Hydrate Capacitor?" Geochem. Geophys. Geosyst., **2**(1), 1037, doi: 10.1029/2000GC000131

Equitable (2004), "Equitable Hires Meter-reading Firm", Pipeline and Gas Technology, June 2004

Fuglestveld, J.S., Jonson, J.E., and Isaksen, I.S.A. (1994), "Effects of Reduction in Stratospheric Ozone on Tropospheric Chemistry Through Changes in Photolysis Rates", Tellus, **46B**, 172–192

Fung, I., John, J., Lerner, J., Mattew, E., Prather, M., Steeler, L.P., and Fraser, P.J. (1991), "Three-Dimensional Model Synthesis of the Global Methane Cycle", J. Geophys. Res., **96**, 13033–13065

Hein, R., Crutzen, P.J., and Heinmann, M. (1997), "An Inverse Modelling Approach to Investigate the Global Atmospheric Methane Cycle", Global Biogeochem. Cy., **11**, 43–76

Hill, Z. (2005), "LNG Will Supply More Than 20% of US Gas by 2025", NAFTA Journal, Zagreb, Croatia, Year 56, No. 6, p. 220

Houweling, S., Kaminski, T., Dentener, F., Lelieveld, J., and Heinmann, M. (1999), "Inverse Modeling of Methane Sources and Sinks Using the Adjoint of a Global Transport Model", J.Geophys. Res., **104**, 26137–26160

IPCC (1991), "Assessment of the Vulnerability of Coastal Areas to Sea Level Rise - A common Methodolog", The First Report of Intergovernmental Panel on Climate Change, Cambridge University Press, Cambridge

Janovic, Z. (2005), "Oil and Petrochemical Processes and Products", Hrvatsko, drustvo za goriva I maziva, Zagreb, Croatia, pp. 1–427

Lelieveld, J., Crutzen, P.J., and Dentener, F.J. (1998), "Changing Concentration, Lifetime and Climate Forcing of Atmospheric Methane", Tellus, **50B**, 128–150

Mote, P.W., Dunkerton, T.J., McIntyre, M.E., Ray, E.A., Haynes, P.H., and Russel, J.M. (1998), "Vertical Velocity, Vertical Diffusion and Dilution by Midlatitude Air in the Tropical Lower Stratosphere", J.Geophys. Res., **103**, 8651–8666

Moyer, E.J., Irion, F.W., Yung, Y.L., and Gunson, M.R. (1996), "ATMO Stratospheric Deuterated Water and Implications for Troposphere-Stratosphere Transport", Geophys. Res. Lett., **23**, 2385–2388

NEGTAP (2001) "Transboundary Air Pollution: Acidification Eutrophication and Ground-Level Ozone in the UK", National Expert Group on Transboundary Air Pollution, Department for Environment, Food and Rural Affairs (DEFRA), UK

NS (2005), "CO_2 Corrosion Rate Calculation model", NORSORK STANDARD, Norwegian Technological standards Institute Oscarsgt. 20, Majorstural, Norway, 287–294

Obanijesu, E.O. and Sonibare, J.A. (2005), "Natural Gas Flaring and Environmental Protection in Nigeria", NAFTA Journal, Croatia, Year 56, No. 7, pp

Obanijesu, E.O., Sonibare, J.A., Bello, O.O., Akeredolu, F.A., and Macaulay, S.R.A. (2006), "The Impact of Pipeline Failures on the Oil and Gas Industry in Nigeria", Eng. J. Univ. Qatar, **19**, 1–12

Obanijesu, E.O. and Macaulay, S.R.A. (2008) "Modeling the H_2S Contribution to Internal Corrosion Rate of Natural Gas Pipeline", Energy Sources Part A: Recovery, Utilization and Environmental Effects, USA. In Press

Olivier, J.G.J., Bouwman, A.E., Berdowski, J.J.M., Veldt, C., Bloos, J.P.J., Visschedijk, A.J.H., van der Maas, C.W.M., and Zasndveld, P.Y.J. (1999), "Sectorial Emission Inventories of Greenhouse Gases for 1990 on a Per Country Basis as Well as on 1x1", Environ. Sci. Policy, **2**, 241–263

Oyekunle, L.O. (1999), "Gas Flaring in Nigeria and Environment Pollution Control", Natural Gas: The Energy for the Next Millennium, Proceedings of the 29th Annual Conference, Nigerian Society of Chemical Engineers, Nigeria

Patin, S. (2004), "Gas Impact on Fish and Other Marine Organisms", http://www.offshore-environment.com/gasimpact.html

Petit, J.R., Jouzel, J., Raynaud, D., Barkov, N.I., Barnola, J.M., Basile, I., Bender, M., Chappellaz, J., Delaygue, D.G., Delmotte, M., Kotlyakov, V.M., Legrand, M., Lipenkov, V.Y., Lorius, C., Pepin, L., Ritz, C., Saltzman, E., and Stievenard, M. (1999), "Climate and Atmospheric History of the Past 420,000 Years from the Vostok Ice Core, Antarctica", Nature, **399**, 429–436

Prather, M. and Ehhalt, D. (2001), "Atmospheric Chemistry and Greenhouse Gases". In: Climate Change 2001 – The Scientific Basis, Contribution of Working Group 1 to the Third Assessment Report of the Intergovernmental Panel on Climate Change, Cambridge University Press, Cambridge, pp. 239–287

Prather, M., Derwent, R., Ehhalt, D., Frazer, P., Sanhueza, E., and Zhou, X. (1995), "Other Tracer Gases and Atmospheric Chemistry". In: Climate Change 1994, edited by Houghton, J.T., MeiraFilho, L.G., Bruce, J., Hoesung L., Callander, B.A., Haites, F., Harris, N., and Maskell, K., Cambridge University Press, Cambridge, pp. 73–126

Rosenlof, K.H., Tuck, A.F., Kelly, K.K., Russel, J.M., and McCormick, M.P. (1997), "Hemispheric Asymmetries in Water Vapor and Interferences About Transport in the Lower Stratosphere", J. Geophys. Res., **102**, 13213–13234

Saipem (2004), "Saipem to Lay Dolphin Line", Pipeline and Gas Technology, June 2004

Schimel, D., Alves, D., Enting, I., Heinmann, M., Joos, F., Raynaud, D., Wigley, T., Prather, M., Derwent, R., Ehhart, D., Frazer, P., Sanhueza, E., Zhou, X., Jonas, P., Charlson, R., Rodha, H., Sadasivan, S., Shine, K.P., Fouquart, Y., Ramaswamy, V., Solomon, S., Srivinasan, J., Albritton, D., Isaksen, I., Lal, M., and Wuebbles, D. (1996), "Chapter 2, Radioactive Forcing of Climate Change", In: Climate Change 1995: The Science of Climate Change, Contribution of Working Group 1 to the Second Assessment Report of the Intergovernmental Panel on Climate Change, Cambridge University Press, Cambridge

Shell (2004), "Shell to Sell Gulf of Mexico System", Pipeline and Gas Technology, June 2004

Singh, H.B., Thakur, A.N., Chen, Y.E., and Kanakidou, M. (1996), "Tetrachloroethylene as an Indicator of Low Cl Atom Concentrations in the Troposphere", Geophys. Res. Lett., **23**, 1529–1532

Sonibare, J.A. and Akeredolu, F.A. (2004), "Natural Gas Domestic Market Development for Total Elimination of Routine Flares in Nigeria's Upstream Petroleum Operations" Energy Policy, UK, Vol. 34(6), 743–753

Steven, E. (2000), "Plan for Degassing Lakes Nyos and Monoun", www.lake nyos\Lake Nyos plan for degassing lakes Nyos and Monoun, Cameroon Gas disaster at Nyos mitigation of a natural hazard at Nyos.htm

Stress (2003), "Pipeline Corrosion", Stress Engineering Services, USA

Ueber, E. and McCall, A. (1982), "The Rise and Fall of the California Sardine Empire", In: Climate Variability, Climate Change and Fisheries, edited by Glantz, M.H., Cambridge University Press, Cambridge

US DOE (Department of Energy) (1985), "Characterization of Information Requirements for Studies of CO_2 Effects: Water Resources, Agriculture, Fisheries, Forests and Human Health", DOE/ER-0236, Carbon Dioxide Research Division, Washington, DC

Watanabe, K., Koyama, H., Tanoguchi, H., Ohma, T., and Himmelblau, D.M. (1993), "Location of Pinholes in a pipeline", Comput. Chem. Eng., 17(1), 61–70

Wooster, W.S. (1982), "King Crab Dethroned", In: Climate Variability, Climate Change and Fisheries, edited by Glantz, M.H., Cambridge University Press, Cambridge

Part III
Water Treatment, Purification and Protection

Activated Carbon for Water Treatment in Nigeria: Problems and Prospects

I.K. Adewumi

Abstract A field survey of activated carbon (AC) users and usage in Southwestern Nigeria was made using both purposive and random sampling methods to pick respondents. The survey was to identify sources of AC used, the cost and specifications and relevant information for users. Using predetermined activation conditions, a locally developed furnace was to produce good quality AC from PKS sourced in three different parts of the Rain forest belt of Southern Nigeria under the same conditions used in a standard furnace modified for the process. The results of the survey showed that almost all AC sold in Nigeria are imported from Europe and Asian countries and they have no specifications. There is no evidence of local production of this high-demand engineering material. There was no locational variation in the quality of the carbon adsorbents currently produced which has comparable quality to imported ones. The AC industry is viable, especially from agricultural farm waste including palm kernel shells, which have a 70% yield of AC. There is the need to regulate or standardize AC imported or produced locally. The prospect for local production in developing countries is high and this will be enhanced by the development of equipment and adoption of appropriate production technology.

Keywords Activated carbon · activation furnace · adsorption · carbon yield · waste management

1 Introduction

Activated carbon (AC) is a valuable engineering material for water treatment and material refining. AC has wide acceptance for use because of its relative cheapness and universal adsorptive capacity for majority of impurities over other favoured

---

I.K. Adewumi(✉)
Department of Civil Engineering, Faculty of Technology, Obaferi Awolowo University, Ile-Ife, Nigeria
e-mail: ife_adewumi@oauife.edu.ng or kenadewumi@yahoo.co.uk

adsorbents such as silica gel and molecular sieves. Despite its wide use and ready source of raw materials for its production, AC is imported by most major industries, in particular, from Europe and Asian countries such as Thailand and Malaysia. The high demand, especially from the growing *packaged water industries* (PWIs) that produce treated water in plastic sachets or bottles, has led to the presentation of ordinary charred materials as AC to unsuspecting buyers. This is because there are no ready methods for the determination of the quality, except in few companies such as the beverage and liquor industries, which mostly receive supplies directly from foreign sources.

Literature is replete with studies showing that AC can easily be manufactured from carbonaceous wastes such as palm kernel shells (PKS), coconut shells, corn cobs and stalks (Ogedengbe et al. 1985; Adewumi 1999, 2006; Guo and Lua 2000a–c) bagasse wastes, and most other agricultural and livestock wastes including blood. What is critical to each raw material used as matrix is the carbonization and activation conditions for the production and the medium for such processing (Adewumi 2006).

Obisesan (2004) has shown that there are many palm oil plantations producing fresh fruit bunches (FFBs). These plantations produce as much as 15–18 t per hectare-year of mean (FFBs) of palm fruits, of which 30–50% are of the improved *Dura* type and 30–65% are of the wild (unimproved) *Dura* end up as PKS. The resulting PKS from these was estimated by Obisesan (2004) to be about 64% of the total biomass produced as fruit bunch.

Adewumi and Ogedengbe (2005) have established conditions for the development of AC from PKS at laboratory scale level using a modified electric muffle furnace that allowed the introduction of activating medium through a pipe into a specially fabricated metal crucible. In a recent study Adewumi (2006) developed a solid fuel fired furnace that took cognizance of erratic electricity energy source and also took advantage of readily available biomass. The furnace was lined with lateritic clay and fuelled with charred PKS with inlet that feeds the activation water medium into the material being processed as recommended by Adewumi (1999). The PKS used in the production of AC in the study were sourced from three different locations Ile-Ife (South Western), Uromi (Delta region) and Port Harcourt (South Eastern) along the Rain forest belt of Southern Nigeria. These varied locations helped to determine locational variation in the quality of the carbon adsorbent. The AC produced with the fuel-fired furnace and the laboratory scale AC types had comparable adsorptive quality with a prominently used AC available on the market (Adewumi and Ogedengbe 2005; Adewumi 2006). The adsorptive capacities of the AC samples were tested according to standard methods and each had a minimum iodine value of 1,000 and a surface area of not less than 1,100 m^2/g of AC which are comparable to specifications given in literature (Adewumi and Ogedengbe 2005; Adewumi 2006). There was also no locational variation and significant difference at 95% confidence level in the quality of the AC produced. What is of paramount importance are the processing conditions and the heating method which for PKS has been determined as specified above.

This paper reports the major problems in the production of AC in Nigeria and the prospects of establishing a viable AC production industry not only in Nigeria but the entire West African rain forest belt sub region.

2 Materials and Methods

2.1 Field Survey

As a preliminary stage of the study on the characterization and specification of AC that can be produced from PKS for water treatment, field survey was made of Producers and Consumers of AC in southwestern Nigeria where most of the industry is concentrated. This involved the use of questionnaires and spot interviews of users and usage of AC. The key industries were first identified and then a decision as to the sampling method was taken based on the number of each type of industry. A list of registered PWIs prepared by the National Agency for Foods and Drugs Administration and Control (NAFDAC) was helpful in identifying industries within the study area.

The questionnaire used elicited information on the source and cost of AC purchased, the quantity required for operation per year and the impurities that the AC is expected to remove from the material being treated or processed. The ready availability of the material, whether or not they know other users and producers of AC, and if the AC bought or available came with specifications as to the quality and capacity, grain size, etc. were determined.

The sampling of the major multinational beverage or pharmaceutical industries was purposive because of the scarcity of plants, while the sampling of small scale enterprise establishments such as the PWIs was randomized using the *Epitab-Info 6* method (WHO 1991) to determine the minimum sample population. The list of registered industries in Osun State was keyed into the Excel package and the *RandF* programme was used to identify names with *RandF* values ≤0.33.

The analysis of the questionnaire was completed and the key impurities respondents expected to be removed were identified and used in the determination of the adsorptive capacity of AC that was produced in the laboratory from PKS apart from the standard tests for specification of carbon adsorbents.

3 Results and Discussion

3.1 Field Survey

The results of the field survey showed that AC is in high demand in Nigeria and the mean cost per kg varies between $1.04 and $1.67(₦ 125 and ₦ 200) (Adewumi 2006; Tables 1 and 2). There were 173 registered PWIs and beverage industries in Osun State

Table 1 Average quantity and cost of activated carbon (AC) used in removing specified impurities in sampled industries in Southwest Nigeria

S/No	Major industries using AC in Nigeria	Average quantity needed annually (purposive) (kg)	The mean (and range) of cost of AC (2002) ($:00/kg)	Average quantity needed annually (random) (kg)	The mean (and range) of cost of AC (2005) ($:00/kg)	*Code of impurities reportedly removed with AC
1	PWIs:		1 (0.83–1.67)	350	1.04 (0.83–1.67)	a–h
2	Beverages	2,000		350		a–h
3	Breweries	2,000	1 (0.5–1.67)		1.67 (0.83–2.33)	a, b, c, e, f
4	Pharmaceutical	750	1.67			a, c, d, f, g, h
5	Steel rolling mills	1,500	0.83			c, g, h
		(7.5 m³)	1.25			

*Code of impurities in water/liquor usually removed with AC:

a – acidity

b – chlorine

c – colour

d – heavy metal ions

e – ⁺microbes

f – (non-filterable) turbidity (NFT)

g – odour

h – taste

⁺Microbes are mainly removed with a combination of the ultraviolet light (UV) units and microfilter units in all the packaged water industries (PWIs) visited.

at the time of the survey. Choosing values of z, d and p as being equal to 95%, 33% and 5% respectively the minimum sample was found to be 55. The summary of the responses to the inquiries in the questionnaire is presented in Table 1.

Both the purposive sampling of large scale industrial establishments in four industrial States (Edo, Lagos, Oyo and Osun States) and the randomized sampling of NAFDAC-approved PWIs in Osun State showed that AC is an essential engineering material needed for physical-chemical treatment of water and other liquids by adsorption. The analyses of the questionnaires showed that most of the TWIs have to replace spent AC within 3–5 months.

It is interesting to note that all the AC imported into Nigeria had no specification of iodine number, phenol value, methylene blue value, ash content, and appropriate medium for its use…Even NAFDAC has no specification yet for the types of AC to be used in the country. The only thing closest to a specification involves the requirements for the regulation of registration of PWIs. This is probably similar to requirements in developing countries. The methods used in quality testing in Table 2 showed that the major users have no scientific means of doing this due to lack of awareness or product education.

Table 2 Availability and specifications of AC in Southwest Nigeria

Type of industry	Availability		Specifications	
	Source of AC by industries	Usual weight of package (kg)	Specifications on package	ªCoded method of quality verification
Packaged water industries:				
(i) Small scale	Open market	25	None	0⁺⁺, 2⁺, 3⁺, 5
(ii) Large scale	Open market	25	None	2, 3, 5
Beverages	Imported	20–25	None	4
Breweries	Imported	20	Grain size, density	4, 6
Liquor	–	–		–
Pharmaceutical	Imported	ᵇNS	–	2, 4
Vegetable oils	–	–	NS	–
Others:			–	
(i) Steel rolling mills	Imported/open market	20–25		4

ªCoded methods of in-house verification of quality of AC in sampled industries:
ᵇNS – not specified in the response to the questionnaire
0 – None
1 – Unspecified laboratory tests
2 – Visual assessment
3 – By hand abrasion
4 – Specified standard tests in in-house laboratory of the company
5 – Analysis by appointed consultant
6 – Claimed centralized analysis at the headquarters (of multinational companies) not known to respondent

Degree of responses: Numeral alone implies one or two respondents; '+' as superscript implies *a few* respondents; and '++' implies *most* respondents.

The study also showed that there are no major producers of AC in Nigeria, and even main users of AC only have information on where they buy the material and probably other users but do not know whether or not such AC comes with specification. Evaluation is made visually or taken at the seller's words which experience and the findings presented in the present study have found to be far from the truth (Table 2).

The need for NAFDAC or other regulatory bodies to make it mandatory for producers or sellers of AC to provide specifications of the AC produced or imported into Nigeria cannot be overemphasized. Specifically, the listing of the iodine number, phenol value, grain size ranges, and medium for use must be put on the packaging for all AC whether imported or produced locally. Such simple procedures would enable quality verification by regulatory bodies and users.

For other farm wastes, such as corn cobs and stalks, the conditions for AC production have to be determined through laboratory analyses. Since these farm wastes are virtually free of charge, their conversion to carbon adsorbent adds value to the crops as it also provides a solution to solid waste management and pollution.

The charred PKS used by Adewumi (2006) was procured at about the cost of $0.16–0.25 (₦ 20–30) per kg whereas the raw PKS would cost much less than this or even nothing to a farmer who wants to integrate AC production with oil processing. If the AC produced is then sold at the current market price of $1.04–1.25 (₦ 125–150) per kg, it means the valorization of the erstwhile waste is both a viable venture and a source of employment. The use of PKS as fuel also reduces the cost of heat generation for the activation process. The cleaner environment and prevention of open burning of the wastes most importantly imply a cleaner production method.

In Adewumi (2006)'s study, the PKS and wood ash wastes from the heating of the PKS was used as raw material for extraction of the lye solution used for both the quenching of the charred shells and the activation of the pulverized shells. The need for chemical purchase for the activation is therefore solved within the same production process. Obisesan (2004) had shown that there is as much as PKS waste of 4.5–9.0 t/ha-year. Such quantity can support local AC production at a sustainable level.

The average AC yield producible from PKS in the optimization of AC production in the 2^3-factorial experiment is as shown in Table 3. For the BBC945 type of AC, which was the best in the study and had a 70% yield implies that at least 700 kg is producible from a metric ton of PKS. This aspect of the work is useful in estimating the quantity of AC that can be produced from a given mass of raw or charred PKS. It is also useful to engineers in the design of the furnace for activation of biomass.

With an estimated AC requirement of about 75 kg per filtration column and 150 t of AC per annum in Osun State alone or a conservative 4,320 t per annum in Nigeria. At a conservative selling price of $1.04/kg this volume of wastes would generate an income of $4.5 million from palm produce waste alone, not including other agricultural wastes. Even if a producer has to buy raw or charred PKS that presently costs less than $0.16 per kg, 1.43 kg or say 1.5 kg of the raw or charred PKS will yield 1 kg of the BBC type AC which is at 70% yield. The operating and maintenance costs are not estimated in this present study but both are not expected to cost more than $0.21 per kg. This will leave a net $0.67 per kg as return on investment.

4 Conclusion

Activated carbon is in high demand in Nigeria by almost all industries. The results presented in this paper showed that although AC is not presently produced in Nigeria due to lack of facilities, the production equipment could be produced locally to meet local demands for AC. The AC materials imported to Nigeria presently have no specification and regulatory bodies on health and food are yet to provide information on the minimum standards for AC quality. The AC industry may be a viable industry that will turn current wastes into useful engineering material and provide producers with high return on investment if well managed. There is the need for regulatory bodies in Nigeria to enforce the listing of the specifications for AC on the packaging to facilitate quality control of such product.

Table 3 The effect of temperature, treatment medium and contact time on the yield of AC produced from PKS

AC code[a]	Mass after carbonization (%)	Average mass after carbonization (%)	Mass after activation (%)	Average mass after activation (%)	Average yield of process (%)
ACC520	90.24				
	79.9				
		85.1	–	–	85.1
BCC520	80.41				
	83.86				
		82.0	–	–	82.0
OCC520	81.71				
	89.03				
		85.5	–	–	85.5
BAC715	90.16		70.00		
	82.62		69.84		
		86.4		70.0	70.0
BAC745	91.84		67.14		
	84.38		66.41		
		88.4		66.8	66.8
BAC915	88.84		65.85		
	90.49		68.73		
		89.7		67.3	67.0
BAC945	89.87		58.14		
	89.01		66.23		
		89.4		62.2	62.0
BBC715	87.17		68.26		
	91.27		72.75		
		89.2		70.5	70.5
BBC745	92.03		72.40		
	85.14		70.00		
		88.58		71.2	71.2
BBC915	89.34		69.00		
	88.48		70.76		
		88.91		70.0	70.0
BBC945	90.63		73.72		
	86.55		65.82		
		88.59		69.78	70.0

[a]The codes ACC520, BCC520, OCC520 are PKS carbonized only and quenched in acidified water, alkaline water or ordinary water respectively; BAC715, BBC715 are BCC520 shells activated in acidified steam or alkaline steam respectively at 700 °C for 15 min; 745 ones are at a duration of 45 min; BAC915, BBC915 are the BCC520 shells activated in acid steam or alkaline steam respectively at 900 °C for 15 min; the 945 ones are at a duration of 45 min.

References

Adewumi, I.K. (1999), Development of a Commercial Grade Activated Charcoal Using Palm Kernel Shells, Unpublished M.Sc. thesis, Obafemi Awolowo University (OAU), Ile-Ife.

Adewumi, I.K. (2006), Characterization and Specification of Activated Charcoal Produced from Palm Kernel Shells for Water Treatment, Ph.D. thesis, Obafemi Awolowo University, Ile-Ife.

Adewumi, I.K. and Ogedengbe, M.O. (2005), Optimizing conditions for activated charcoal production from palm kernel shells, *Pakistan Journal of Applied Sciences*, 8(6):1082–1087.

Grayson, M. (ed) (1985), Kirk-Othmer Concise Encyclopaedia of Chemical Technology, Wiley, New York, pp. 97–105.

Guo, J. and Lua, A.C. (2000a), Preparation and characterization of adsorbents from oil palm fruit solid wastes, *Journal of Oil Palm Research*, 12:64–70.

Guo, J. and Lua, A.C. (2000b), Effect of surface chemistry on gas-phase adsorption by activated carbon prepared from oil-palm stone with pre-impregnation, *Separation and Purification Technology*, 18:47–55.

Guo, J. and Lua, A.C. (2000c), Effect of heating temperature on the properties of chars and activated carbons prepared from oil palm stones, *Journal of Thermal Analysis and Calorimetry*, 60:417–425.

Obisesan, I.O. (2004), Yield, the Ultimate in Crop Improvement, Inaugural Lecture Series No. 168, Obafemi Awolowo University, Ile-Ife.

Ogedengbe, O., Oriaje, A.T. and Tella, A. (1985), Carbonization and activation of palm kernel shells for household water filters, *Water International*, 10:132–138.

World Health Organization (WHO) (1991), An Epidemiologic Approach to Reproductive Health, Wingo, P.A., Higgins, J.E., Rubin, G.L. and Zahnizer, S.C. (eds), WHO, Geneva.

Impact of Feedwater Salinity on Energy Requirements of a Small-Scale Membrane Filtration System

B.S. Richards, L. Masson, and A.I. Schäfer

Abstract Many remote communities in both developed and developing countries lack electricity and clean drinking water. One solution, for such communities that rely on brackish groundwater, is a photovoltaic (PV) powered hybrid ultrafiltration (UF)/nanofiltration (NF) or reverse osmosis (RO) membrane filtration system. The system prototype described here can produce between 150–280 L of clean water for each peak sunshine hour, depending on the salinity of the feedwater (1–5 g/L of total dissolved solids (TDS)) and membrane choice. The best specific energy consumption (SEC) for achieving drinking water quality with a salinity of less than 0.5 g/L TDS from 1, 2.5 and 5 g/L salinity feedwater was 1.1, 1.8 and 2.6 kWh/m^3, respectively. Slightly higher feedwaters (7.5 g/L) can be treated with one of the membranes tested, and as long as sufficient power is available for providing an adequate transmembrane pressure. Higher salinities cannot be treated effectively with the current system due to pressure limitations. Energy recovery would need to be investigated in order to achieve a competitive SEC for such high salinity feedwaters.

Keywords Photovoltaic · solar energy · desalination · membranes · nanofiltration and reverse osmosis · submerged ultrafiltration pretreatment

---

B.S. Richards (✉)
School of Engineering and Physical Sciences, Heriot-Watt University, Edinburgh, EH14 4AS, United Kingdom
Centre for Sustainable Energy Systems, Australian National University, Canberra, ACT 0200, Australia
e-mail: B.S.Richards@hw.ac.uk

L. Masson
ESIGEC, Université de Savoie, Le Bourget du Lac 73376, France
Environmental Engineering, University of Wollongong, Wollongong NSW 2522, Australia

A.I. Schäfer
School of Engineering and Electronics, The University of Edinburgh, Edinburgh, EH9 3JL, United Kingdom

1 Introduction

Small-scale membrane filtration systems that are able to be powered from a renewable energy source are of great interest for remote communities in both developed and developing countries. This is underpinned by the fact that an estimated 1.3 billion people do not have access to clean drinking water, while a further 2 billion are living without electricity (UNDP 1998). The overlap between these two groups – living with neither electricity nor clean water – has been estimated at 1 billion people, or 17% of the world's population (Parodi et al. 2000).

The research presented here is an initial effort to characterise the range of feedwater salinities that can be effectively treated using a small-scale membrane filtration system using a variety of nanofiltration (NF) and reverse osmosis (RO) membranes. The hybrid system described here is novel in that it includes an ultrafiltration (UF) pre-treatment stage in order to reduce problems of membrane fouling that have occurred in other small-scale desalination systems (Mathew et al. 2000). The applicability such a similar system to be powered via a renewable energy source – in this case photovoltaic (PV) solar energy – has been demonstrated during a 6 week field trip in Central Australia (Schäfer et al. 2007). While other small-scale PV-powered systems have been presented in the literature and even commercialised (Laborde et al. 2001; Alajlan and Smiai 1996; Keefer et al. 1985; Mathew et al. 2000; Weiner et al. 2001; Al Suleimani and Nair 2000), until now results reporting the performance of such small-scale systems over a wide range of feedwater salinities, ranging from 1 g/L total dissolved solids (TDS) to 35 g/L TDS have not been presented. The application of this technology to portable water supplies as required in various emergency situations demands an understanding of system performance over a range of water supplies as opposed to stationary installations where variations in feed water do occur but not over such a wide range. However, situations where conditions have changed drastically due to a disaster have also occurred, forexample, following a tsunami that has lead to seawater intrusion into the normally brackish groundwater (Vrba and Verhagen 2006).

Furthermore, correlations between the experimental results presented here and the modelling results from the literature (Laborde et al. 2001) can be drawn. Of particular importance when considering renewable energy as a source of power is the specific energy consumption (SEC; units kWh/m^3), which defines the energy consumption required to produce 1 m^3 of clean drinking water. This SEC translates directly into the requires solar panel area and hence capital cost, while in non-renewable energy powered systems energy would be a operating cost. Further performance parameters of interest that are presented in this paper include flux and permeate salinity, recovery and retention, as well as power consumption.

The aim of this project was to investigate the applicability of using photovoltaic (PV) modules for powering a RO desalination system. PV was chosen given its propensity to operate well for over 20 years in harsh, remote environments. The majority of PV-powered RO systems are designed to desalinate seawater, which has a salinity of about 35 g/L TDS, and therefore require very high pressure (40–80 bar) pumps to overcome the natural osmotic pressure of the feedwater. However, a synergistic relationship often exists between lack of fresh surface water (e.g. rivers, rainfall) and the abundance of both solar irradiation and groundwater that is of marginal (0.5–1.5 g/L TDS) or brackish (1.5–5.0 g/L TDS) quality (Schäfer et al. 2007). Therefore, the objective was to determine the performance of PV-membrane system that is designed to desalinate groundwater of marginal and brackish quality without the need for electrical storage (batteries). Due to fluctuations in solar radiation, ROSI needs to be tested over a wide range of operating conditions, and this was performed in the laboratory with water of widely varying salinity.

2 Equipment

Since 2001, this project has developed five prototypes and is in progress of commercialization (Schäfer and Richards 2007). The system described here is from the third project stage prior to the development of a customized pump, which is often the weakest link in such technologies. Figure 1 shows a schematic diagram of the components in the small-scale hybrid membrane desalination system.

2.1 Power Generation and Electronics

Electricity can either be produced by the PV array or converted to DC from an AC power source such as electricity grid or a backup generator providing $230\,V_{AC}$ power. In this paper the pump was powered by DC power supply to provide a stable energy source for experiments, however the system has also been powered by PV panels (Schäfer et al. 2007).

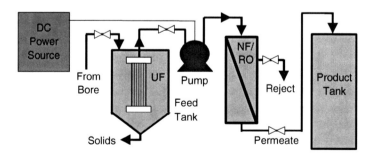

Fig. 1 Schematic diagram of the small-scale UF/NF hybrid membrane desalination system that is designed to be powered by a renewable energy source. The thick lines are water flows, while the thin lines are electrical connections

2.2 Pumps

Initially, feedwater needs to be pumped from the groundwater bore into a 200-L feedwater tank. Both the bore and its associated pump are assumed to already exist at the intended location and the energy requirements of such a pump are not included in the analysis in this paper. Six ultrafiltration (UF) submerged membrane modules (see below) are immersed in the feed tank and water is drawn through the membranes by the progressive cavity pump shown in Fig. 1 (LF502, Mono Pumps, Australia). This pump, which also subsequently feeds the water through the NF or low pressure RO membranes, is of the same type that has been successfully used in PV-powered water pumping systems. The pressure is determined by the length of the rotor and stator, while the flow rate is determined by the motor speed. The LF502 pump is capable of high pressures (up to 24 bar maximum) and delivers a maximum flow of 470 L/h at 15 bar pressure. The pressure was varied from 4 to 15 bar in 1 bar increments at constant feed flow, which naturally results in a increase in recovery and power consumption.

2.3 Membranes

The six UF membrane modules are connected in parallel and submerged in the feedwater tank to form the pre-treatment system (see Fig. 2a). A gentle suction (about 0.5 bar) is applied to draw the permeate through the hollow fibres, which have a nominal pore size of 0.04 μm (Côté et al. 2001). Figure 2b shows of photo of the third prototype of the small-scale hybrid membrane treatment system, prior to its mounting onto a trailer equipped with PV panels and a solar tracker (Schäfer et al. 2007). Experiments were performed with three different low-pressure NF/RO membranes (all 4″ in diameter and 40″ in length) to identify the optimum membrane choice for different feedwater salinities: Dow Filmtec NF90 and BW30 (Dow Filmtec Membranes 2007a, b), and Koch (Fluid Systems) TFC 4920-S (Koch Membrane Systems 2007). All membranes have a maximum flow of 320–380 L/h, and the recommended operating pressure of the two Filmtec membranes was 15.5 bar, while the Fluid Systems membrane was 5.5 bar. According to the manufacturers the NF90, BW30 and TFC 4920-S membranes are able to retain up to 97%, 99.5% and 85% of sodium chloride (NaCl), respectively.

There are several advantages to the hybrid UF and NF/RO membrane configuration:

1. Since the UF membrane is suspended in the feed tank, heavier particulates can sink to the bottom of the tank rather than accumulating on the membrane surface. The feed tank can be periodically emptied to remove any solids.
2. By effectively removing all turbidity and bacteria, the UF membranes assist in extending the life of the NF/RO membrane by supplying it with clean water that contains only salts and trace elements, thus reducing maintenance requirements, fouling, and water cost.

Fig. 2a Six Zenon ZeeWeed 10 UF pre-treatment membranes connected in parallel

Fig. 2b Components of the third prototype of the UF/NF hybrid membrane desalination system: UF membranes in feed tank; LF502 pump; NF membrane; PV array; motor controller; meters for water (pH, conductivity, and temperature) and solar radiation measurements; and computer for data acquisition

3. The UF membrane is able to remove bacteria, cysts, most viruses and other microbiological contents. Thus, for the treatment of low-salinity surface waters where disinfection is the main concern – such as a polluted river – the UF membrane alone may be sufficient. Where desalination is required, the UF and NF/RO membranes are a second barrier to pathogens and this stage effectively removes all viruses.
4. The pressure drop across the UF membrane is small and the same pump that delivers water to the NF/RO module can be used to suck water through the UF module.

5. The UF membranes can be cleaned quite simply, either by a backflush (flow reversal) or by the injection of compressed air at the base of the membranes in order to scour the membrane surface, both requiring further investigations. In terms of chemical cleaning household bleach can be used which is readily available and reasonably cheap.
6. Where applicable, a NF membrane that is designed to operate at lower pressures (5–10 bar) than a RO membrane can be used to reduce the power requirements of the system. However, the salt rejection of the NF membrane is likely to only be sufficient to achieve clean drinking water from brackish water sources.

Further advantages of a hybrid UF and NF/RO membrane configuration were summarized by Redondo (2001) as being: modularity; relative insensitivity to changes in feedwater quality; and lower whole-of-life pre-treatment costs compared to conventional pre-treatment.

2.4 Data Acquisition

Samples were collected at each interval from the feed tank, UF permeate, NF/RO permeate and NF/RO concentrate. Turbidity, pH, temperature, and conductivity were measured immediately, while 25 mL of sample was collected for later elemental analysis. Permeate flux, voltage and current were also measured. Both permeate and concentrate were recirculated into the feed tank.

3 Results

The selection of an appropriate RO or NF membrane depends on the feedwater quality and targeted contaminants that have to be retained. This paper explores a wide range of feedwater salinities, ranging from 1–35 g/L TDS, in essence the range from surface to seawater. Systematic experiments were performed under different conditions to evaluate the performance of ROSI using the three different NF/RO membranes with an emphasis on specific energy consumption. The feedwater used for all experiments in this section was 200 L of tap water with the addition of 1, 2.5, 5, 7.5, 10, 15, 25 or 35 g/L TDS of swimming pool salt, to simulate marginal, brackish and saline water. The first set of experiments consisted of maintaining the feed flow rate constant at 150 L/h and running the system with different operating pressure (4–15 bar) and salt concentrations (1 to 35 g/L). The second set of experiments monitored the performance of the system again, but this time operating at a constant operating pressure (10 bar) and a variable feed flow rate of 150–500 L/h and salinity of 5 g/L TDS. Those variations of flowrate and pressure simulate the energy fluctuations in such a system when operated with renewable energy that is linked directly to the pump (in absence of battery or converter).

Figure 3 plots the performance of ROSI with the Filmtec BW30 membrane, which is described by the manufacturer as a RO element for desalination of brackish water. Several trends are observed in the flux (Fig. 3a) performance as a function of pressure. Firstly, flux increases linearly with the applied pressure for all salt concentrations at low operating pressures. However, at high pressures there is a slight departure from linearity, which can be explained with concentration polarisation (Masson et al. 2005). Secondly, the permeate flux is also shown to decrease with increasing feedwater salinity. This is due to higher salt concentrations resulting in more concentration polarization and hence larger osmotic pressures, which work against the applied pressures, and hence the effective transmembrane pressure is smaller and provides a smaller driving force across the membrane.

The recovery (Fig. 3b) is shown to increase as a function of pressure and to decrease with increasing feedwater salinity, as expected. The retention of the BW30 membrane (Fig. 3c) of all feedwater salinities less than 5 g/L TDS is greater than 90%. This results in a permeate salinity (Fig. 3d) that meets the Australian Drinking Water Guideline (ADWG) value of 500 mg/L TDS (NHMRC 2004). For higher salinity feedwaters retention declines due to the pressure limitation and concentration polarization and in consequence the ADWG can no longer be complied with. Feedwaters of 7.5 g/L require a pressure of at least 8 bar to produce a drinkable product. The progressive cavity pump used in the system has a linear relationship between transmembrane pressure and power consumption (Fig. 3e). Once the power consumption and flux are known, then the SEC (Fig. 3f) can be calculated as the ratio of these two values (units: kWh/m^3). The SEC for the BW30 membrane shows a minimum at low salinity feedwaters due to the lower osmotic pressure and decreases with increasing transmembrane pressures due to higher flux and hence increased recovery. The minimum SEC for 2.5 g/L TDS is 2.1 kWh/m^3 at a pressure of 15 bar and a flow of 280 L/h and producing a flux of 24.4 m^{3}m^{-2}s^{-1}, while 5.0 g/L TDS feedwater had a SEC of and 2.6 kWh/m^3 at a pressure of 15 bar and a flow of 290 L/h and producing a flux of 19.5 m^{3}m^{-2}s^{-1}. Higher salinity feedwater (7.5 g/L TDS) can be treated as long as at least 300 W of power is available to provide a transmembrane pressure of at least 8 bar, however the SEC is increasing significantly due to the limited flux at higher salinities. Figure 4 shows the performance of the system with the Filmtec NF90 nanofiltration membrane. The trends observed for the NF90 membrane are similar to those for the BW30 membrane. Notable differences are the increased flux (Fig. 4a), and hence increased recovery (Fig. 4b), reduced retention at high salinities (Fig. 4c), and hence increased permeate salinity (Fig. 4d) and, due to the higher productivity at identical pressure, reduced SEC (Fig. 4f). When equipped with the NF90 membrane, the system can still satisfactorily desalinate feedwaters with salinities of up to 5 g/L TDS, but a water of 7.5 g/L can no longer be treated. The SEC (2.2 kWh/m^3 at a pressure of 15 bar and a flow of 276 L/h and producing a flux of 25.4 m^{3}m^{-2}s^{-1}) than for the BW30 membrane (2.6 kWh/m^3) which reflects a saving of 20% due to membrane choice. The trends

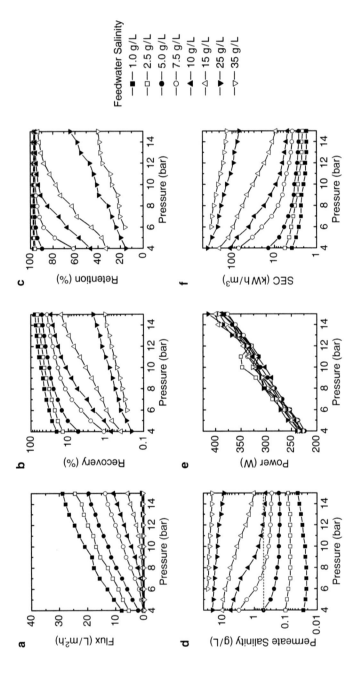

Fig. 3 Performance of ROSI system with UF pre-filters and BW30 membrane for a wide range of feedwater salinities: (a) flux; (b) recovery; (c) retention; (d) the resulting salinity in the permeate stream; (e) power; and (f) specific energy consumption. The Australian Drinking Water Guideline (ADWG) of 0.5 g/L TDS is also plotted in grey in Fig. 3d

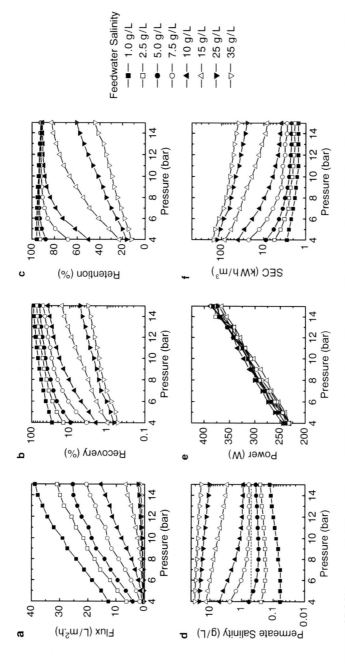

Fig. 4 Performance of ROSI system with UF pre-filters and NF90 membrane for a wide range of feedwater salinities: (a) flux; (b) recovery; (c) retention; (d) the resulting salinity in the permeate stream; (e) power; and (f) specific energy consumption. The ADWG of 0.5 g/L TDS is also plotted in grey in Fig. 4d

observed in system performance with the NF90 membrane become much more obvious with the experiments conducted with the Fluid Systems TFC-4920-S "softening" membrane (Fig. 5). The flux (Fig. 5a) and recovery (Fig. 5b) are both significantly higher, however so is the permeate salinity (Fig. 5d). This results in the system only being able to meet ADWG values for feedwater salinities of 1 g/L TDS which makes this an ideal surface water membrane. The SEC of such waters is then only 1.2 kWh/m^3 at a pressure of 13 bar and a flow of 290 L/h and producing a flux of 39.2 m^{3}m^{-2}s^{-1}. For the experiments where the ROSI system was operated at a constant pressure of 10 bar and feedwater salinity of 5 g/L TDS, the performance of all three membranes is plotted in Fig. 6a–f as a function of feedwater flow.

4 Discussion

The performance of all three membranes used in these experiments was specified by the manufacturer at a recovery of 15% (Dow Filmtec Membranes 2007a, b; Koch Membrane Systems 2007). Often small-scale membrane filtration systems are designed such that the recovery would remain less than about 25% (Laborde et al. 2001; Alajlan and Smiai 1996; Keefer et al. 1985; Mathew et al. 2000; Weiner et al. 2001; Al Suleimani and Nair 2000) in order to prevent concentration polarization and increased fouling of the membrane (Laborde et al. 2001). In addition, recovery depends on the other factors such as, firstly, feedwater temperature, as the flux may increase by 2.7%/°C rise in temperature (Alajlan and Smiai 1996) and, secondly, the age and cleanliness of the membrane. Therefore, even in a system operated at constant pressure and flow, recovery will vary with time. As seen above, the recovery of the membrane modules used in this system varied significantly, due to the nature of the experiments in that pressure was varied in accordance with pump characteristics, and results show that a much broader range of recoveries can be tolerated in terms of water quality. Actual limitations are feedwater dependent as with very high recovery the risk of scaling may increase.

The BW30 membrane has the highest SEC for all feedwater salinities, the NF90 membrane the next lowest, while the TFC-S membrane exhibits the lowest SEC. This reflects the permeabilities of those membranes. However, a low SEC alone is not a good indication of satisfactory system performance, as the permeate salinity needs to be monitored as well. The BW30 membrane is able to desalinate 1–5 g/L TDS feedwater at any operating pressure, while 7.5 g/L TDS feedwater can only be desalinated to within the ADWG value of 0.5 g/L for operating pressures of 8 bar or greater. The NF90 can also satisfactorily desalinate 1–5 g/L TDS feedwater at any pressure, however higher feedwater salinities cannot be treated to meet the ADWG. Finally, while having the lowest SEC, the TFC-S membrane is only able to desalinate the lowest salinity feedwater (1 g/L) at pressures of less than 12 bar – all other feedwaters, the product water still contain more than 0.5 g/L TDS. Therefore, the lower SEC achieved for higher feedwater salinities

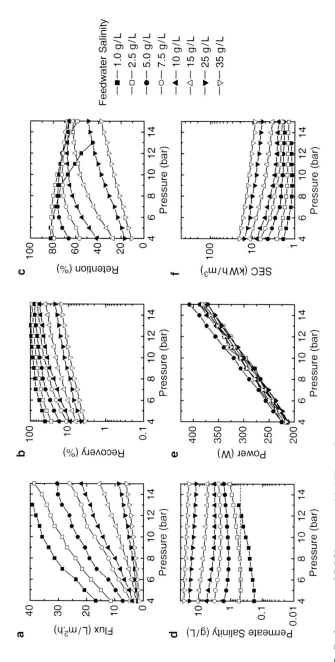

Fig. 5 Performance of ROSI system with UF pre-filters and TFC-4920-S membrane for a wide range of feedwater salinities: (a) flux; (b) recovery; (c) retention; (d) the resulting salinity in the permeate stream; (e) power; and (f) specific energy consumption. The ADWG of 0.5 g/L TDS is also plotted in grey in Fig. 5d

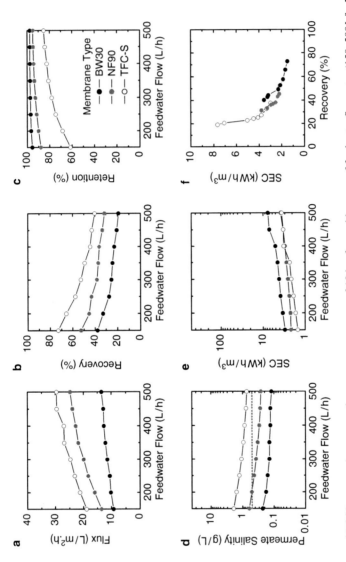

Fig. 6 Performance of ROSI system when operated at a constant pressure of 10 bar for a wide range of feedwater flow rates (150–500 L/h, 5 g/L TDS salinity), which is reflected in the recovery. The different curves correspond to the BW30 (.), NF90 (.), and TFC-4920-S (–) membranes: (a) flux; (b) recovery; (c) retention; (d) the resulting salinity in the permeate stream; (e) specific energy consumption. In Fig. 6f the variation of SEC is plotted as a function of recovery. The ADWG of 0.5 g/L TDS is also plotted in grey in Fig. 6d

– for example, 1.8 kWh/m³ for 5 g/L TDS – is not suitable unless a multi-stage membrane filtration process is considered (Crutcher et al. 1982; Alajlan and Smiai 1996; Al Suleimani and Nair 2000), which is unrealistic in a developing country application thus the TFC-S membrane is most suitable for surface water treatment.

The SEC increases with higher feed salt concentration, while it decreases with increasing applied pressure. There are two competing mechanisms at work here. On one hand, the SEC increases for higher salt concentrations as this increases the osmotic pressure, and thus more energy is required to overcome this natural pressure and to desalinate the water. On the other hand, the SEC is seen to decrease at higher applied pressures. The major effect here is that the permeate flux is increasing more rapidly than the power consumption, while the slight reduction of the salt retention plays a minor role (Masson et al. 2005). This suggests that, the system should be able to be operated at a pressure which produces a maximum permeate flux and minimum SEC, and indeed this is seen for the TFC-S membrane for the lowest salinity feedwater. In general, the SEC of this system is acceptable for feedwater salinities of 7.5 g/L TDS or less.

The increased permeate salinity can be attributed to reduced retention at higher transmembrane pressures, caused by high feedwater fluxes leading to reduced a crossflow velocity as recoveries approach 100%. The increased boundary layer thickness that forms at high feedwater fluxes results in an increased diffusion of salt across the membrane. As discussed previously, operating the membranes at high recoveries is outside the normal operating specification for the membranes, and this may have to be better controlled in the future.

For the experiments performed at constant pressure of 10 bar and feedwater salinity of 5 g/L TDS, the SEC (Fig. 6e) increases at higher feedwater flows. This is understandable, as although a slightly higher flux is achieved at higher feedwater flows, much less water is recovered (see Fig. 6b) and more the power consumption is many times greater (not plotted) and it is highest for the BW30 membrane and lowest for the TFC-4920-S membrane. The good overlap between the three curves in Fig. 6f indicates that the SEC is primarily a function of membrane recovery and the system itself, rather than other effects such as temperature. A further interesting result that can now be seen from Fig. 6d is that the NF90 membrane is only able to meet the ADWG value for permeate salinity at a feedwater flow of greater than 250 L/h. However, it is believed that reduced flux at higher recoveries will result in the permeate salinity *in the product tank* maintaining an average that meets the ADWG value of 0.5 g/L TDS. In contrast, the BW30 membrane is able to produce good quality permeate from the 5 g/L feedwater at any recovery ratio (feedwater flow rate), albeit at a much higher SEC. Therefore, there is a trade-off in the optimum recovery for system design, with lower recoveries being favoured to avoid excessive concentration polarisation (Laborde et al. 2001), however this achieved only at a higher SEC. The results from this work suggest that a small-scale system equipped with the NF90 membrane is quite suitable to being operated autonomously from a fluctuating power source as it keeps the recovery ratio in the range 31–53% for those membranes tested.

5 Conclusions

Extensive optimization was performed on the ROSI system prototype in the laboratory for a wide range of feedwater salinities. It was found that the most appropriate membranes to treat different feedwater salinities are as follows: BW30 membrane for higher salinities (7.5 g/L TDS), the NF90 membrane for medium salinity brackish water (range 2.5–5 g/L TDS), and the TFC-4920-S membrane for marginal (1 g/L TDS) feedwater. Feedwater salinities greater than 7.5 g/L cannot be satisfactorily treated with this small-scale membrane treatment system due to pressure limitations of the pump). The best SEC for 5 g/L TDS feedwater was 2.2 kWh/m^3 using the NF90 membrane (at a pressure of 15 bar and a flow of 276 L/h and producing a flux of 25.4 m^{3}m^{-2}s^{-1}). This value is significantly lower than previous prototypes of the ROSI system (Richards and Schäfer 2002, 2003). It should be noted that no energy recovery system has been used in this system, however this could further reduce the SEC and should be investigated.

Acknowledgments Zenon Environmental (Canada), Dow Chemicals (Australia) and Koch Membrane Systems (USA) are thanked for the provision of membrane modules and/or technical support. Funding for this project is provided through the Australian Research Council Linkage Projects scheme (LP0349322). The authors would like to thank Andrew Moore from Mono Pumps (Australia) for useful discussions and project support.

References

Alajlan, S.A., Smiai, M.S., 1996. Performance and development of PV-plant for water pumping and desalination for remote areas in Saudi Arabia. In: Proc. of World Renewable Energy Congress IV, Denver, Colorado, USA, pp. 441–446.

Al Suleimani, Z., Nair, V.R., 2000. Desalination by solar-powered reverse osmosis in a remote area of the Sultanate of Oman. Applied Energy 65, 367–380.

Block, D.L., Melody, I.I., 1989. Assessment of solar desalination processes. In: Proc. of Solar'89, American Solar Energy Society, Denver, USA, pp. 62–71.

Côté, P., Coburn, J., Munro, A., 2001. A new immersed membrane for pretreatment to reverse osmosis. Desalination 139, 229–236.

Crutcher, J.L., Norbedo, A.J., Cummings, A.B., 1982. Photovoltaic powered seawater desalination systems: Experience with two installations. In: Proc. of 16th IEEE Photovoltaics Specialists Conference, San Diego, CA, pp. 1400–1404.

Dow Filmtec Membranes, 2007a. BW30 technical information. URL: http://www.dow.com/PublishedLiterature/dh_058e/09002f138058eb00.pdf (last accessed 28 Feb 2007).

Dow Filmtec Membranes, 2007b. NF90 technical information. URL: http://www.dow.com/PublishedLiterature/dh_0603/09002f13806035f5.pdf (last accessed 28 Feb 2007).

Keefer, B.G., Hembree, R.D., Schrack, F.C., 1985. Optimized matching of solar photovoltaic power with reverse osmosis desalination. Desalination 54, 89–103.

Koch Membrane Systems, 2007. TFC-S technical information. URL: http://www.kochmembrane.com/pdf/8492000spiral.pdf (last accessed 28 Feb 2007).

Laborde, H.M., França, K.B., Neff, H., Lima, A.M.N., 2001, Optimization strategy for small-scale reverse osmosis water desalination system based on solar energy. Desalination 133, 1–12.

Masson, L., Richards, B.S., Schäfer, A.I., 2005. System design and performance testing of a hybrid membrane photovoltaic desalination system. Desalination 179, 51–59.

Mathew, K., Dallas, S., Ho, G.E., Anda, M., 2000. A solar-powered village water supply system from brackish water. In: Proc. of World Renewable Energy Congress VI, Brighton, UK. pp. 2061–2064.

National Health and Medical Research Council (NHMRC), 2004. Australian Drinking Water Guidelines, National Health and Medical Research Council, National Water Quality Management Strategy.

Parodi, O., Preiser, K., Schweizer-Ries, P., 2000. Clean water with clean energy: A huge market for PV – but not yet explored. In: Proc. of 16th European Photovoltaic Solar Energy Conference, pp. 2997–3001.

Redondo, J.A., 2001. Brackish-, sea- and wastewater desalination. Desalination 138, 29–40.

Richards, B.S., Schäfer, A.I., 2002. Design considerations for a solar-powered desalination system for remote communities in Australia. Desalination 144, 193–199.

Richards, B.S., Schäfer, A.I., 2003. Photovoltaic-powered desalination system for remote Australian communities. Renewable Energy 28, 2013–2022.

Schäfer, A.I., Richards, B.S., 2007. From concept to commercialization: Student learning in a sustainable engineering innovation project. European Journal of Engineering Education 32(2), 1–23.

Schäfer, A.I., Broeckmann, A., Richards, B.S., 2007. Renewable energy powered membrane technology. 1. Development and characterization of a photovoltaic hybrid membrane system. Environmental Science & Technology, 41, 998–1003.

United Nations Development (UNDP) Report, 1998. Human Development Report, Goetzky, Bonn, Germany.

Vrba, J., Verhagen, B.T., 2006. Groundwater for emergency situations – a framework document. International Hydrological Programme VI, Series on Groundwater No. 12, UNESCO, Paris, URL: http://unesdoc.unesco.org/images/0014/001427/142762e.pdf.

Weiner, D., Fisher, D., Moses, E.J., Katz, B., Meron, G., 2001. Operation experience of a solar- and wind-powered desalination demonstration plant. Desalination 137, 7–13.

Modern Technology for Wastewater Treatment and Its Application in Africa

D. Thierno and S. Asplund

Abstract Water covers 71% of the earth's surface and makes up 65% of our bodies. We all want clean water to drink, for recreation, and just to enjoy looking at. If water becomes polluted, it loses its value economically and aesthetically, and can become a threat to our health and to the survival of the aquatic being living in it and the wildlife that depends on it.

The main reason of water pollution is human activities. In Africa water is used daily in homes and industries; about 700 m^3 per person/year mainly from rivers and boreholes. This is well below the internationally fixed water poverty level of 1,000 m^3 per year per person. Industrial demand for water by the year 2017 is expected to reach 5.5 billion m^3 per year. After it is used and contaminated most of it returns to the closes river. The used water of a community is called wastewater, or sewage. If it is not treated before being discharged into waterways, serious pollution and persistence of malaria and related sicknesses are the results. Water pollution also occurs when rain water runoff from urban and industrial areas and from agricultural land and mining operations making its way back to receiving waters and into the ground.

In this article common water pollutants in Africa are considered and their environmental and health impacts are shortly discussed. This will be followed by a fast review of the water condition in the countries. Finally modern technology for wastewater treatment using both end-of-pipe solutions and Cleaner Production approaches will be considered, and examples of their technical and economic performance will be given.

Keywords Modern · waste · water · biofuel

D. Thierno (✉)
BBA and BB in Systems of Telecom, Sales Manager Preseco Oy, Africa and Middle East
e-mail: thierno.diallo@preseco.eu

S. Asplund
Ph Dr Physics, Mg. D. Preseco Pomiltek Oy
e-mail: Staffan.asplund@preseco.eu

1 Introduction

The main reason of water pollution is human activities. In Africa water is used daily in homes and industries; about 700 m^3 per person/year mainly from rivers and boreholes. This is well below the internationally fixed water poverty level of 1,000 m^3 per year per person. Industrial demand for water by the year 2017 is expected to reach 5.5 billion m^3 per year. After it is used and contaminated most of it returns to the closes river. The used water of a community is called wastewater, or sewage. If it is not treated before being discharged into waterways, serious pollution and persistence of malaria and related sicknesses are the results. Water pollution also occurs when rain water runoff from urban and industrial areas and from agricultural land and mining operations making its way back to receiving waters and into the ground.

Current economical development plans of certain African countries have shown that more residential land will be reclaimed and more industrial areas will be constructed. Consequently, this will add more drinking water scarcity and pollutants to the existing non-treated system if the effluents from these areas are not treated. The sick condition of the Nile, the Niger, the Senegal, the Zambezi, the Orange, the Volta, the Congo rivers etc. will be quickly transferred to those who are using its water, and a full range of water-borne diseases are a result from this situation. Dumping raw sewage in these rivers (or irrigation canals) can result in diarrhea, cholera, yellow fever, and is the major cause of infant mortality in the developing world. Viral A hepatitis, and of even more danger Viral E hepatitis, can also result from contamination of surface and ground waters. Furthermore, contamination with industrial effluent can result in many different forms of cancers including leukaemia and hepatomas, which have been on the rise in Africa in the past few years,[1] to mention but a few consequences.

The Nile as an example: Preliminary assessment results of a study on costs and water quality of the Nile claimed that treatment at-source will not be cost-effective if made only for some of the point sources along rivers. On the other hand, if made for all point sources, treatment at-source will be very costly.[2] On this basis, the authors suggest that the standards for the pollutants' concentrations in the Nile River may be reviewed, allowing the Nile to be more polluted. Obviously, this conclusion is difficult to accept as it gives green line to industries to pollute and to authorities to relax their measures and efforts to enforce the regulations. The costs used in the study are also questionable, as they depend on the classic end of-the-pipe solutions available at that time.

In order to arrive at reliable and applicable results, we have to use modern technology, particularly those depending on Cleaner Production, and a cost-benefit analysis instead of the cost-effective approaches the authors mentioned. The statement that 'treatment at the source will be very costly' should be questioned by asking 'with relative to what?' to the life of the people being endangered by the pollution?

After considering the types of pollution and the current state of pollution of the River Nile, this article will provide examples of using cost-effective modern technologies for sustainable equipment.

1.1 Definition of Water Pollution

Water pollution is "contamination of water by undesirable foreign matter." It has impacts on our sea, our surface, and our underground water. Pollution comes in many forms some conventional as ammonia, BOD, nitrogen and nitrate, pathogens, phosphorus and suspended solids, and other toxics as cadmium, copper, lead, mercury, phenol and total residual chloride. A more convenient way is to classify the pollutants with respect to the method of treatment.

Thus the type of pollution could be:

1.1.1 Microbiological

Disease-causing (pathogenic) micro-organisms, like bacteria, viruses and protozoa that can cause swimmers to get sick. Fish and shellfish can become contaminated and lead to some serious diseases like polio and cholera.

1.1.2 Chemical

A whole variety of chemicals from industries, such as metals and solvents, and even chemicals which are formed from the breakdown of natural wastes (ammonia, for instance) are poisonous to all aquatic life. Pesticides used in agriculture and around the home insecticides for controlling insects and herbicides for controlling weeds are another type of toxic chemical.

1.1.3 Oxygen-Depleting Substances

Many wastes are *biodegradable*, that is, they can be broken down and used as food by micro-organisms like bacteria. We tend to think of biodegradable wastes as being preferable to non-biodegradable ones, because they will be broken down and not remain in the environment for very long times. Too much biodegradable material, though, can cause the serious problem of *oxygen depletion* in receiving waters.

Like fish, *aerobic* bacteria that live in water use oxygen which is *dissolved* in the water when they consume their "food". But, oxygen is not very soluble in water. Even when the water is saturated with dissolved oxygen, it contains only about 1/25 the concentration that is present in air. So if there is too much "food" in the water, the bacteria that are consuming it can easily use up all of the dissolved oxygen, leaving none for the fish, which will die of suffocation. Once the oxygen is gone (depleted), other bacteria that do not need dissolved oxygen take over. But while *aerobic* micro-organisms those which use dissolved oxygen convert the nitrogen, sulphur, and carbon compounds that are present in the wastewater into odourless and relatively harmless *oxygenated* forms like nitrates, sulphites and carbonates, these *anaerobic* micro-organisms produce toxic and smelly ammonia, amines, and sulphides, and flammable methane (bio gas).

1.1.4 Nutrients

The elements phosphorus and nitrogen are necessary for plant growth, and are plentiful in untreated wastewater. Added to lakes and streams, they cause nuisance growth of aquatic weeds, as well as "blooms" of algae, which are microscopic plants. This can cause several problems.

Suspended substances increase the turbidity and decrease the clarity of water. Pollutants, referred to as *particulate* matter, consist of very small particles which are just *suspended* in the water. Although they may be kept in suspension by turbulence, once in the receiving water, they will eventually settle out and form sediment or sludge at the bottom. These *sediments* can decrease the depth of the body of water. If there is a lot of biodegradable organic material in the sediment, it will become anaerobic and contribute to problems mentioned above. Toxic materials can also accumulate in the sediment and affect the organisms which live there and can build up in fish that feed on them, and so be passed up the food chain, causing problems all along the way. Also, some of the particulate matter may be grease – or be coated with grease, which is lighter than water, and float to the top, creating an aesthetic nuisance.

1.2 Case Study

Good numbers of African countries have not sewerage network, movable wastewater treatment plant and event don't know what does it mean treating wastewater. It will be then very difficult to study the sources of pollution in all African countries in this paper work at once, thus as a very advanced country in this area and heavily industrialised among many African countries, Egypt can be taken as a case study.

1.3 Sources of Water Pollution in Egypt

Degradation of Water quality is a major issue in Egypt. The severity of present water quality problems in Egypt varies among different water bodies depending on: flow, use pattern, population density, extent of industrialization, availability of sanitation systems and the social and economic conditions existing in the area of the water source. Discharge of untreated or partially treated industrial and domestic wastewater, leaching of pesticides and residues of fertilizers; and navigation are often factors that affect the quality of water. Generally, the poorest water quality exists in the rural areas.

1.3.1 Industry

Industries discharge a variety of toxic compounds and heavy metals, and industrial process wastewater may also be too hot or too low in dissolved oxygen to support life. Silt-bearing runoff from construction sites and farms can inhibit the penetration of sunlight through the water column, and hampers water organisms in their search for food.

The industrial sector in Egypt is an important user of natural resources and a contributor to pollution of water and soil. There are estimated to be some 24,000 industrial enterprises in Egypt, about 700 of which are major industrial facilities. The manufacturing facilities are often located within the boundaries of major cities, in areas with readily available utilities and supporting services. In general, the majority of heavy industry is concentrated in Greater Cairo and Alexandria.

1.3.2 Municipal Wastewater

Based on the population studies and rates of water consumption, the total wastewater flows generated by all governorates, assuming full coverage by wastewater facilities is estimated to be 3.5 km^3/year. Approximately 1.6 km^3/year receives treatment. By the year 2017, an additional capacity of treatment plants equivalent to 1.7 km^3/year is targeted.[3] In Greater Cairo and other cities, the sewerage systems also serve industrial and commercial activities. Therefore, instances of high levels of toxic substances in wastewater have been reported. Improper sludge disposal and/or reuse may lead to contamination of surface and ground water. In general, the bulk of treated and untreated domestic wastewater is discharged into agricultural drains. Total coliform bacteria reach 10^6 MPN/100 ml as recorded in some drains of Eastern Delta.[4]

Wastewater treatment is always cost-effective on the social basis, if we take into consideration the cost of health care and environment degradation. Even on the individual scale of the industrial establishment, wastewater treatment is cost-effective when we apply modern technology solution as Cleaner Production (CP). This will be demonstrated in the following table.

Comparative cost to health problems caused by wastewater

Population × 1,000	Quantity of waste m^3/day × 100	Type of health problem	Cost of treatment plant/000€	Cost of medical care
10	15	Cholera, leukaemia, hepatomas, malaria, cancer, etc.	20–300	€ + death
20	30	Cholera, leukaemia, hepatomas, malaria, cancer, etc.	50–1,000	€ + death
100	150	Cholera, leukaemia, hepatomas, malaria, cancer, etc.	1,000–2,000	€ + death
200	300	Cholera, leukaemia, hepatomas, malaria, cancer, etc.	250–5,000	€ + death

1.4 Modern Wastewater Treatment Technologyv

1.4.1 Prevention Measures

The first key word in modern treatment technology of wastewater is prevention of pollution before it occurs. This applies to both communal and to industrial wastewater and could be accomplished through upgrading of water usage policies and industrial processes, replacement of polluting materials and chemicals by other substances, which have more friendly impact on the environment and using more efficient equipment which conserve water consumption. It should be remembered that decreasing the amount of water used does not only decrease the cost of treatment, but it decreases also the pumping cost, hence smaller electricity bill.

Industries have implemented a wide variety of pollution prevention measures, and a large number of successes have been documented in recent years. According World Bank report published in 1997, the adoption of cleaner production technologies in industry can reduce or even eliminate the need for investment in end-of-pipe treatment technology. As a rough guide, 20% to 30% reductions in pollution can often be achieved without requiring any capital investment, and additional reductions of 20% or more can be achieved with investments that have a payback period of only a few months. Such efforts do, however, require continuous managerial attention.

Sometimes all what is needed in CP application is changing the practice. A good example is the case of the slaughterhouse process changes to reduce biological oxygen demand (BOD) in Bosnia and Herzegovina.[5] The main environmental issues associated with meat processing are high consumption of water and discharge of effluents with high pollutant loads containing blood, fat, manure, undigested stomach contents, meat and meat extracts, dirt and cleaning agents. New practices included extension of bleeding time, construction of a blood collection system, and the introduction of controlled composting of manure, instead of release into the river. Cattle were not fed before slaughter to reduce undigested food. Hoses were fit with nozzles and drains were fitted with traps to prevent solids from entering the effluent. Annual Savings in this case were 15% reduction in water consumption ($274\,m^3$/year), 60% reduction in salt consumption (1.8 t/year), and 42% reduction in BOD ($1,468\,mg\ O_2$/l). The annual savings were US$ 913/year; whereas the investments did not exceed US$56/year. The payback period was thus less than 1 month.

1.4.2 On-site Treatment Combined with CP Application

After having ensured that pollution is prevented to the largest possible extent, on-site treatment would be often needed. The type of the on-site treatment depends on the wastewater quality and the type of pollutants. On the other hand the most common CP applications that are combined with this type of treatment is recycling of the water and/or the pollutants. Recycling allows water to be reused, which reduces costs of purchasing, treating, and disposing of water. Recycling systems also collect contaminants such as oil and heavy metals so that they do not enter soil and ground water.

In addition, water recycling for industry and agricultural uses could sometimes result in unexpected savings. A typical example is the case of paper industry, food industries in Finland, following the application of water regulations. These industries had to decrease the suspended matter (SM) in the wastewater from 10% to 3%. This increased the productivity by 10%, and at the end meant economic profit in spite of the installation of the wastewater treatment facility.

Before recycling one should use water conservation measures, to determine how to reduce the amount of wastewater generated by making changes in the operating procedures. You can save money by conserving water. A variety of systems are available, utilizing processes such as flotation, ultra filtration, electro coagulation, and carbon filtration. Most systems also include gravity settling, oil-water skimming, and chlorination and/or aeration.

As specialist in the field of environment, Preseco Oy offers new and cost effective technologies for pollution abatement, among others: Bio Gas Plan, Land Fill Technologies, Composting Technologies, Soil Treatment Technologies, Water and Wastewater treatment Technologies, Bio Fuel Technologies.

2 Waste Handling

2.1 The Process of the Preseco Biogas Plant

2.1.1 Waste Reception, Pre-processing and Sorting

Mixed waste, for example, is separated from organic waste and the appropriate pulp for the bioreactor is obtained. Separated waste can be collected, if required.

2.1.2 Hygienization

The pulp is heated in order to destroy microbes that are harmful to human beings. This also breaks up the structure of the pulp and facilitates the activity of the anaerobic microbes in the bioreactor.

2.1.3 The Bioreactor and ADA

These together build the core of the process. Here, anaerobic microbes break down the biowaste and turn it into biogas and nutrient-rich sludge. The sludge in the bioreactor is recycled between itself and the pressurised ADA-reactor in order to achieve the optimal process balance.

2.1.4 Biogas Treatment and Energy Production

The biogas is dried and cleaned of harmful gases. After storage, the clean gas is pumped to be used either as heating fuel or for an electric generator. Also, heat is always obtained as a by-product of electricity production. The digestion residue is obtained straight from the bioreactor. This very nutrient-rich fluid is dried mechanically. The separated solid humus is transferred into aerobic end stabilization and the water is partly returned to the pre-processing phase. The excess water is treated in a waste water treatment plant that Preseco supply.

2.1.5 The Preseco Accelerated Drum Composting Process

When the biowaste reaches the plant, support material is added to it. This support material is needed to add carbon to the process and to ensure an optimal structure of the mixture in terms of air and humidity. The mixing of waste and support material is an automated process. The biowaste pulp is fed into an aerated composting drum, where temperature rises to 50–70 °C as a result of microbial activity. Air is constantly blown into the drum, and its amount is regulated by monitoring exhaust gases and process temperature. The composted, ready pulp is transferred automatically into the storage unit. The process is continuous. The heat generated in the process can be recovered and used for heating, drying or producing hot water, for example. The end product is clean and hygienic. During storage it will mature and cool down becoming odour-free. It is suitable as a soil conditioner, among other applications.

2.1.6 The Process of the Preseco Landfill Technologies

(a) Closing and post-treatment
After active operation landfill sites are closed and post-treated in order to prevent harmful emissions into atmosphere and discharges into soil, surface waters and groundwater.

(b) Capping
Landfill sites are capped with impermeable soil cover or plastic liner in order to minimize precipitations entering into the landfill body and leachate production. In addition, top cover helps to keep the recovery rate of landfill gas high and odour problems away. Suitable top cover is also needed for vegetation. There will be a wastewater treatment plant installed in order to complete the ecosystem recovery. Landfill gas production continues decades after closing a landfill site.

Modern Technology for Wastewater Treatment

3   Preseco Water and Wastewater

3.1   *Preseco Wastewater Treatment Process*

Man-made water purification is always a combination of mechanical, biological and chemical treatments. As knowledge and know-how develop, biological purification gains ever more importance and plays an increasingly significant role. Just like in nature and with as few chemicals as possible if not with none.

3.1.1   Process of Wastewater Treatment

The first phase of the process is to remove solid matter from the incoming sewage or water. It is removed with a mechanical screen that has a screening dimension that is small than the size of the solids. From the screen the water is led to sand and grease separation. The screen options are:

- Step screen
- Bar screen
- Drum screen
- Screw screen

Solid matter separated by the screen is washed and dewatered to remove bad odours. In the screening press the water is separated from the washed solids by pressure. The water is led back to the process. The screening washing and drying options are:

- Screening washer
- Screening press

In sand and grease separation the sand settles to the bottom and the grease is floated to the surface by aeration. The sand is pumped to sand classifier and the grease is scraped from the surface and taken to grease/water separation unit. The separated water is led to the beginning of the process. The options for the construction are:

- Aerated sand removal unit of concrete
- Aerated sand removal unit of steel structure

The sand and water mixture from the sand removal is separated in the sand classifier equipped with inclined screw conveyor. The sand settles on the bottom of the classifiers and is pulled up by the screw along the through. The water is led to the beginning of the process. Equipment used:
- Sand classifier

Phosphorus is removed by dosing aluminium salt, iron salt or lime to precipitate the phosphate and separating the insoluble compound by sedimentation, filtration or flotation. Chemical solution preparation equipment:

- Automatic dosing systems
- Containers for reagents
- Silos for reagents

Organic matter related to BOD and COD is removed in the activated sludge process.

Nitrogen is removed in nitrification denitrification process. Phosphorus is also removed in biological P-removal process.

The incoming waste water is mixed with return activated sludge and is lead to mixed anaerobic part, if biological phosphorus removal is required, then to mixed anoxic denitrification. In anoxic phase the microbes oxidize ammonium-N with help of organic carbon from incoming water and recycled nitrates from aerated phase to nitrogen gas and the nitrates are also reduced to nitrogen. Nitrogen as gaseous substance is released to air.

In the aeration phase bacteria oxidize ammonium nitrogen first into nitrites and thereafter to nitrates. The aeration keeps the dissolved oxygen level required by aerobic bacteria.

Organic substrate is utilized by the bacteria and even colloidal and soluble material is rendered in solid form as bacteria mass that is able to be separated from the water by sedimentation or flotation and filtration. The biomass collected from the clarification is taken back to the beginning of the process as return activated sludge and part of it is taken to sludge handling as excess sludge. The biologically removed P is taken out of the process with excess sludge. Equipment used:

- Circular clarifiers
- Rectangular clarifiers
- Flotation unit

3.1.2 Tertiary Process

If effective removal of phosphorus is required, it is made by using chemical precipitation as tertiary phase. The reagents are dosed, mixed and flocculated and the resulted insoluble phosphorus (P) precipitate is removed either in tertiary clarifier, flotation or sand filter. The removed sludge is taken from to sludge handling. Reject water is taken to the beginning of the plant.

- Chemical dosing equipment
- Sand filters, conventional and continuous models
- Flotation units
- Tertiary clarifier with flocculation unit

3.1.3 Sludge Handling

The excess sludge is first thickened in gravity thickener or as conditioned with polyelectrolyte in pre-dewatering drum. The conditioned thickened sludge is dewatered in filter belt press to solid sludge cake and the reject water is led to the beginning of the water process. The sludge cake is further able to be utilized as soil improvement material, humus source in landscaping after composting. Thickening equipment:

- Pre-dewatering drum
- Filter belt press
- Sludge conveyors
- Sludge silos

(a) Latest development on flotation process

The use of Preseco's floatation in the University of Kuopio in Finland has been proven to be able to *remove 99.9% of cells and viruses such as enteric bacteria and coliphase viruses etc.*

3.2 Clean Water with Reversed Osmosis

Water is forced under pressure through a membrane in the opposite direction from the salty side to the salt-free side. Reversed osmosis is mostly utilised on industrial scale in the purification of municipal and industrial wastewater, drinking water, landfill runoff and other heavily loaded waters. As a result we get drinkable water. It has a modular structure, consisting of small units. The equipment can be fitted into a building or installed in an easily movable container. Only water molecules can fit through the organic membranes that we use.

3.3 Biodiesel

Biodiesel (FAME/Fatty Acid Methyl Ester) can be produced from all vegetable oils and nearly all animal oils. Even used cooking oils can be utilized for biodiesel production.

FAME is formed from oil, methanol and catalyst by transesterification reaction and this is called Chemical Conversion of Biodiesel.

Biodiesel is a clearly more environmentally friendly and less pollutant alternative compared to fossil diesel fuel. Using biodiesel reduces significantly unburn hydrocarbons (−93%) as well as carbon monoxide (−50%) and particulate (−30%) emissions. Vehicle Sulphur emissions are eliminated totally and cancer causing PAH and nitrated PAH compounds are reduced up to 90%.

4 Conclusion

When several polluting institutions establishments are close to each others as the case in Industrial cities, agricultural areas, residential areas the wastewater from these entities could be collected and treated together in one plant. This reduces the total costs substantially, provided that wastewaters are of similar qualities. The reuse of valuable waste substances and treated water will reduce the total cost of the treatment plants.

If adjacent plants are planed simultaneously, the possibility of constructing integrated systems is increased. In this case industries are selected I n such a way that the wastes of one industry could be used as input material to another industry. In rural Asia, integrated systems are an old concept that has been applied for hundreds or probably even thousands of years. In China, for example, there are huge farms that are almost completely self-sufficient in terms of energy and nutrients because of the effective recycling of their waste streams. The application of integrated concepts provides a good balance between resource use and reuse and environmental protection.

References

1. Nadia El-Awady, The Nile and Its People: What Goes Around Comes Around. Islamonline, 27 May 2004.
2. Said, A. et al., Analysis of Nile water pollution control strategies: a case study using the decision support system for water quality management. Proceedings of the 2nd Inter-Regional Conference on Environment-Water, Lauzanne, Switzerland, September 1–3, 1999.
3. National Water Research Center of Egypt (NWRC), Research Institute for Groundwater (RIGW), 2002. "Groundwater Quality Status" Baseline Report.
4. Ministry of Water Resources and Irrigation of Egypt, "Survey of Nile System Pollution Sources" Report No. 64. September 2002.
5. Thad Mermer, "Slaughterhouse process changes to reduce biological oxygen demand (BOD)". Bosnia and Herzegovina, 1998. http://www.cema-sa.org
6. Preseco Oy, 2002, Finland. www.preseco.eu. (info@preseco.eu)

Ultrafiltration to Supply Drinking Water in International Development: A Review of Opportunities

J. Davey and A.I. Schäfer

Abstract One of humanities biggest problems at present are millions of preventable deaths in developing countries. Most of those deaths are caused by microorganisms, often from sewage contaminated drinking water. Hence, technology to remove such contaminants is a first step to solving the problem. One such technology is ultrafiltration (UF). UF is a membrane filtration process in which water is pushed through a physical filter with a transmembrane pressure supplied by a pump or gravity. The pore size of such membranes is such that bacteria and most viruses can be effectively retained. As a consequence, this process has the ability to disinfect water physically and hence prevent water related disease and death from microorganisms. In this paper the performance of existing UF membranes and systems will be reviewed in terms of pathogen removal, water productivity (system capacity and flux), specific energy consumption per volume of water produced, which affect cost. Specific needs of systems to be installed and operated in developing countries as well as opportunities for the global community will be outlined.

Keywords Ultrafiltration · developing countries · decentralised treatment · renewable energy · pathogen removal · water supply

1 Introduction

Water related problems are increasing around the globe, with regard to both quantity and quality. In the current international Decade for Action 'Water for Life' (2005–2015) (UN 2006) and the United Nations Millennium Goals (UN 2005)

J. Davey and A.I. Schäfer(✉)
School of Engineering and Electronics, The University of Edinburgh, William Rankine Building, The King's Buildings, Edinburgh EH9 3JL, United Kingdom
e-mail: Andrea.Schaefer@ed.ac.uk

policy makers, practitioners and researchers are searching for ways to address the problems of millions of water related deaths each year, most of which affect children (Gleick 2002). In developing countries water and sanitation infrastructure is still lacking which is the major cause for this problem. While technology has long been available, common obstacles are 'lack of investment, lack of political will, and difficulty in maintaining services' (Montgomery and Elimelech 2007). Many communities in developing countries are vastly lacking infrastructure such as water pipes, an electricity grid, or access to a knowledge base to design, build and maintain treatment facilities, this poses a significant challenge. In now developed countries similar conditions prevailed prior to industrialisation and the construction of a sewer and water supply infrastructure, commonly referred to as public health engineering (Strang 2004). This infrastructure remains in service today, although not without problems and in many remote locations in developed countries (such as aboriginal communities in Central Australia, Islands in the Mediterranean, some rural areas in Europe) remain without access to safe drinking water and or rely on expensive water delivery by truck or boat.

A secondary problem after microbiological pollution is the presence of natural or man-made contaminants such as arsenic, selenium, uranium, fluoride, nitrate or boron, which often result in crippling health effects or death due to chronic exposure (Schwarzenbach et al. 2006). Those compounds require advanced treatment technologies for reliable continuous removal as opposed to conventional methods such as sand filtration.

In developed countries, micropollutants such as endocrine disrupting chemicals and pharmaceuticals have become a concern, often also due to sewage contamination of drinking water sources. This has resulted in a questioning of the suitability of existing infrastructure (Weber 2006) as all water is being treated to a very high drinking water standard and then used for many applications such as toilet flushing, cleaning, car washing, and irrigation which do not require such high standards. While it is clear that the debate on this topic is ongoing, it invites the opportunity to take a different approach in developing countries and, at least in remote locations, trial a decentralised approach. UF lends itself as a very suitable technology for such an approach and it is in this spirit the technology is reviewed with regards to performance and likely energy demands of such systems.

2 Ultrafiltration for Provision of Clean Drinking Water

2.1 Principle of Ultrafiltration (UF)

Membrane systems achieve a physical disinfection of water by physically sieving the waterborne organisms that are larger than the smallest pore size. Any material that is smaller passes through the membrane with the clean water or permeate. Subsequently, disinfection can take place without the need for chemicals.

The assessment of the pore size and their distribution within a membrane are very important in the microorganism removal potential of the membrane (Jacangelo et al. 1991). Because of the primary application of UF to retain macromolecules, the 'pore size' of a UF membrane is usually expressed as molecular weight cut-off (MWCO) in Dalton (Da; g/mol). This molecular weight cut of can be converted into an actual pore size, which is commonly used for membranes with larger pores and UF is in the range of 0.005 to 0.04 μm (von Gottberg and Persechino 2000). Therefore, viruses, which have a larger diameter (0.01–0.1 μm) can be retained, although smaller species may pass through the more open membranes. Bacteria are much larger than the pores (1–10 μm) and their retention is hence not a problem as long as the filter is intact. A comparison of UF membrane pore size with the size of bacteria, viruses, water molecules and ions is illustrated in Fig. 1. An overview of some common microorganisms and their sizes is given in Table 1.

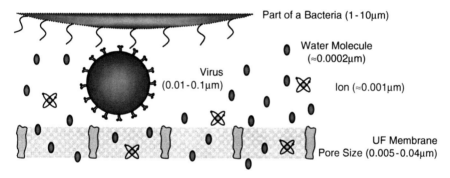

Fig. 1 Equivalent diameters of a number of microorganisms and typical membrane pore sizes, adapted from Herath et al. (1998)

Table 1 Sizes of microorganisms used to model viruses (Adapted from Herath et al. 1998)

Microorganism	Microorganism size (μm)		
	Eq. diameter	Length	Width
Qβ	0.025	–	–
MS2	0.025	–	–
Fr	0.019	–	–
T4	0.08	2.25	–
Psudomonas diminuta	1.05	2.68	0.56
Psudomonas putida	1.63	4.77	0.80
E-coli B	1.41	2.33	0.96
E-coli K12 A/λ	1.80	4.22	1.00
E-coli K12 C15	1.82	3.84	1.08
E-coli R	1.62	2.75	1.09
Alcaligenes eutro (β)	1.00	1.34	0.79

Membranes have been used in sanitation and sterilization processes due to their ability to perform this physical disinfection. While microfiltration (MF, pore size 0.1–0.45 μm) have been shown to retain small bacteria such as *Pseudomonas diminuta* (Waterhouse and Hall 1995), MF cannot usually retain viruses. However, the pore size classifications of those two membranes overlap somewhat and often depend on manufacturer preferences. In consequence, it is important to know the pore diameters of specific materials and apply a healthy level of caution when selecting a membrane.

The transmembrane pressure, which is the driving force in UF, varies from -0.5 bar to about 3–4 bar. A negative pressure (suction pressure or vacuum) is applied in submerged systems, in which the membrane is immersed in a water tank and the product water then sucked through the membranes to the permeate side. This configuration is common in high solids applications such as membrane bioreactors for wastewater treatment (DiGiano et al. 2004) or high turbidity surface waters as the process allows simultaneous sedimentation of particles. In general, pressure increases as the pore size decreases due to the higher membrane resistance at a given productivity (flux). Flux is the permeate flow normalised per membrane area and ranges for UF from as low as $50 L/m^2.h$ to as high as $1,000 L/m^2$ h. While high fluxes reduce the cost of membranes due to the reduced need in membrane area, fouling and hence cleaning and maintenance costs tend to increase. Pilot trials with specific waters are recommended to determine realistic flux values and establish pre-treatment needs.

2.2 Ultrafiltration Applications

UF is increasingly used to produce safe drinking water (Laîné et al. 2000), which can partially be attributed to increasing problems with microbiological contamination (in particular *giardia* and *cryptosporidium*). In fact, low pressure membrane processes such as UF and MF are capable of producing drinking water that falls in line with the standards set by the Surface Water Treatment Rule (SWTR) (Jacangelo et al. 1991). This combined with the decreasing costs of membranes (between 1994 and 2000 membrane costs fell by approximately 70%) has resulted in an increased adaptation of the technology (Laîné et al. 2000). However, in 2003, the majority of large scale plants are installed in the US (about 50%) while only 3 of 450 large scale (>400 m^3/d) applications are in Africa (Adham et al. 2005). It is evident that a significant market potential is yet to be satisfied. Less information is available on small scale systems and hence this paper aims at summarizing the most prominent small scale UF systems.

A number of other important applications of MF and UF are

1. Coupling with chemical additives such as coagulants or powdered activated carbon if smaller contaminants such as natural organic matter or micropollutants are to be removed.

2. Pre-treatment for nanofiltration (NF) and reverse osmosis (RO) to minimise fouling and prolong optimal operation conditions and membrane lifespan (Vedavyasan 2007). This is an important consideration for international development applications where long term operation of systems with minimal maintenance is important (Schäfer et al. 2007).
3. Wastewater treatment and reuse, especially membrane bioreactors (MBR). Increased concern about contaminants in wastewater discharge and in some cases water recycling applications are driving such change combined with the advantage of a smaller footprint of membrane technology compared to conventional treatment. This area is a significant opportunity for international development for the control of wastewater and hence the contamination of water supplies (DiGiano et al. 2004).

2.3 Retention of Pathogens (Viruses and Bacteria)

Removal of bacteria and viruses is describes as log removal (LR) which is defined in Equation (1),

$$LR = -Log\left(\frac{n_{final}}{n_{initial}}\right) \qquad (1)$$

where n_{final} is the final virus count and $n_{initial}$ is the initial virus count. Given the varying shapes and sizes of bacteria and viruses (see Table 1), retention is often specific to the type of microorganisms provided its size is similar to the pore size. Tests are rarely performed with viruses and hence little data is available for specific species but common model viruses rather.

Microorganism removal with predominant 'conventional' water treatment technologies such as slow sand filtration or biological filtration is generally low and unreliable as virus retention in such filters is often dependent on the establishment of a biofilm. For example, sand filters achieved a 2 log removal (99%) for the MS-2 (0.028 µm) and 3 log (99.9%) for the PRD-1 (0.065 µm) bacteriophages (Yahya et al. 1993). The level of microorganisms must be reduced or inactivated to an extent so that they do not pose a treat to health when consumed and these results fall short of the guidelines set by the SWTR where, to be compliant, a 3 log removal or inactivation of *Giardia* needs to be achieved and a 4 log (99.99%) removal of viruses (Yahya et al. 1993).

The removal of viruses and bacteria from water supplies using low-pressure membranes has been well documented (Jacangelo 1995). Due to the nominal pore size of a UF membrane the majority of bacteria (>6 log) and viruses (>4 log) can be removed from a contaminated drinking water source.

Tables 2 and 3 show the LR of viruses for a number of commercially available UF systems. Reported virus removal averages about 4 log (99.99%) in a reported range

Table 2 Table of selected commercial UF systems with published data (max. capacity, typical transmembrane pressure [TMP], log virus removal) and published/calculated specific energy consumption (SEC) (based on operation at max. capacity)

#	Company	System	Membrane	Max. capacity (m³/h)	TMP (kPa)	SEC (Wh/m³)[a]	Log virus removal	References
1	KMS	Konsolidator 78	Koch FEG tubular membrane	2.38	70	2,333	4	Koch Membrane Systems (1) (2004)
2	KMS	HF-4	Koch PMPW hollow fiber	13.3	240	410	4	Koch Membrane Systems (2) (2005)
3	KMS	HF-6	Koch PMPW hollow fiber	25.4	240	615	4	Koch Membrane Systems (2) (2005)
4	US Filter (Memcor)	EFC-400	—	5.5	—	—	—	Siemens Water Technologies Corp (2007)
5	US Filter (Memcor)	EFC-424	—	11.0	—	—	—	Siemens Water Technologies Corp (2007)
6	US Filter (Memcor)	EFC-1200	—	11.0	—	—	—	Siemens Water Technologies Corp (2007)
7	Separation Dynamics	Extran Model "E"	Extran A2A	0.78	150	641	—	Separation dynamics
8	Zenon	Z-Box-S, S6	ZeeWeed 1000	11.4	—	—	3.5	Deakin (2007), Zenon (1)
9	Zenon	Homespring	—	2.52	—	—	5	Deakin (2007), Zenon (1)
10	Pall Corporation	AP-1	Microza hollow fiber membrane	7.0	250	214	4.5–6[b]	Pall Corporation (1)
11	Pall Corporation	AP-2	Microza hollow fiber membrane	12.0	250	855	4.5–6[b]	Pall Corporation (1)
12	Pall Corporation	Septra	Septra CB	11.3	280	1,197	4	Pall Corporation (1), Pall Corporation (3)
13	Norit	Perfector	X-Flow Capfil Aquaflex	2.0	300	513	4	Norit (1); Norit (2); Norit (3)
14	Gamma Filtration	Microlab 130S	Patterson Candy International (PCI-BX6) membrane	20.0	300	2,308	—	Drouiche et al. (2001)
15	Norit	Lineguard UF	X-flow S-30	6.0	200	86	4	Norit (4)
16	Solco	Skyhydrant	—	0.83	—	—	—	Solco (2004)
17	Mono Pumps	WPS	Zenon	144.0	60	26	3	Deakin (2007), Moore (2006)

[a] Reliable data for SEC determination is to date difficult to access and varies significantly depending on controls and operating parameters as well as assumption in estimation. This topic requires a more in depth analysis. For this reason data from Table 3 for membranes only was used in this paper.
[b] Based upon coagulation processes before filtration.

Table 3 Table of commercially available membranes used in various systems (max. Capacity, membrane area, TMP, log removal) and published/calculated data (flux, pore diameter, SEC)

#	Manufacturer	Name	Membrane	Max. capacity (m^3/h)	Area (m^2)	TMP (kPa)	Flux (L/m^2h)	Pore diameter (µm)	Log virus removal	SEC (Wh/m^3)[a]	References
1	KMS	PMPW	HF 8-48-35-PMPW	4.6	32.1	210	143	0.018 (MWCO 100kDa)	4	90	Koch Membrane Systems (1) (2004)
2	KMS	PMPW	HF 8-72-35-PMPW	7.3	50.5	210	145	0.018	4	90	Koch Membrane Systems (1) (2004)
3	KMS	PMPW	HF 10-72-35-PMPW	11.6	80.9	240	143	0.018	4	103	Koch Membrane Systems (1) (2004)
4	KMS	PMPW	HF 10-48-35-PMPW	7.4	51.5	240	144	0.018	4	103	Koch Membrane Systems (1) (2004)
5	Pall Corp.	Microza	OLT-3206	3.5	10.7	300	327	0.005	3	128	Pall Corporation (1); Pall Corporation (2)
6	Pall Corp.	Microza	OLT-5026	3.5	23	300	152	0.005	3	128	Pall Corporation (1); Pall Corporation (2)
7	Pall Corp.	Microza	OLT-5026G	7.5	23	300	326	0.005	3	128	(Pall Corporation (1))
8	Pall Corp.	Microza	OLT-6036	16	34	300	471	0.004	3	128	Pall Corporation (1); Pall Corporation (2)
9	Pall Corp.	Microza	LGV-3010	–	–	–	–	0.006	4	–	Pall Corporation (4)
10	Pall Corp.	Microza	LGV-5210	–	–	–	–	0.006	4	–	Pall Corporation (4)
11	Pall Corp.	Septra	–	11.3	13.9	280	813	0.02	4	–	Pall Corporation (1); Pall Corporation (3)
12	Hydranautics	Hydracap	40″	4.6	30	152	100	0.02	4	65	(Hydranautics)
13	Hydranautics	Hydracap	60″	6.8	46	152	148	0.10	4	65	Hydranautic
14	Inge	Dizzer 3000	–	3	30	80	100	0.05	–	34	Inge
15	Inge	Dizzer 5000	–	5	50	80	100	0.037	–	34	Inge
16	Liqui-flux	Membrana	–	4.8	61	70	120	0.016	–	–	Liqui-flux
17	Zenon	Zeeweed	500	1.8[b]	41[b]	200	51	0.035	2	–	Deakin (2007), Zenon (1)

18	Zenon	Zeeweed	1,000	3.6[b]	56[b]	200	65–85	0.018	3.5	—	Deakin (2007), Zenon (1)
19	Norit	Aquaflex	S-225 FSFC PVC	17.3	35	400	494	0.026	4	171	Norit (2); Norit (3)
20	Norit	Xiga	—	3.5	35	400	100	0.026	4	171	Norit (4)
21	Norit	X-Flow	S30	4.7	6.2	300	758	0.026	4	128	Norit (4)
22	USFilter	—	—	—	—	—	110	0.10	0.5	—	Siemens Water Technologies Corp (2007)
23	USFilter	—	—	—	—	—	85	0.10	0.5	—	Siemens Water Technologies Corp (2007)
24	Nitto Denko	NTU-3306-K6R	—	15	30	300	500	0.004	—	128	Nitto Denko
25	Nitto Denko	NTU-3306-K4R	—	7	14	300	500	0.004	—	128	Nitto Denko

[a]SEC calculations are based on published typical performance data which is likely to deviate from real performance. It is essential that more data on power consumption be made available to assess the actual performance of such membranes.
[b]Results are based on a coagulation pretreatment prior to the membrane filtration.

of 0.5 (MF) to 6 log with coagulation pre-treatment. Drouiche et al. (2001) have reported a 6 log removal of bacteria and a 4 log removal of viruses using small UF units. Given the size of bacteria, retention by an *intact* UF membrane is guaranteed.

However, viruses vary significantly in size and shape and in consequence validity of retention needs to be investigated with care. For example the the poliovirus, which is one of the smaller viruses, is spherical with a diameter of 25 nm (Andrews and Pereira 1964) which is smaller than a typical UF pore. There are hundreds of enteric viruses found in the faeces of humans and animals, each of them having very different shapes and sizes. Therefore any membrane that is investigated needs to be able to handle the range of viruses that are likely to occur in the water to be treated.

Given the difficulty of working with infectious viruses, phages (viruses that affect bacteria not humans) are widely used in membrane investigations to determine the efficiency of removal. Phages are organisms that are excreted by a certain proportion of the population (animal or human) at all times, and these individuals are non-infected, while viruses are excreted by infected individuals for a short period of time (Grabow et al. 1999). Coliphages are widely used as model viruses. They are easy to work with, give acceptable results and, most importantly, are non hazardous. Their size is also advantageous as it is very similar to pathogenic viruses (Otaki et al. 1998). For example, Herath et al. (1998) used RNA coliphages to replicate pathogenic viruses, their size and shape is very similar to that of pathogenic enteroviruses, along with four strains of *E-coli* phages, T4, QB, MS2 and fr. The T4 E-coli was used as it has a very irregular shape while the other three have icosahedron shapes, which is similar in shape to a sphere. *E-coli* is a thermophilic coliform which indicates faecal pollution from warm blooded animals (Ashbolt et al. 2001). Table 1 shows the size of micro-organisms used to model viruses.

Looking at the sizes presented in Table 1 one can see that many microorganisms will not be challenging to remove. Removing the smaller viruses or phages effectively is more difficult, however. Figure 2 shows the difference in size between the sizes of selected micro-organisms and the pore sizes of typical UF and MF membranes. Organisms larger than the membrane cut-off lines are retained. The area shown of specific interest in Fig. 2 is the region where MF cannot remove viruses effectively (see Fig. 3 for more detail). However, in this domain some UF membranes can successfully remove these viruses based on size exclusion.

In summary, based on bacteria and virus removal data available in the literature, UF is an ideal process to alleviate the problem of water related disease and death in developing countries, although further work to specifically investigate the retention of some common viruses in the field, system maintenance, membrane integrity and component failures would be beneficial.

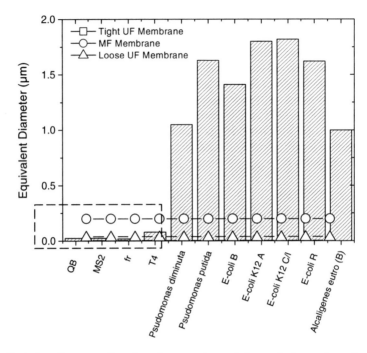

Fig. 2 Equivalent diameters of smaller micro-organism and the nominal pore sizes of a number of commercially available membranes

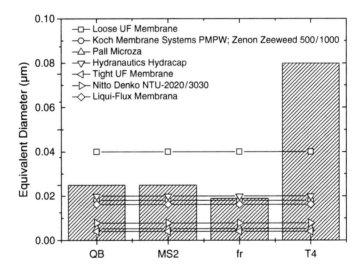

Fig. 3 Equivalent diameters of smaller micro-organism and the nominal pore sizes of a number of commercially available membranes

3 UF Systems in International Development

Despite the obvious potential of UF to solve the problem of safe water in developing countries, to date very few such technology are making their way into the developing world on a long-term basis. UF systems are being used in humanitarian situations, an example being the donation of a number of small-scale UF/MF units for emergency disaster relief, such as the Sumatran Tsunami in 2004 (Zenon (2) 2005). During natural disasters damage is often caused to the sanitary infrastructure. Therefore, surface waters can become rapidly contaminated with faecal microbiota. If this water is to be used for drinking and is left untreated then an epidemic is likely to break out. As a result, UF can be used to remove these microorganisms to a suitable level and provide a sustainable alternative to bottled water. Unit operation in such disaster situations is generally short term and hence provides an opportunity for technologies that are not yet proven in terms of long term technical sustainability. Oxfam, for example, is seeking technologies for disaster relief that can operate for period of 1 year and that are 'disposable' meaning that they will not normally be recovered after a period of service (Bastable 2007).

Water supply systems sought for international development are typically 'low-tech'. Naturally, countries are poor, disposable incomes, if any, low and the lack of 'public health infrastructure' calls for the development of a sewerage and water distribution infrastructure as in the developed world. The required investment for such infrastructure is enormous and in consequence little progress is made. An alternative is to implement novel concepts such as decentralized UF systems, particularly in areas where distances from urban areas are large and infrastructure (water distribution systems and electricity grid installations) are expensive. This provides an area of potential investigation for the development of autonomous low-pressure membrane systems powered by renewable energies. To evaluate feasibility and potential costs it is of interest to review existing UF systems, their performance and energy consumption, in particular.

3.1 Existing Ultrafiltration Systems

Table 2 summarises a number of UF and MF system that are commercially available. Those systems are not necessarily suited to international development applications and most require existence of infrastructure such as an electricity grid. Some are designed for sewage rather than drinking water treatment. However, most of those systems can potentially be adapted into renewably powered systems for remote applications. A positive side-effect of the high quality water produced by UF is that any remaining viruses or regrowth of microorganisms in this water can be effectively controlled by disinfection processes as turbidity, which often interferes with disinfection processes, has been removed. This ultimately reduces costs of such post-treatment and the generation of a residual disinfectant concentration.

3.2 Performance of Membranes in Terms of Productivity and Energy Consumption

In order to determine power requirements of a UF system using a specific membrane needs to know the transmembrane pressure and feed flow. The power consumption P can be calculated as (Mulder 1996)

$$P = \frac{Q_F \Delta P}{\eta} \quad (2)$$

Where Q_F is the volumetric feed flowrate (m³/s), ΔP is the transmembrane pressure (N/m²) and is the pump efficiency (typically 0.5–0.8), giving a power value in Watts (Mulder 1996). The majority of the energy usage of such membranes systems is due to the applied transmembrane pressure, while monitoring and control equipment as well as fouling control also need to be considered. Energy requirements of control systems have been neglected in calculations for the purpose of this paper.

The specific energy consumption (SEC) of a system is defined as the energy consumed in the production of a unit volume of water. This is based on the power consumption of the system, in this case simplified to the dominating unit which is the pump. SEC is defined as

$$SEC = \frac{IU}{Q} \quad (3)$$

Where I is the current of the pump (A), U is the voltage (V) and Q is the permeate flowrate (m³/h). However, since P = IU, power can be substituted into Equation (3)

$$SEC = \frac{P}{Q} \quad (4)$$

For calculation in this paper, flux data (if not published) was calculated using Equation (5) with the maximum capacity of the membrane under question in order to give the maximum productivity of the system. This results inevitably in the minimum SEC value.

$$J = \frac{1}{A}\frac{dV}{dt} = \frac{Q}{A} \quad (5)$$

When the membrane was classified with a MWCO it was converted to µm by using the Einstein-Stokes Equation. Einstein-Stokes Equation is defined in Equation (6), adapted from Worch (1993).

$$d = 4.074 \times 10^{-11} \cdot M^{0.53} \quad (6)$$

where M is the molecular weight cut off expressed (Da, g/mol) and the d is the equivalent pore diametre (m). The Einstein-Stokes Equation assumes that the molecules are spherical and in consequence that pores are cylindrical.

Table 3 shows the performance of a number of commercially available membranes determined using the above relationships assuming maximum capacity operation and typical transmembrane pressure as published by manufacturers.

Figure 4 shows, as expected, that SEC decreases with increasing pore size and hence increased permeability. Flux data is somewhat misleading as it is determined by transmembrane pressure and hence does not increase with pore size as permeability.

Figure 5 shows log virus removal as a function of pore diameter. Based on the data collected in Tables 3 and 4, it is evident that the performance of commercial membranes at full scale is lower than experimental studies. However, no clear distinction is visible between virus retention of MF and UF membranes. While MF membranes with pores >0.05 μm cannot remove more than 3 log viruses, results for UF are scattered over a <2 log to >8 log range. This emphasises the need to perform tests for specific membranes and specific viruses to be removed.

One can further elucidate from the UF performance in Figs. 4 and 5 that higher virus removal is not necessarily resulting in a higher SEC as the more open UF membranes show a lower virus retention at higher SEC than the tighter UF membranes. This warrants further investigation. Overall a substantial 4 log virus removal can be achieved with 'maximum capacity' SECs as low as 100 Wh/m^3, which is about 10–20 times lower than for a small scale brackish water desalination system using UF as pretreatment (Schäfer et al. 2007).

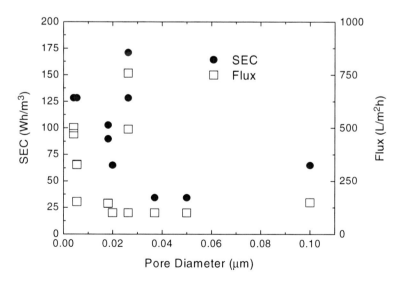

Fig. 4 Specific energy consumption and flux versus pore diameter (from Table 3)

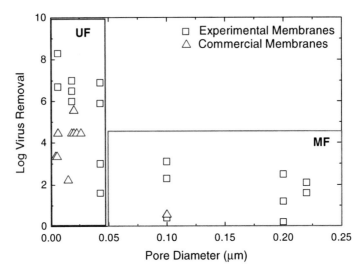

Fig. 5 Log virus removal as a function of membrane pore diameter (commercial [see Table 3] and experimental studies [see Table 4])

Table 4 Experimental investigations of membrane performance (Adapted from Ueda 2001)

Membrane Type	Pore Diameter (µm)	Microorganisms Removed	Log Removal	References
MF	0.22	Poliovirus	1.6	Madaeni et al. (1995)
MF	0.22	Poliovirus	2.1	Madaeni et al. (1995)
MF	0.2	Coliphage MS2	0.2–1.2	Jacangelo (1995)
MF	0.2	Coliphage MS2	0.2–2.5	Jacangelo (1995)
MF	0.1	Coliphage Qβ	2.3	Urase et al. (1993)
MF	0.1	Coliphage Qβ	3.1	Urase et al. (1993)
MF	0.1	Coliphage MS2	0.4	Jacangelo (1995)
UF	0.043	Coliphage MS2	1.6–6.9	Jacangelo (1995)
UF	0.043	Coliphage MS2	3.0–5.9	Jacangelo (1995)
UF	0.018	Coliphage MS2	6.0–7.0	Jacangelo (1995)
UF	0.018	Coliphage Qβ	6.5	Jacangelo (1995)
UF	0.006	Coliphage MS2	8.3	Otaki et al. (1998)
UF	0.006	Poliovirus	6.7	Otaki et al. (1998)

3.3 Comparison of Specific Energy Consumption With Desalination Systems

Since UF is a low pressure process (typically 1.0–5 bar), the required energy is significantly lower than that in nanofiltration and reverse osmosis units that operate in the range 5–20 bar and 10–100 bar, respectively (Mulder 1996). UF is ideal for the treatment of surface waters, while NF or RO are required if salinity or trace contaminants are of concern.

Schäfer and Richards have designed an autonomous desalination system that uses UF as a pretreatment step (Richards and Schäfer 2002, 2003; Schäfer et al. 2001, 2005, 2007; Schäfer and Richards 2005) driven by renewable energy for use in remote areas. There are a number of systems, which have coupled renewable energy, such as photovoltaics and wind power, in desalination applications to varying degrees of success. The performances of these systems with regards to specific energy consumption (SEC) vary significantly ranging from $1-2\,kWh/m^3$ (Schäfer et al. 2007) to $26\,kWh/m^3$ (Joyce et al. 2001). Since NF/RO operates at far higher pressure than UF, the SEC of UF would naturally be significantly lower. As result UF has a huge potential to be used cost effectively as a decentralised system in remote communities and in international development.

4 Conclusions

The review in this paper confirms that UF is a very appropriate choice for resolving the issue of lack of access to safe drinking water in developing countries as well as in disaster relief. The nature of this process in that it is reliable, simple, and energy efficient, makes it suitable for small decentralised systems. The energy requirements can be met in various ways using grid power (where available), renewable resources, generators, handpumps or sometimes gravity. If desalination is required such a system provides an excellent pre-treatment and acts as a double barrier for micro-organisms.

The implementation and maintenance of such systems and their technical sustainability remains to be dealt with. A possible approach would be for some companies focusing investment on selected countries in which, with the aid of donors or aid agencies, many similar systems are installed. With a large number of systems in certain areas it becomes workable to have maintenance personnel based in the area which then can assist with training local operators and offering initiatives such as a 'mobile membrane cleaning' service.

> *We are the first generation that can look extreme and stupid poverty in the eye, look across the water to Africa and elsewhere and say this and mean it: we have the cash, we have the drugs, we have the science – but do we have the will? Do we have the will to make poverty history?*
>
> Bono (2005)

We indeed have the technology and we have the knowledge to solve water problems in developing countries.

Acknowledgements The contributions of many colleagues and companies to the queries looking for small scale UF systems are acknowledged.

References

Adham, S., Chiu, K.-P., Gramith, K., and Oppenheimer, J. (2005). Development of a Microfiltration and Ultrafiltation Knowledge Base. Awwa Research Foundation, U.S. Department of the Interior, Bureau of Reclamation, Pasadena.

Andrews, C. and Pereira, H. G. (1964). Viruses of Vertebrates. Williams and Wilkins, London.

Ashbolt, N. J., Grabow, W. O. K., and Snozzi, M. (2001). Indicators of microbial water quality. In: Water Quality: Guidelines, Standards and Health (L. Fewtrell and J. Bartram, eds.), IWA Publishing, London.

Bastable, A. (2007). Personal Communication (A. I. Schäfer, ed.).

Deakin, C. (2007). Personal Communication (A. I. Schäfer, ed.), Edinburgh.

DiGiano, F. A., Andreottola, G., Adham, S., Buckley, C., Cornel, P., Daigger, G. T., Fane, A. G., Galil, N., Jacangelo, J., Pollice, A., Rittmann, B. E., Rozzi, A., Stephenson, T., and Ujang, Z. (2004). Safe water for everyone: Membrane bioreactor technology. Science in Africa June. http://www.scienceinafrica.co.za/2004/june/membrane.htm

Drouiche, M., Lounici, H., Belhocine, D., Grib, H., Piron, D., and Mameri, N. (2001). Economic study of the treatment of surface water by small ultrafiltration units. South African Water Research Commission 27, 199–204.

Gleick, P. H. (2002). Dirty water: estimated deaths from water related diseases 2000–2020. Pacific Institute for Studies in Development, Environment, and Security.

Grabow, W. O. K., Botma, K. L., de Villiers, J. C., Clay, C. G., and Erasmus, B. (1999). Assessment of cell culture and polymerase chain reaction procedures for the detection of polioviruses in wastewater. Bulletin of the World Health Organization 77, 973–980.

Herath, G., Yamamoto, K., and Urase, T. (1998). Mechanism of bacterial and viral transport through microfiltration membranes. Water Science and Technology 38, 489–496.

Hydranautics. HYDRAcap®. From Website: http://www.hydranautics.com/index.php?pagename=hydracap.

Inge. Dizzer 220/450. From Website: http://www.inge.ag/en/produkte/dizzer_220_450.html.

Jacangelo, J. G. (1995). Removal of cryptosporidium parvum, giardia muris and MS2 bacteriophage by hollow fibre microfiltration and ultrafiltration. In: Drinking Water from Sewage Using Membranes Symposium UNSW, Sydney, Australia.

Jacangelo, J. G., Laine, J.-M., Carns, K. E., Cummings, E. W., and Mallevialle, J. (1991). Low-pressure membrane filtration for removing giardia and microbial indicators. Journal of American Water Works Association 38, 97–106.

Joyce, A., Loureiro, D., Rodrigues, C., and Castro, S. (2001). Small reverse osmosis units using PV systems for water purification in rural places. Desalination 137, 39–44.

Koch Membrane Systems (1) (2004). Ultrafiltration – filtration overview. From Website: http://www.kochmembrane.com/sep_uf.html, Vol. 2006.

Koch Membrane Systems (2) (2005). Packaged water treatment ultrafiltration system. From Website: http://www.kochmembrane.com/pdf/Brochures/FINAL_PWBRO_FIN3-1.pdf.

Laîné, J.-M., Vial, D., and Moulart, P. (2000). Status after 10 years of operation – overview of UF technology today. Desalination 131, 17–25.

Madaeni, S. S., Fane, A. G., and Grohmann, G. S. (1995). Virus removal from water and wastewater using membranes. Journal of Membrane Science 102, 65–75.

Montgomery, M. A. and Elimelech, M. (2007). Water and sanitation in developing countries: Including health in the equation. Environmental Science and Technology 41, 17–24.

Moore, A. (2006). Personal Communication (J. Davey).

Mulder, M. (1996). Basic Principles Of Membrane Technology. Kluwer, Dordrecht.

Nitto Denko. Membrane products. From Website: http://www.nitto.com/product/datasheet/menbren/, Vol. 2007.

Norit (1). X-flow introduces the Perfector-E. From Website: http://www.norit.nl/?RubriekID=2216. 15th April 2007.

Norit (2). CAPFIL – capillary membrane products. From Website: http://www.norit.com/p3.php?RubriekID=2160.
Norit (3). Potable water. From Website: http://www.norit.com/import/assetmanager/5/8815/NORI260069%20Potable%20Water.pdf.
Norit (4). POE systems – Filtrix® LineGuard. From Website: http://www.norit.com/p3.php?RubriekID=2923.
Otaki, M., Yano, K., and Ohgaki, S. (1998). Virus removal in a membrane separation process. Water Science and Technology 37, 107–116.
Pall Corporation (1). Power generation products. From Website: http://www.pall.com/power_product_list.asp. 15th April 2007.
Pall Corporation (2). Microza UF modules, OLT series. From Website: http://www.pall.com/datasheet_MicroE_2801.asp?sectionid=specifications.
Pall Corporation (3). Septra™ CB filtration systems. From Website: http://www.pall.com/datasheet_FoodandBev_41058.asp?sectionid=description.
Pall Corporation (4). Microza membranes. From Website: http://www.pallcorp.com/pdf/198_Microza.pdf.
Richards, B. S. and Schäfer, A. I. (2002). Design considerations for a solar-powered desalination system for remote communities in Australia. Desalination 144, 193–199.
Richards, B. S. and Schäfer, A. I. (2003). Photovoltaic-powered desalination systems for remote Australian communities. Renewable Energy 28, 2013–2022.
Schäfer, A. I. and Richards, B. S. (2005). Testing of a hybrid membrane system for groundwater desalination in an Australian national park. Desalination 183, 55–62.
Schäfer, A. I., Fane, A. G., and Waite, T. D. (2001). Cost factors and chemical pretreatment effects in the membrane filtration of waters containing natural organic matter. Water Research 35, 1509–1517.
Schäfer, A. I., Broeckmann, A., and Richards, B. S. (2005). Membranes and renewable energy – a new era of sustainable development for developing countries. Membrane Technology 2005(11), 6–10.
Schäfer, A. I., Broeckmann, A., and Richards, B. S. (2007). Renewable energy powered membrane technology. 1. Development and characterization of a photovoltaic hybrid membrane system. Environmental Science and Technology 41, 998–1003.
Schwarzenbach, R. P., Escher, B. I., Fenner, K., Hofstetter, T. B., Johnson, C. A., von Gunten, U., and Wehrli, B. (2006). The challenge of micropollutants in aquatic systems. Science 313, 1072–1077.
Separation Dynamics. Extran ultrafiltration system (EUS), Model "E". From Website: http://www.separationdynamics.com/html/esystem.pdf.
Siemens Water Technologies Corp (2007). Water technologies: Advanced membrane filtration. From Website: http://www.usfilter.com/NR/rdonlyres/A83CE626-788E-4521-A7A2-09D8FE8475D3/0/IWSAMFBR110412.pdf.
Solco (2004). Skyhydrant. From Website: http://www.solco.com.au/products/solco_water_solutions/skyhydrant.
Strang, V. (2004). The meaning of water. Berg.
Ueda, T. (2001). Removal of microorganisms from wastewater with a membrane bioreactor. Bulletin of the National Institute of Agricultural Sciences (Japan) 40, 1–94.
UN (2005). United Nations millennium goals. From Website: http://www.un.org/millenniumgoals/.
UN (2006). Water, a shared responsibility. United Nations World Water Development Report 2.
Urase, T., Yamamoto, K., and Ohgaki, S. (1993). Evaluation of virus removal in membrane separation processes using coliphage Q beta. Water Science and Technology 28, 9–15.
Vedavyasan, C. V. (2007). Pretreatment trends – an overview. Desalination 203, 296–299.
von Gottberg, A. J. M. and Persechino, J. M. (2000). Using membrane filtration as pretreatment for reverse osmosis to improve system performance. In North American Biennial Conference of the American Desalting Association. Ionics Technical Paper.
Waterhouse, S. and Hall, G. M. (1995). The validation of sterilising grade microfiltration membranes with Psedomonas diminuta. Journal of Membrane Science 104, 1–9.

Weber, W. J. J. (2006). Distributed optimal technology networks: An integrated concept for water reuse. Desalination 188, 163–168.

Worch, E. (1993). Eine neue Gleichung zur Berechnung von Diffusionskoeffizienten gelöster Stoffe. Vom Wasser 81, 289–297.

Yahya, M. T., Cluff, C. B., and Gerba, C. P. (1993). Virus removal by slow sand filtration and nanofiltration. Water Science and Technology 27, 445–448.

Zenon (1). Membrane water treatment systems. From Website: http://www.zenon.com/products/. 15th April 2007.

Zenon (2) (2005). Hurricane victims in Gulf Coast will receive clean water from water treatment systems donated by ZENON and Maytag. From Website: http://www.zenon.com/about/corporate_goodwill/hurricane_katrina.shtml.

Groundwater Pollution in Shallow Wells in Southern Malawi and a Potential Indigenous Method of Water Purification

M. Pritchard, T. Mkandawire, and J.G. O'Neill

Abstract The provision of safe drinking water is a fundamental right of basic health and an extremely high priority of the Malawi Poverty Reduction Strategy. Only 40% of the people in Malawi have access to safe drinking water at any one time. Conventional water purification systems are prohibitively expensive for developing countries. The bulk of research work carried out in developing countries has concentrated on surface and borehole water quality with barely any work on monitoring water quality from shallow wells. The extent of pollution in shallow wells together with innovative, sustainable and economical solutions for rural villagers needs to be developed.

This research work has focused on establishing data on water quality from shallow wells in southern Malawi with the view to developing a technology that uses indigenous plant extracts to purify the groundwater. An in-situ water testing kit was used to determine the water quality. The majority of the physico-chemical parameters were found to be within the recommended limits; however, microbiological water quality results showed that the water can be grossly polluted with faecal matter and the likely presence of disease causing microorganisms. Preliminary laboratory tests on a powdered extract from the common indigenous plant *Moringa oleifera* are sufficiently encouraging for microbiological purification for further more detailed work to be planned.

M. Pritchard(✉)
Leeds Metropolitan University, School of the Built Environment, Leeds, LS2 8AJ, United Kingdom
e-mail: m.pritchard@leedsmet.ac.uk

T. Mkandawire
University of Malawi, The Polytechnic, Department of Civil Engineering, Private Bag 303, Chichiri, Blantyre 3, Malawi
e-mail: tmkandawire@poly.ac.mw

J.G. O'Neill
Centre for Research in Environment and Health, 6 Abbotsway, York, YO31 9LD, United Kingdom
e-mail: garyoneill@crehyork.co.uk

Keywords Groundwater quality · Malawi · *Moringa oleifera* · shallow wells

1 Introduction

About 1.1 billion people in the developing world are compelled to use contaminated water for drinking/cooking (UNICEF 2004). In Africa, over half of the population is without safe drinking water and over 6 million children are believed to die every year from water-related illnesses (DeGabriele 2002; Fernando 2005). In Malawi, this equates to roughly one in five children dying before they reach the age of five (UNICEF 2003). Malawi has been ranked the worst water manager in Southern Africa Development Community (SADC) by the World Water Council (Banda 2003). A large part of the sub-region of Malawi is characterised by small towns, villages and dispersed rural settlements. As a result, access to reticulated surface water resources has been limited because of the high costs and long distances that need to be covered in order to establish infrastructure for formal water services. As a result people use untreated groundwater from borehole/shallow wells, which pose a threat to their health.

Groundwater is the main source of water for about 60% of both rural and urban residents throughout Southern Africa, (UNEP 2002). About 63% of the people in Malawi use groundwater (Staines 2002). Groundwater is mainly supplied through boreholes or shallow wells. Boreholes are mechanically drilled holes between 20 and 80 m deep, with diameters ranging from 0.1 to 0.2 m. In comparison shallow wells have larger diameter holes (>1 m), which are either hand dug or drilled, with depths <15 m. There are typically two types of shallow wells, either open (unprotected) or covered (protected) as illustrated in Fig. 1. The main forms of contamination in boreholes normally emanate from chemical elements, whilst in shallow wells stem from bacteriological and physical constituents (Chilton and Smith-Crington 1984). The detrimental impact on human health of chemical contamination normally requires many decades of exposure before it can be recognised. Where life spans are short due to high incidence of infectious diseases (e.g. cholera) emanating from bacteriological contamination, it is this form of contamination

Fig. 1 Open (unprotected) and covered (protected) shallow wells in Malawi

that needs to be addressed first in groundwater purification techniques for Southern Africa, namely Malawi. The majority of research work carried out in developing countries has concentrated on surface and borehole water quality. This research work was undertaken to obtain data on the biological, physical and chemical water quality from shallow wells. The research work also conducted preliminary testing of the effectiveness of using powder extract from *Moringa oleifera* seeds as a water purifier.

2 Water Quality from Shallow Wells

2.1 Sampling/Monitoring

The research work was undertaken in three districts (Blantyre, Chiradzulu, and Mulanje) in the southern region of Malawi; collectively containing 350 shallow wells, where a large number of water-related diseases have been reported. The water quality from the shallow wells was monitored at selected times within a typical year to represent seasonal variations i.e. twice in the dry season and twice in the wet season, with the average seasonal value reported. Monitoring was undertaken using an in-situ water testing kit, which enabled microbiological, physical and chemical water quality of the wells water to be tested in line with the World Health Organisation (WHO) standards (2006), Malawi Bureau of standards (MBS) (1990, 2005) and Ministry of Water Development standards (MoWD) (2003).

The numbers of total and faecal coliforms were established using a membrane filtration technique carried out in-situ to ensure that the sample did not deteriorate with storage. Bacteria retained on the membranes were incubated at 37°C and 44°C for total and faecal coliforms respectively for a period of 24h. Bacteria that were present produced visible colonies that were counted and converted to represent a count per 100ml. Duplicate samples were taken for consistency during all testing stages. Test meters together with reagents were used to determine: turbidity, pH, temperature, total dissolved solids, electrical conductivity ammonia, arsenic, nitrite, nitrate, sulphate and hardness.

2.2 Results and Discussion

The water quality parameters obtained from the shallow wells were evaluated against the WHO (2006); MBS (1990, 2005); and MoWD (2003) guidelines to determine the amount and fluctuation in the level of contaminates between the dry and wet season. The standard acceptable guideline values are given at the top of Table 1 followed by the on-site microbiological and physical analyses for the three sampling districts in Malawi. In a similar format, the chemical standards are presented in Table 2 together with the chemical data for the wells. The mean of values obtained for both the dry and wet seasons are presented in the data tables.

Table 1 Microbiological and physical drinking water parameters obtained from shallow wells in Malawi

			Microbiological				Physical									
			Total coliforms (per/100 ml)		Faecal coliforms (per/100 ml)		Turbidity (NTU)		TDS (mg/l)		Electrical (µS)		pH		Temperature (°C)	
	Parameter	Season	Dry	Wet	Dry	Wet	Dry	Wet	Dry	Wet	Dry	Wet	Dry	Wet	Dry	Wet
Standards values	WHO		0		0				1,000					6.5–8.5			
	MBS		0		0		5		1,000					6.5–8.5			
	MoWD		50		50		5		2,000		3,500			6.0–9.5			
Districts/Villages Blantyre	Cedric		161	545	2	164	0.47	0.00	273	229	455.0	382.0	7.08	7.02	31.2	147.5
	Chemusa		780	2,555	113	205	1.76	0.40	342	308	570.0	513.0	7.41	7.05	26.2	25.9
	Fred (1)		80	152	3	74	0.83	0.01	220	205	365.0	341.5	b	6.75	27.5	26.8
	Fred (2)		a	419	a	113	a	1.21	a	184	a	306.0	a	7.16	a	27.2
	Kumazale		175	1,420	8	320	8.45	5.14	237	218	394.5	363.0	8.81	6.83	26.0	27.1
	Kumponda		1,4675	1,860	c	108	1.26	0.00	195	179	325.5	299.0	7.36	6.56	28.1	27.0
	Kumponda		1,4675	17,175	5,650	15,225	61.76	8.99	206	176	343.5	294.0	7.10	7.18	27.7	25.6
	Pasani		767	325	45	435	0.35	0.00	330	282	550.5	469.0	b	6.74	25.1	27.4
	Saili		80	1,350	10	435	0.51	0.74	203	183	339.0	305.5	b	7.20	25.3	26.1
Chiradzulu	Chelewani		212	d	37	854	0.96	3.01	114	111	190.7	184.9	6.44	6.43	24.2	23.4
	Chelewani		a	2,2850	a	818	c	26.36	c	113	c	188.6	c	6.42	c	22.3
	Makawa		4,350	2,165	250	373	2.13	0.00	132	125	219.5	207.5	6.65	6.47	25.3	24.6
	Mlandani		433	d	58	d	9.42	26.50	173	161	289.5	268.5	7.47	6.81	25.2	23.7
	Mtembo		1,970	864	30	603	2.87	14.38	172	71	286.5	117.8	7.31	6.69	24.3	25.0
	Mtembo		a	22,675	a	4,550	a	30.29	a	64	a	105.9	a	7.37	a	26.9
	Ng'omba		c	935	c	245	c	0.00	c	162	c	270.0	c	6.69	c	26.3
	Nlukla		708	5,070	237	823	2.03	19.32	112	109	188.7	181.0	6.56	6.83	24.4	24.1
	Nyasa		50	3,265	13	733	2.84	1.87	140	114	235.0	189.8	6.94	6.25	26.1	24.3

Method of Water Purification

	Village												
	Chipoka	c	c	c	c	c	c	c	c	c	c	c	c
	Mulola	e	985	e	560	e	2.88	34	c	58.1	c	c	c
	Naluso	550	1,905	102	968	2.39	20.93	161	84	c	140.6	6.38	b	26.4	26.4
	Namaja	148	d	10	3,060	0.77	320.00	138	157	268.0	261.5	6.92	b	27.7	25.7
	Namazoma (1)	163	1,955	15	715	2.52	10.67	51	85	229.8	142.3	5.71	b	26.1	26.5
Mulanje	*Namazoma (2)*	*11,590*	*12,450*	*1,630*	*350*	*10.76*	*3.56*	*40*	*40*	*27*	*86.1*	*67.0*	*5.56*	*26.4*	*26.9*
	Nande	c	c	c	c	c	c	c	c	37.4	45.8	c	c	c	25.1
	Nyimbiri	118	1,425	103	1,500	0.63	44.90	107	95	66.1	c	6.69	b	26.4	26.9
										177.6	157.8			28.1	

[a] Well was dry.
[b] Failure of equipment.
[c] Well has fallen into disrepair.
[d] Result was nullified (membrane was stuck to Petri dish).
[e] Unable to gain access to village well.
Italics indicate open/unprotected wells.
Grey shading indicates the wells which have failed to meet the MoWD (2003) standard.

Table 2 Chemical drinking water parameters obtained from shallow wells in Malawi

	Parameter	Chlorine, Free (mg/l)		Chlorine, Total (mg/l)		Sulphate, SO$_4$ (mg/l)		Hardness CaCO$_3$ (mg/l)		Chemical Nitrate, N (mg/l)		Ammonia, N (mg/l)		Arsenic (mg/l)		Nitrite, NO$_2$ (mg/l)	
	Season	Dry	Wet	Dry	Wet	Dry	Wet	Dry	Wet	Dry	Wet	Dry	Wet	Dry	Wet	Dry	Wet
Standards values	WHO			0.6–1.0			250		500		50		1.5		0.01		3
	MBS						400		500		10					0.05		
	MoWD						800		800		100					0.05		
Districts/villages Blantyre	Cedric	0.01	0.09	0.01	0.09	1.5	4.0	115	170	0.01	0.01	0.02	0.01	≤0.003	≤0.003	0.02	0.00
	Chemusa	0.03	0.02	0.03	0.05	71.5	79.0	141	220	0.00	0.03	0.42	0.11	≤0.003	≤0.003	0.01	0.03
	Fred (1)	0.01	0.03	0.01	0.03	2.5	1.5	88	145	0.09	0.14	0.50	0.05	≤0.003	≤0.003	0.00	0.00
	Fred (2)	a	0.02	a	0.03	a	3.5	a	140	a	0.74	a	0.03	a	≤0.003	a	0.01
	Kumazale	0.03	0.01	0.03	0.02	5.5	4.0	105	143	0.00	1.92	0.19	0.02	≤0.003	≤0.003	0.00	0.01
	Kumponda	0.01	0.03	0.04	0.04	5.0	2.5	57	138	0.61	0.05	0.00	0.03	≤0.003	≤0.003	0.00	0.00
	Kumponda	0.14	0.00	0.18	0.00	8.0	0.0	75	120	0.07	0.16	0.21	0.09	≤0.003	≤0.003	0.06	0.01
	Pasani	0.01	0.02	0.01	0.02	8.5	18.5	170	210	0.00	0.06	0.00	0.20	≤0.003	≤0.003	0.00	0.01
	Saili	0.00	0.04	0.00	0.04	10.5	6.0	56	118	0.02	0.09	0.00	0.05	≤0.003	≤0.003	0.00	0.00

Method of Water Purification 175

District	Village																
Chiradzulu	Chelewani	0.02	0.00	0.02	0.00	2.5	5.0	5	57	0.00	0.00	0.00	0.01	≤0.003	≤0.003	0.54	1.23
	Chelewani	c	0.12	c	0.15	c	7.5	c	40	c	0.01	c	0.51	≤0.003	≤0.003	c	0.02
	Makawa	0.00	0.01	0.00	0.01	7.5	6.5	39	50	0.00	0.00	0.00	0.01	≤0.003	≤0.003	0.00	0.02
	Mlandani	0.02	0.08	0.03	0.09	3.0	8.5	65	120	0.01	0.02	0.00	0.05	≤0.003	≤0.003	0.07	0.05
	Mtembo	0.02	0.03	0.03	0.03	1.5	8.0	61	30	0.02	0.01	0.02	0.01	≤0.003	≤0.003	0.15	b
	Mtembo	a	0.03	a	0.08	a	6.5	a	30	a	0.03	a	0.19	a	≤0.003	a	b
	Ng'omba	c	0.05	c	0.05	c	4.0	c	106	c	0.00	c	0.01	c	≤0.003	c	1.76
	Nlukla	0.01	0.09	0.01	0.10	4.0	11.0	22	64	0.00	0.01	0.00	0.03	≤0.003	≤0.003	2.45	2.00
	Nyasa	0.02	0.06	0.02	0.06	0.0	6.0	32	61	0.00	0.00	0.00	0.02	≤0.003	≤0.003	0.22	0.72
Mulanje	Chipoka	c	c	c	c	c	c	c	c	c	c	c	c	c	c	c	c
	Mulola	d	0.07	d	0.07	d	2.5	d	63	d	0.00	d	0.03	d	≤0.003	d	0.18
	Naluso	0.02	0.12	0.03	0.12	2.5	0.0	c	200	0.00	0.00	0.01	0.05	≤0.003	≤0.003	1.45	1.68
	Namaja	0.01	0.51	0.02	0.57	2.0	5.5	23	200	0.00	0.01	0.00	0.09	≤0.003	≤0.003	0.00	0.39
	Namazoma (1)	0.02	0.04	0.02	0.05	1.5	0.0	0	46	0.00	0.00	0.02	0.02	≤0.003	≤0.003	1.39	0.98
	Namazoma (2)	0.05	0.06	0.05	0.06	3.0	5.0	0	14	0.00	0.00	0.00	0.03	≤0.003	≤0.003	1.12	1.19
	Nande	c	c	c	c	c	c	c	c	c	c	c	c	c	c	c	c
	Nyimbiri	0.03	0.165	0.04	0.17	5.0	2.5	11	148	0.00	0.00	0.01	0.10	≤0.003	≤0.003	1.13	0.55

[a] Well was dry.
[b] Failure of equipment.
[c] Well has fallen into disrepair.
[d] Unable to gain access to village well.
Italics indicate open/unprotected wells.

In terms of total coliform, all wells tested in both the dry and wet seasons did not meet the temporary drinking water guidelines set by the MoWD (2003), of a maximum of 50 TC/100ml for untreated water. Approximately 45% of the wells failed to meet the faecal coliform drinking water guideline of 50 FC/100ml in the dry season while this figure had increased to 100% of the wells failing to meet the standard in the wet season. There was a prominent rise in the number of coliforms present in the wet season compared to the dry season. Such an increase can be expected as a result of the increase in the mobility of pollutants during the rains. There is a large variation between some wells, with some containing a very high level of faecal contamination – probably a result of livestock roaming free and poor sanitation facilities in close proximity to these wells. Water quality from open wells is considerably inferior to that of covered wells, and by far exceeds guideline values.

In terms of physical and chemical pollution, shallow well water does not pose any real potential health risk i.e. the majority of the physico-chemical parameters were within the recommended guideline values. However, 12.5% of water samples did not meet the MoWD (2003) minimum pH standard of 6.0 in the dry season. About 5% of the samples did not meet the MoWD (2003) standard value for turbidity in the dry season. In the wet season this figure had increased to 21% not meeting the turbidity standard value.

3 Water Purification

The government of Malawi through the Ministry of Water Development aims at ensuring that sufficient quantities and acceptable qualities of water is equitably accessible to every individual on a sustainable basis (Government of Malawi 2003). Attempts to overcome water quality problems have been undertaken but at an unaffordable cost by rural livelihoods. There is a vital need to develop sustainable cost effective technologies to remove biological contamination from shallow well water for rural people who live below the poverty line.

3.1 Plant Extracts

Natural plant extracts have been used for water purification for many centuries. For example, *Strychnos potatorum* was noted as being used as a clarifier between the 14th and 15th centuries BC; *Zea mays* was used as a settling agent by sailors in the 16th and 17th centuries. *Tigonella foenum; Cyamopsis psoraloides; Hibiscus sabdariffa; Lens esculenta*, have also been used, in history, to aid water purification (Schulz and Okun 1984). However, the science and civil engineering application of the use of plant extracts have not, really, been developed beyond its embryonic stage. This is largely due to the fact that money has been spent in developed countries, on more technologically advanced water purification processes.

One of the most promising plant extracts identified to date is the powder obtained from the seeds of the *Moringa oleifera* tree. In Sudan, rural women use dry *Moringa oleifera* seeds powder to treat highly turbid Nile water (Muyibi and Evison 1995). The University of Leicester in the UK has used the *Moringa oleifera* seeds within a contact flocculation filtration process for the treatment of low turbidity water. Sutherland et al. (1994) reported that the seeds have been effective in the removal of suspended solids in the order of 80–99.5% from surface waters containing medium to high initial turbidities.

3.2 Moringa oleifera as a Water Purifier – Preliminary Laboratory Trial

To assess the coagulant potential of the *Moringa oleifera* powder as a purifier for water from shallow wells to be used on a local level, a jar test was undertaken using different amounts of the powder, ranging from 0.1 to 20 g/l. A series of 11 beakers containing the contaminated water was first placed on the flocculator; the mixing paddles were inserted and at a slow speed (20 revs/min) the selected amount of the powder was added to the beaker. The samples were then stirred at 250 revs/min (high speed) for 1 min to ensure complete dispersion after which the speed of the mixing paddles was reduced to 20 revs/min for 15 min as recommended by Peavy et al. (1985) to aid in the formation of flocs. The flocculator was then switched off and samples were allowed to settle for 15 min to simulate the settling time at a treatment plant. This was undertaken to allow optimum conditions to be established. Each sample was then filtered through double 'Whatman #4' filter papers and the remaining 'purified' water was sent to a commercial testing laboratory for analysis.

It can be seen from Table 3 that approximately a 50–100% improvement was obtained in terms of the elimination of total coliforms. A 100% improvement was obtained in terms of E. coli at all concentrations. Colour appears to be readily removed and to some extent turbidity. The optimum dose appears relatively low – in the region of 1 g/l. Although the situation is not entirely clear there is certainly sufficient data for further more detailed work to establish the optimum dose and any link with factors such as pH.

4 Conclusions and Recommendations

The study showed that shallow wells yield water of unacceptable microbiological quality and that the situation is significantly worse in the wet season; almost certainly due to the mobility of this type of pollutant increasing. The physical and chemical parameters were, in general, within standard values and did not fluctuate significantly with season.

Table 3 Raw water quality and the change in quality with *Moringa oleifera* powder

Sample reference	Total coliforms per 100 ml	E.Coli per 100 ml	Colonies 2 days 37 °C No/ml	Colonies 3 days 22 °C No/ml	pH	Conductivity 20 °C (µS/cm)	Turbidity (FTU)	Colour (Hazen)
Raw water	9	9	28	40	5.9	79	5.84	130
Moringa oleifera 1.0 g/l	0	0	3	4	5	95	2.25	17.2
Moringa oleifera 2.5 g/l	1	0	1	3	4.5	168	4.61	12.8
Moringa oleifera 5.0 g/l	5	0	15	16	4.3	219	16.6	11.4
Moringa oleifera 10.0 g/l	4	0	2	3	4.1	369	46.8	14.1
Moringa oleifera 15.0 g/l	5	0	6	2	3.9	463	75.2	17
Moringa oleifera 20.0 g/l	4	0	7	6	3.8	557	167	18.6

The study also showed that *Moringa oleifera* powder can significantly reduce the number of both total and faecal coliforms, turbidity and colour for certain contaminated waters. As coliforms are indicators of faecal contamination and hence disease causing bacteria, such as *Vibrio choleri* which causes cholera; any capability for their removal from drinking water should significantly contribute to an improvement in public health.

References

Chilton, P.J. and Smith-Crington, A.K. (1984). "Characteristics of the Weathered Basement Aquifer in Malawi in Relation to Rural Water Supplies". In: Challenges in African Hydrology and Water Resources (Proc. Harare Symp., July, 1984, pp. 235–248. IAHS publ. no. 144).

Banda, M. (2003). "Malawi Ranked Worst Water Manager in SADC", Daily Times Newspaper, National News, April 2, p. 3.

DeGabriele, J. (2002). "Improving Community Based Management of Boreholes: A Case Study from Malawi". Available from: http://www.ies.wisc.edu/ltc/live/basprog9.pdf [Accessed 3/1/08].

Fernando, T.S. (2005) "Murunga: The Ultimate Answer to Polluted Water". Available from: http://www.infolanka.com/org/diary/13.html [Accessed 3/1/08].

Government of Malawi (2003) "Devolution of Functions to Assemblies: Guidelines and Standards", Ministry of Water Development (unpublished official document from the Regional Water Officer-South).

Malawi Bureau of Standards – MBS (1990). 'Potable Water Standards for Treated Water Supply Schemes.' (MBS 214:1990).

Malawi Bureau of Standards – MBS (2005). 'Malawi Standard; Drinking water – Specification' Malawi Standards Board; MS 214:2005, ICS 13.030.40 (First revision).

Ministry of Water Development – MoWD (2003). 'Government of Malawi; Devolution of Functions of Assemblies: Guidelines and Standards.

Muyibi, S.A., and Evison, L.M. (1995). "Moringa oleifera seeds for softening hard water," Water Research, Vol. 29, No. 4, pp. 1099–1105.

Peavy, H.S., Rowe, D.R., and Tchobanoglous, G. (1985) "Environmental Engineering," McGraw-Hill, New York.

Schulz, C.R. and Okun, D.A. (1984) "Surface Water Treatment for Communities in Developing Countries", Intermediate Technology Publications, Great Britain.

Staines, M. (2002). "Water/Wastewater Problems and Solutions in Rural Malawi." M.Phil. thesis, University of Strathclyde, Glasgow.

Sutherland, J.P., Folkard, G.K., Mtawali, M.A., and Grant, W.D. (1994). "Moringa oleifera as a Natural Coagulant," 20th WEDC Conference, Affordable Water Supply and Sanitation, Colombo, Sri Lanka, pp. 297–299.

UNEP (2002). "Past, Present and Future Perspectives", Africa Environment Outlook, United Nations Environment Programme, Nairobi, Kenya.

UNICEF (2003). "The State of the World's Children 2004". UNICEF, New York. Available from: http://www.unicef.org/sowc04/sowc04_contents.html [Accessed 3/1/08].

UNICEF (2004). "Meeting the MDG Drinking Water and Sanitation Target – A Mid-term Assessment of Progress", UNICEF, New York, Available from: http://www.unicef.org/wes/files/who_unicef_watsan_midterm_rev.pdf [Accessed 3/1/08].

World Health Organisation – WHO. (2006). Guidelines for Drinking-Water Quality, First Addendum to Third Edition, Volume 1 Recommendations. Available from: http://www.who.int/water_sanitation_health/dwq/gdwq0506.pdf [Accessed 3/1/08].

Photoelectrocatalytic Removal of Color from Water Using TiO_2 and TiO_2/Cu_2O Thin Film Electrodes Under Low Light Intensity

Z. Feleke, R. van de Krol, and P.W. Appel

Abstract This work describes, photoelectrocatalytic degradation of organic pollutants by using methyl orange (an azo dye) as a model compound. The TiO_2 thin film and TiO_2/Cu_2O composite electrodes were used as semiconductor photo electrodes. Photo catalysis by UV light Corresponding to the light intensity range of the solar light was employed with the aim of using renewable and pollution-free energy. Result showed that the rate of removal of color was enhanced when potential bias of 1.5 V was applied. The degradation rate was also increased either in acidic (pH 2) or alkaline (pH 10) conditions. The application of a positive potential higher than the flat-band potential on the TiO_2 electrode decreases the rapid charge recombination process, and enhanced the degradation of organic compound. When the TiO_2/Cu_2O thin film electrode was used, more efficient electron and hole separation was observed in the composite system under very low potential. It is considered that the photo-generated holes migrate towards the interface while the electrons migrate towards TiO_2 and then to the back contact transparent fluorine doped tin-oxide-coated glass (TCO), making the behavior of the composite film analogous to that of an n-type semiconductor. In all cases, the kinetics of the photo catalytic oxidation of methyl orange followed a pseudo first order model and the apparent rate constant may depend on several factors such as, the nature and concentration of the organic compound, radiant flux, the solution pH and the presence of other organic substances.

Keywords Photoelectrocatalysis · titanium dioxide · cuprous oxide · composite thin film · photo electrode

Z. Feleke (✉)
Department of Chemistry, Faculty of Science, Addis Ababa University, P.O. Box 1176, Addis Ababa, Ethiopia
e-mail: zewge@chem.aau.edu.et or fbeshah@yahoo.com

R. van de Krol and P.W. Appel
Department of Chemical Technology, Faculty of Applied Science, Delft University of Technology, Julianalaan 136, 2628 BL Delft, The Netherlands

1 Introduction

Photo catalytic processes to degrade organic pollutants in water by oxidation utilizing TiO_2 as a catalyst in the form of suspended TiO_2 powder have been the subjects of extensive research recently [1–4]. In practice, the rapid unfavourable electron-hole recombination reaction in TiO_2 compared with the relatively slow redox reactions of organic compounds resulted in low quantum yield [5, 6].

Modifications of semiconductor surfaces such as addition of metals as dopant or combinations with other semiconductors have been considered to decrease the electron and hole recombination rate and thereby increase the quantum yield of photocatalytic process [7–9]. Coupling of two semiconductor particles with different band gap energy levels has also been considered either to minimize electron-hole recombination or to modify the catalyst so as to absorb light from the sun in the visible region. A variety of semiconductor heterojunctions such as CdS/TiO_2, CdS/PbS, CdS/ZnO, CdS/AgI, Cd_3P_2/TiO_2 and Cd_3P_2/ZnO and AgI/Ag_2S have been investigated [10–16]. Further improvements have been reported by employing photoelectrochemical approach [17]. However, most of the semiconductor materials with relatively shorter band gap have very low stability for the purpose of water purification.

Among the numerous transition metal semiconducting oxides, cuprous oxide (Cu_2O) is of intense interests and has been extensively investigated for its distinctive properties. Cu_2O is a reddish p-type semiconductor with a direct band gap of 2.0–2.2 eV [18, 19]. Cu_2O has been tested as photo electrode in electrochemical cells and water-splitting materials [20–21], investigated the photocatalytic degradation of cyanide in water using titanium dioxide/copper oxide composite catalyst system in powder form. Vigil et al. [22] Studied the incorporation of copper oxide to nanoporous TiO_2 from an aqueous solution of copper formate with the purpose of sensitizing the TiO_2 to wavelengths in the visible range for the application of heterojunctions in solar energy. Based on the direction and wavelength dependence of photocurrent they show that photons are absorbed by the copper oxide and electrons injected to the TiO_2. Most of the studies reported in the literature use high intensity UV lamp, which makes it difficult for wide applications.

In the present work, photoelectrocatalytic approach was considered using both TiO_2 thin film electrode and TiO_2/Cu_2O composite electrode. The objectives of the study were to investigate the photoelectrocatalytic activity of titanium dioxide (TiO_2) thin film, and electrodes titanium oxide (TiO_2)/cuprous oxide (Cu_2O)/ composite thin film electrode for the destruction of organic contaminants in water under low light intensity by using methyl orange as a model organic compound. We have focused our attention on the low-intensity range of UV illumination, from 1 to 2 mW/cm in our studies, since the ultraviolet light in direct sunlight is generally 2–4 mW/cm depending on the geographical location.

2 Experimental

2.1 Preparation of Cu_2O Thin Film

Cu_2O was electrodeposited on a transparent conducting substrate TCO, from alkaline Cu (II) lactate aqueous solution [23, 24] in a conventional two-electrode cell. Prior to electrodeposition, the TCO was sonicated in ethanol solution and dried in air. Copper wire was glued to the conducting side using aqueous based graphite conductive adhesive. The pH of the solution was adjusted between 9 and 12 by the addition of sodium hydroxide. The solution temperature was kept constant during deposition by a thermostat (IKA. ETS-D4 fuzzy, The Netherlands). Deposition temperature was varied in the range from 25 °C to 60 °C. The deposition potential was maintained at −0.4 V and the current density varied from 0.1 to 0.8 mA/cm². The duration of deposition ranging from 2 to 30 min was used to obtain films of various thickness. All potentials are reported versus the SCE reference electrode. Electrochemical deposition was controlled by an EG&G 283 potentiostat. Cuprous oxide films were annealed in air at 300 °C for 1 h or used without annealing. The structure of the films was identified by X-ray diffraction (XRD) under θ–2θ mode (AXS D8 Advance, Bruker, England).

2.2 Preparation of TiO_2 and TiO_2/Cu_2O Composite Thin Film Electrodes

TiO_2 films were prepared from pure Anatase Paste (ECN, The Netherlands). Two types of paste were used to prepare the TiO_2 electrode with different particle size 9 and 100 nm. The electrodes are represented as E1 (9 nm) and E2 (100 nm). Then, the films were annealed in air at 450 °C for 3 h in a furnace equipped with temperature control unit. The TiO_2/Cu_2O composite film was prepared by electrodepositing the Cu_2O on to the TiO_2 film from alkaline copper lactate solution at bath temperature of 60 °C. To vary the amount of deposited Cu_2O, the time of deposition was varied from 2 to 30 min.

2.3 Measurement of Photoelectrochemical Response

The photoelectrochemical measurements were performed in a single compartment Teflon cell having 1 cm diameter circular quartz three-electrode system. The light source was a light emitting diode (LED) with maximum emission of 375 nm (FWHM~15 nm, Roithner) array that was aligned for proper sample illumination. The potential was controlled with the potentiostat (EG&G 283, The Netherlands).

Current–time and current-voltage measurements were made in 0.5 M K_2SO_4 solution. For quantitative photoelectrochemical characterization at different wavelength, white light from a halogen lamp (200 W, Osram, The Netherlands) in an Oriel housing was passed through a monochromator (Spectropro150, Acton). The light intensity was measured with a UV photodiode (PD300, Ophir, Germany). The spectral photo response was measured with a potentiostat (EG&G 283, The Netherlands) equipped with shutter (UNIBLITZ, Model T132, France). The photo response was measured with and without oxygen bubbling.

2.4 Photo catalysis and Photoelectrocatalysis Experiments

The photocatalytic activity of the TiO_2 thin film electrodes and the Cu_2O/TiO_2 composite thin film electrode was evaluated using -[4-(dimethylamino)phenylazo] benzenesulfonic acid (methyl orange) as a model compound (Fig. 1). Screening photo catalysis experiments were carried out using rectangular 3 mL capacity quartz cell. The light source was a deuterium lamp with light intensity of 2 mW/cm² (Model C4545, Germany). Further photocatalytic (PC) and photoelectrocatalytic (PEC) degradation experiments were performed in a single-compartment Teflon cell (40 mL) reactor having a circular quartz window of 20-mm diameter (Fig. 2). A circular opening in the cell opposite to the quartz window allowed exposure of 3.14 cm² of the working electrode (TiO_2 or TiO_2/Cu_2O) to UV illumination. The platinum coil counter electrode was placed just in front of the working electrode. A saturated calomel electrode (SCE), used as a reference, was placed close to the working electrode. The experiments were carried out in the pH range from 2 to 10 in 0.1 M K_2SO_4 or 0.1 KCl

Fig. 1 [4-(Dimethylamino)phenylazo]benzenesulfonic acid, sodium salt (methyl orange)

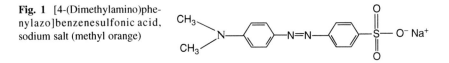

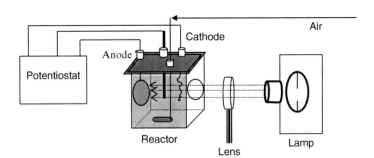

Fig. 2 Schematic diagram of photocatalysis and photelectrocatalysis reactor system

M electrolyte solutions. A potentiostat model EG&G 243 was used to bias the working electrode. Unless noted otherwise, the concentration of dyes used was 20 mg/L. Aqueous solution of methyl orange (40 mL) was photolysed under various operating conditions. The pH of the solution was adjusted to desired value with sodium hydroxide and sulfuric acid solution. The potential was controlled with the potentiostat (EG&G 243). The photo degradation of methyl orange was monitored with UV-Vis spectrophotometer (VARIAN) at its wavelength of maximum absorption (464 nm).

3 Results and Discussion

3.1 Film Compositions and Structure

The XRD spectrum of film sample, deposited on tin oxide conducting glass (TCO) substrate from lactate solution under alkaline medium is shown in Fig. 3. The film deposited is polycrystalline and chemically pure Cu_2O with no traces of CuO. The XRD spectrum indicates a strong Cu_2O peak with (200) preferential orientation. This is in good agreement with the results of other studies [23, 25, 26].

3.2 Photo Response

The dark and photocurrent of the Cu_2O electrode at -0.4 V vs. Ag/AgCl is represented in Fig. 4. A typical p-type photocurrent versus time behavior is observed. In

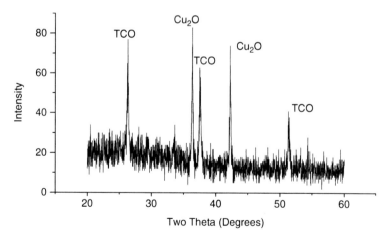

Fig. 3 X-ray diffraction patterns of copper oxides deposited at −0.4 V on TCO from alkaline lactate solution pH 10

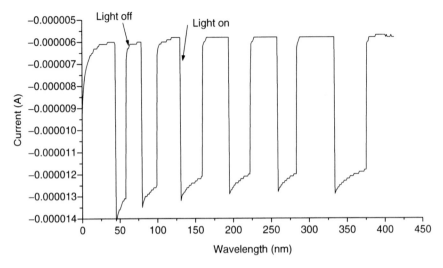

Fig. 4 Photo response Cu$_2$O electrodeposited from alkaline lactate solution on TCO conduction substrate under oxygen bubbling (0.5 M K$_2$SO$_4$, 0.2 mA/cm^2, −0.4 V)

addition, Photocurrent was stable over longer time range. The photo response was not changed after repeated use of the electrode and similar photo response was obtained for films deposited for 2, 5, 10, 20, and 30 min (data not shown). This indicates that the overall process of light absorption by the semiconductor, generation and separation of electron hole pairs, and transfer of electrons, is efficient even for thick films.

The cyclic voltammogram of the Cu$_2$O thin film electrode in 0.5 M K$_2$SO$_4$ solution in the dark and under illumination (375 nm) is shown in Fig. 5. The sweep starts at open-circuit potential and is scanned cathodically at 10 mV/s. At potential more negative than −0.8 V vs. SCE, a fast increasing cathodic current was observed due to the reduction of Cu$_2$O to Cu, which is also reported by other researchers [24]. The deposition currents are in a potential window between −0.3 and −0.7 V, which is about 400 mV window. The potential range corresponds to the Pourbaix diagram for the stability of Cu, Cu$_2$O, and CuO system [27]. Other studies on the electrochemical deposition of Cu$_2$O oxide films showed that the potential window is between −0.35 and −0.55 V [23]. The deposition current in Fig. 5 is very stable indicating that the p-type electrode containing Cu$_2$O as main component is resistant against photo corrosion. Figure 6 shows the spectral photo response of the electrodes prepared from alkaline lactate solution at pH 10. The photoresonse region of Cu$_2$O lies in the UV and visible region (300–600 nm) with absorption maximum around 450 nm. The maximum instantaneous photocurrent efficiency (IPCE) calculated from Equation (1) at 375 nm is 7%. This value is comparable to the value reported by Jongh et al. [24] for absorption in the visible region, but lower for absorption in the UV region. The photocurrent measurement in this study was carried out without air bubbling. The spectral photo response of the TiO$_2$/Cu$_2$O

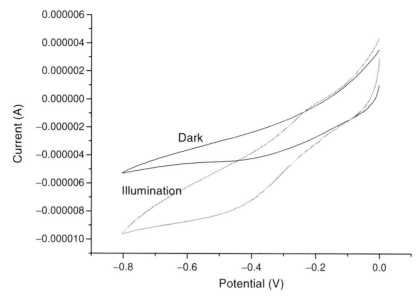

Fig. 5 Cyclic voltammogram of Cu_2O thin film deposited at −0.4 V on TCO from alkaline lactate solution pH 10 (scan rate 10 mV/s, 0.5 M K_2SO_4 electrolyte)

composite film is shifted to a maximum around 400 nm (Fig. 6). The pattern of photo response is similar to that of an n-type semiconductor photo catalyst. It is interesting to note that the maximum IPCE calculated using Equation (1) is 6% at wavelength around 375 nm. The relatively high photo response property of cuprous oxide might be useful for photo catalytic decontamination of water under low light intensity if it is combined with titanium oxide (TiO_2).

$$IPCE = \frac{i}{P \times F} \times 100 \qquad (1)$$

where i = photocurrent density, F = Faraday constant, P = light intensity (E s^{-1} cm^{-2}).

3.3 Photo Catalysis and Photoelectrocatalysis with TiO_2 Thin Film

Figure 7 shows the removal of color by photolysis (P), photo catalysis (PC), electrolysis (EC), and photoelectrocatalysis (PEC) using TiO_2 electrode as a function of time. No apparent change in methyl orange absorbance was observed by electrolysis alone. Whereas about 10% decrease of the maximum absorption peak of methyl orange was observed under photolysis and 40% decrease was observed under photo catalysis in 24 h reaction time. However, by applying a potential bias of +1.0 V, 100% decrease in

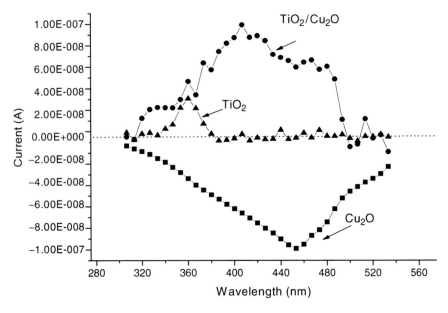

Fig. 6 Spectral photo response of the electrodes prepared from alkaline lactate solution at pH 10

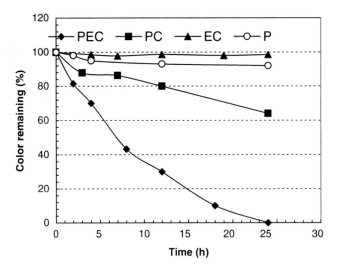

Fig. 7 Removal of color by photolysis, photo catalysis, electrolysis, and photoelectrocatalysis (+1.0 V) using TiO_2 electrode in photoelectrochemical reactor

absorbance was achieved for the same duration of reaction time. The results obtained at pH 6.5 illustrate that under applied potential there is an improvement of the photocatalytic decomposition of the dye. It has been reported that by applying an anodic bias potential higher than the flat-band potential of the semiconductor, the competitive reactions of charge recombination (e⁻/h⁺) can be minimized [17, 28–31]. A potential gradient within the photo catalyst film is provided at potentiostatic conditions to efficiently force the electrons to reach the counter electrode. Therefore photo-generated holes can be trapped in the surface by H_2O/OH^- species to give rise to $OH^{\bullet}$ radicals, which are essential for promoting the efficient degradation of dyes. It has been reported that the photoelectrocatalytic reaction may depend on applied potential, solution pH, and the type of electrolyte used [30–32].

3.4 Effect of pH on Photoelectrocatalytic Degradation Rate

The investigation on the kinetics of methyl orange degradation in aqueous solution under light irradiation was performed and absorbance reduction rate was assumed to follow a pseudo first-order model according to the following Equation (4):

$$\left(\ln \frac{A_t}{A_o}\right) = -Kkt = -k_{app}t \qquad (2)$$

where A_o is the initial absorbance, A_t is absorbance at time t, k is the reaction rate constant, and K a constant associated with light intensity, adsorption behaviour of methyl orange, and the electric field.

The absorbance had an exponential relationship to reaction time and the apparent first-order rate constant k_{app} could be calculated from Equation (2). A plot of $\ln(A_t/A_0)$ versus time represents a straight line as shown in Fig. 8, the slope of which upon linear regression equals the apparent first-order rate constant k_{app}, which is a function of light intensity, adsorption behaviour of methyl orange, and the electric field. The degradation rate was also increased either in acidic (pH 2) or alkaline (pH 10) conditions compared to the neutral (pH 6.5) solution. The reaction rate was decreased when the initial concentration of methyl orange increased at a given pH.

The pH is related to the ionization state of the surface as well as to that of reactants, and to the change of flat band potential of the semiconductor electrode. As illustrated in Fig. 9, both the acidic and alkaline conditions favored color removal and almost complete removal of color was realized in electrochemically assisted photocatalytic degradation process at 6h in acidic media. In neutral and alkaline media, the powerful electron-donating dimethylamino group generates a π-π^* transition which lies in the visible region of the spectrum (λ_{max} = 464 nm) with typical orange color. The molar extinction coefficient of this charge transfer transition is low (λ_{max} [464] = 267 M^{-1} cm^{-1}). The wavelength of maximum absorption shifts to a higher value when the pH is decreased due to protonation of the azo group to form the azonium ion [33]. It is known that pH values can influence the adsorption of dye molecules onto the catalyst surfaces [30, 32, 34]. The flat band potential E_{fb} is a well-known function of the electrode characteristics and the pH of the solution ($E_{fb} = E^0_{fb} + 0.05915$ pH) and follows Nernstian behaviour with a slope of −59 mV. The applied potential is positive to the flat band potential (−0.808 V). Therefore, there is always a potential gradient over the titania film, resulting in an electric field, which keeps photogenerated charges apart.

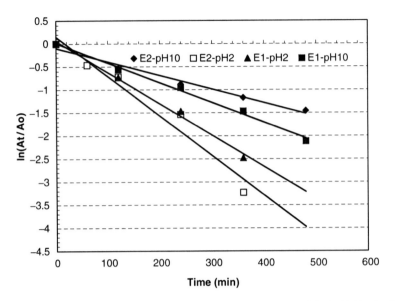

Fig. 8 Effect of pH on photoelectrocatalytic color removal at bias potential of +1.5 V using TiO_2 thin film electrodes

Under alkaline condition, the color removal rate may be affected by an increase in the density of TiO^- groups on the semiconductor electrode surface. Due to the coulombic repulsion, no direct electrostatic interaction between methyl orange and TiO_2 particles can occur. However, since the anode is a positively charged electrode, the influence of electric field becomes predominant on the electrode surface. As a consequence, the interaction of the methyl orange with the electrode surface is both potential and pH-dependent. As illustrated in Fig. 9, the apparent rate constant dramatically increases at higher pH values. This might suggest that different mechanisms take place. It is assumed that the electric field due to applied potential control the interaction of methyl orange and TiO_2 electrode.

In high alkaline solutions (pH > 10), since hydroxyl radicals are easier to be generated by oxidizing more hydroxide ions available on TiO_2 surface according to reactions (3–4), the efficiency of the process is logically enhanced. The minority charge carriers photo generated upon illumination on the photo anode can oxidize the H_2O/OH^- producing $OH^·$ radicals that adsorb on the photo electrode surface with the release of H^+ ions to the solution.

$$TiO_2 + hv \rightarrow TiO_2 - e^-_{cb} + TiO_2 - h^+_{cb} \quad (3)$$

$$TiO_2 - h^+_{cb} + H_2O_s \rightarrow TiO_2 - OH_s^· + H^+ \quad (4)$$

$$TiO_2 - h^+_{cb} + HO_s^- \rightarrow TiO_2 - OH_s \quad (5)$$

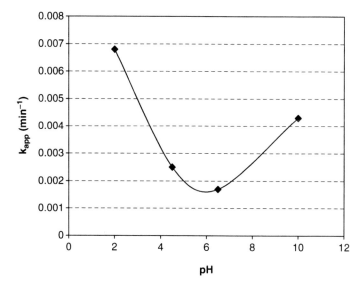

Fig. 9 Variation of apparent reaction rate constant as a function of pH at bias potential of +1.5 V using TiO_2 thin film electrode

On the other hand, the drastic increase in the discoloration rate at lower pH may follow a different mechanism. Methyl orange is an anion in aqueous solution, whose absorption spectrum remains invariant in the pH range 4–14, but changes below pH 4, as expected for protonation of the nitrogen atom of the azo linkage. Whereas the unprotonated compounds is an electron donor, the protonated species has electron-withdrawing character. At the same time, the TiO_2 particles are positively charged at low pH values and the semiconductor electrode is also positively charged. The improved photocurrent behavior in the presence of oxygen suggests that the oxygen may pick up the conduction band electron in TiO_2 particles, leading to the production of peroxy radical according to the reactions (6–7) (data not shown).

$$O_2 + e^-_{cb} \rightarrow O_2^{\cdot -} \tag{6}$$

$$O_2^{\cdot -} + H^+ \rightarrow OOH \tag{7}$$

Indeed experiments conducted without oxygen supply at low pH indicated lower reaction rate. Peroxide radicals are reported to be responsible for the heterogeneous TiO_2 photodecomposition of organic substrates such as dyes [35]. Further, since methyl orange has a sulfuric group in its molecular structure, which is negatively charged, the acidic solution favors adsorption of dye onto the photo catalyst surface, leading to the improvement of the degradation efficiency to some extent. More investigation is required in this regard to elucidate the predominant reaction mechanisms.

Investigation of the influence of the type of electrolyte on the removal of color with 20 mg/L of methyl orange in 0.1 M K_2SO_4 and KCl solutions (Fig. 10) showed that there is no significant difference of color removal under the two supporting electrolyte conditions. This indicates that the photo catalytic oxidation reaction of chloride and sulfate or electrochemical oxidation of chloride ion to chlorine at potentials around +1.3 V versus Ag/AgCl is not significant under the experimental conditions used in this study, as shown previously by others [36, 37].

3.5 Effect of Applied Potential

The effect of applied potential on the photoelectrocatalytic degradation rate of methyl orange was investigated by comparing the color removal during oxidation of 20 mg/L dye in 0.1 M K_2SO_4 at pH 10 in order to check for the importance of selecting the best potential for this process. Potentials ranged between −0.6 and +2.5 V was applied and the decrease in absorbance at 464 nm was monitored. The results in Fig. 11 show that the apparent rate constant for the removal of color increases progressively with increase in bias potential. On the other hand, the current produced was increased as the applied potential increased (data not shown). This indicates the influence of the applied potential to force the transport of photoelectrons across TiO_2 film and also the photo generated electrons are the charge carriers. As anodic potential increased, a large amount of current carrier (photoelectrons) passed through the TiO_2 film. Additionally, photogenerated holes were consumed by methyl orange in

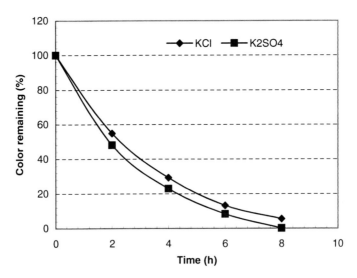

Fig. 10 Color removal as a function of irradiation time at pH 2 with TiO_2 thin film electrodes (E1) at bias potential of 1.5 V in KCl and K_2SO_4 supporting electrolytes

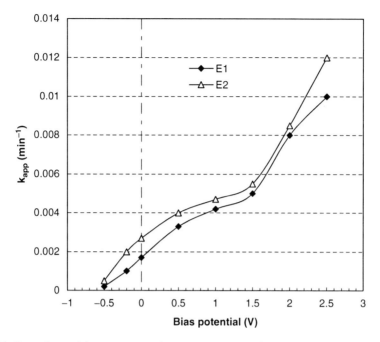

Fig. 11 Dependence of the apparent reaction rate constant (min^{-1}) as function of time at different potential in K_2SO_4 at pH 10 using TiO_2 thin film electrodes

the solution, which is reflected in the decrease of absorbance. The standard electrode potential values for the couple OH•, H$^+$/H$_2$O and O$_3$, 2H$^+$ / O$_2$, H$_2$O are +2.7 and 2.076 V, respectively [37]. The potential applied across the electrodes was 1.5 V in this study and OH• radicals could not be formed by anodic oxidation under the conditions examined. The result shows that color removal was mainly attributed to photocatalytic degradation in the experiments involved.

When the potential exceed 1.5 V, the reaction rate become faster and deviation from linearity was observed. The deviation from linearity at higher potential is associated with the combined effect of the generation of more oxidants at the anode. Overall, the application of a positive potential higher than the flat-band potential on the TiO$_2$ electrode decreases the charge recombination process shown by reaction (8).

$$TiO_2 - h^+_{ch} + TiO_2 - e^-_{ch} \rightarrow Heat \qquad (8)$$

It is interesting to note that the apparent rate constant approach zero as the applied potential approach the flat band potential of TiO$_2$ semiconductor (pure anatase) at pH 10 for the two electrodes. This indicates that charge recombination reaction become predominant if the applied potential is not sufficient for the separation of charge carriers. A slightly higher reaction rate was observed for the electrode made of 100 nm particles (E2).

3.6 TiO$_2$/Cu$_2$O Composite Electrode System

The apparent rate constant for the removal of color by TiO$_2$/Cu$_2$O thin film electrode as a function of bias potential is shown in Fig. 12. Preliminary screening tests (data not shown) indicated that the Cu$_2$O electrodeposited for 5 min at galvanostatic condition (0.2 mA/cm^2) showed better catalytic activity. Good removal efficiency was observed in the potential range −0.6 to −0.2 V *vs.* SCE. This potential range corresponds to the stability range of the Cu$_2$O [27].

At potential below −0.6 V, formation of reduction to Cu was observed and the electrode surface was gradually changed to black color. The color removal continued at reduced rate, possibly due to the catalytic activity of TiO$_2$/Cu system. On the other hand, the reduction in catalytic activity was more pronounced at positive bias potential. This indicates that only Cu$_2$O has strong contribution to the removal of color and further oxidation to other copper oxide species may inhibit the catalytic activity of TiO$_2$. It is interesting to note that apparent rate constant obtained at −0.4 V for the composite thin film (Fig. 12) is comparable to that of pure TiO$_2$ film at bias potential of +1.5 V.

The enhanced catalytic effect by Cu$_2$O at this low potential value is possibly due to more efficient electron and hole separation in the composite system. The potential gradient at the interface promotes the electrons and holes to flow in opposite directions, and the photo generated holes migrate towards the interface while the electrons migrate towards TiO$_2$ and then to TCO, making the behavior of the composite film analogous to that of an n-type semiconductor as shown schematically in Fig. 12. It can be concluded that the composite semiconductor electrode may have

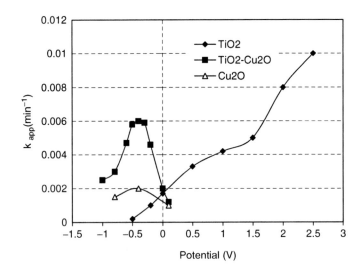

Fig. 12 Dependence of the apparent reaction rate constant (min^{-1}) as function of time at different potential in K$_2$SO$_4$ at pH 10 using Cu$_2$O, TiO$_2$, and TiO$_2$/Cu$_2$O composite thin-film electrodes

better harvesting efficiency of light due to observed shift in absorption spectrum as compared to that of pure TiO_2.

Cu_2O in the composite thin films have a beneficial role in improving charge separation and may also extend the light absorption spectrum of TiO_2.

4 Conclusions

The p-type semiconductor Cu_2O electrode prepared from the cupric lactate solution under alkaline condition showed a stable photocurrent over longer time range. The photo response was not changed after repeated use of the electrode. The rate of removal of color by TiO_2 thin film electrode was enhanced when potential bias of 1.5 V was applied. The TiO_2/Cu_2O composite thin film electrode showed an n-type semiconductor photocurrent characteristic. When the TiO_2/Cu_2O thin film electrode was used, the color removal rate reach maximum, at potential bias of −0.4 V. Further increase or decrease of bias potential caused decreased reaction rate. It is considered that the photo-generated holes migrate towards the interface while the electrons migrate towards TiO_2 and then to the back contact (TCO), making the behavior of the composite film analogous to that of an n-type semiconductor. Cu_2O in the composite thin films have a beneficial role in improving charge separation and may also extend the light absorption spectrum of TiO_2 to the visible region of the electromagnetic spectrum. In all cases, the kinetics of the photocatalytic oxidation of methyl orange followed a pseudo first order model and the apparent rate constant may depend on several factors such as, the nature and concentration of the organic compound, radiant flux, the solution pH and the presence of other organic substances.

Acknowledgements This research was financially supported by the Organization for the Prohibition of Chemical Weapons (OPCW), The Hague, The Netherlands, and the Netherlands Government.

References

1. Serpone, N.; Pelizzetti, E. Photocatalysis Fundamentals and Application. 1st ed., Wiley, New York, **1989**.
2. Hagfeldt, A.; Gratzel, M. *Chem. Rev.* **1995**, 95, 49–68.
3. Karkmaz, M.; Puzenat, E.; Guillard, C.; Herrmann, J.M. *Appl. Catal. B: Environ.* **2004**, 51, 183–194.
4. Chen, C.; Lu, C.;Chung, Y. *J. Photochem. Photobiol. A: Chem.* **2006**, 181, 120–125.
5. Fox, M.A.; Dulay, M. T. *Chem. Rev.* **1993**, 93, 341–357.
6. Fujishima, A.; Rao, T.N.; Tryk, D.A. *J. Photochem. Photobiol. C: Photochem.* **2000**, 1, 1–21.
7. Cordoba, G.; Viniegra, M.; Fierro, J.L.G.; Padilla, J.; Arroyo, R. *J. Solid State Chem.* **1998**, 138, 1–6.
8. Fujitsu, S.; Hamada, T. *J. Am. Ceram. Soc.* **1994**, 77(12), 3281–3283.
9. Wilke, K.; Breuer, H.D. *J. Photochem. Photobiol. A: Chem.* **1999**, 121, 49–53.

10. Spanhel, L.;Weller, H.; Henglein, A. *J. Am. Chem. Soc.* **1987**, 109(22), 6632–6635.
11. Zhou, H.S.; Honma, I.; Komiyama, H.; Haus, J.W. *J. Phys. Chem.* **1993**, 97(4), 895–901.
12. Hotchandani, S.; Kamat, P.V. *J. Phys. Chem.* **1992**, 96(16), 6834–6839.
13. Bedja, I.; Kamat, P.V. *J. Phys. Chem.* **1995**, 99(22), 9182–9188.
14. Tristao, J.C.; Magalhaes, F.; Corio, P.; Sansiviero, C. *J. Photochem. Photobiol. A: Chem.* **2006**, 181, 152–157.
15. Wang, W.; Silva, C.G.; Faria J.L. *Appl. Catal. B: Environ.* **2006**, in press.
16. Xiaodan, Y.; Qingyin, W.; Shicheng, J.; Yihang, G. *Mater. Charact.* **2006**, in press.
17. Hepel, M.; Hazelton, S. *Electrochim. Acta* **2005**, 50, 5278–5291.
18. Grozdanov, I. *Mater. Lett.* **1994**, 19, 281–285.
19. Mizuno, K.; Izaki, M.; Murase, K.; Shinagawa, T.; Chigane, M.; Inaba, M.; Akimasa, T.; Awakura, Y. *J. Electrochem. Soc.* **2005**, 152(4), 179–182.
20. Khan, K.A. *Appl. Energ.* **2000**, 65, 59–66.
21. Hara, M.; Kondo, T.; Komoda, M.; Ikeda, S.; Kondo, J.N.; Domen, K., Shinohara, K.; Tanaka, A. *Chem. Commun.* **1998**, 4, 357–358.
22. Vigil, E.; Gonza'lez, B.; Zumeta, I.; Domingo, C.; Dome'nech, X.; Ayllo'n, J.A. *Thin Solid Films* **2005**, 489, 50–55.
23. Golden, T.D.; Shumsky, M.G.; Zhou, Y.; VanderWerf, R.A.; Van Leeuwen, R.A.; Switzer, J.A. *Chem. Mater.* **1996**, 8, 2499–2504.
24. Jongh, P.E.; Vanmaekelbergh, D.; Kelly, J.J. *J. Electrochem. Soc.* **2000**, 147(2), 486–489.
25. Bohannan, E.W.; Shumsky, M.G.; Switzer, J.A. *Chem. Mater.* **1999**, 11, 2289–2291.
26. Mahalingam, T.; Chitra, J.S.P.; Rajendran, S.; Jayachandran, M.; Chockalingam, M.J. *J. Cryst. Growth* **2000**, 216, 304–310.
27. Pourbaix, M. Atlas of Electrochemical Equilibria in Aqueous Solutions, 2nd ed., Pergamon Press, New York, **1966**.
28. Jorge, M.A.; Sene, J.J.; Florentino, A.O. *J. Photochem. Photobiol. A: Chem.* **2005**, 174, 71–75.
29. Zainal, Z.; Lee, C.Y. *J. Sol-Gel Sci. Tech.* **2006**, 37, 19–25.
30. Li, G.; Qu, J.; Zhang, X.; Ge, *J. J. Mol. Catal. A: Chem.* **2006**, 259, 238–244.
31. Zanoni, M.V.B.; Sene, J.J.; Anderson, M.A. *J. Photochem. Photobiol. A: Chem.* **2003**, 157, 55–63.
32. Candal, R.J.; Zeltner, W.A.; Anderson, M.A. *Environ. Sci. Technol.* **2000**, 34, 3443–3451.
33. Oakes, J.; Gratton, P. *J. Chem. Soc. Perk. Trans.* **1998**, 2, 2563–2568.
34. Galindo, C.; Jacques, P.; Kalt, A. *J. Photochem. Photobiol. A: Chem.* **2000**, 130, 35–47.
35. Hoffmann, M.R.; Martin, S.T.; Choi, W.; Bahnemann, D.W. *Chem. Rev.* **1995**, 95, 69–96.
36. Mills, A; Hunte, S. *J. Photochem. Photobiol. A: Chem.* **1997**, 108, 1–35.
37. Konstantinou, I.K.; Albanis, T.A. *Appl. Catal. B: Environ.* **2004**, 49, 1–14.

Part IV
Mining and Environment

Management of Acid Mine Drainage at Tarkwa, Ghana

V.E. Asamoah, E.K. Asiam, and J.S. Kuma

Abstract This study assesses a natural wetland in Tarkwa, south west Ghana to identify the most sustainable way to remediate acid mine drainage (AMD). The investigation involved mineralogical and bacteria analysis among others. The study traced the AMD occurrence to the presence of acid producing sulphides (pyrite) in a waste rock and the occurrence was found to be catalysed by sulphur oxidising bacteria *Thiobacillus ferro-oxidans*. The pH of the AMD was found to be raised from 4.4 to 7.4 in the natural wetland by *Desulfovibrio desulfurican* bacteria whose activity generally results in producing hydro-oxides. The restoration of pH by diverting part of a River into the wetland to dilute and neutralise the acidic effluent may not be efficient particularly during dry seasons because the flow of the dilutant is much less than the required flow to effect dilution. The study concluded that the sustainable management of AMD can be achieved through the enhancement of *Desulfovibrio desulfurican* activity in a constructed anaerobic wetland.

Keywords Acid mine drainage (AMD) · acidic effluent · natural wetland · microbial activity and dilution

1 Introduction

Acid Mine Drainage (AMD) is one of the biggest environmental problems facing the mining industry today. AMD is very difficult to control or treat once it starts because it is persistent and costly, and tends to be a liability for mines long after decommission.

AMD occurs when iron sulphide minerals such as pyrite and pyrrhotite are exposed to water and air, resulting in the chemical and/or biological oxidation of

V.E. Asamoah (✉)
University of Mines and Technology, Tarkwa, Ghana;
e-mails: victoasamoah@yahoo.com; ekasiam@yahoo.com; jerry_kuma@yahoo.co.uk

the iron sulphide mineral. This produces sulphuric acid, iron hydroxide and dissolved heavy metals, which contaminate both streams and ground water, lowering their pH in the process (Warhurst and Noronha 2000). The overall chemical reaction is presented in Equation (1).

$$4\,FeS_2 + 15\,O_2 + 14\,H_2O = 4\,Fe(OH)_3 + 8\,H_2SO_4 \qquad (1)$$

AMD is becoming a new environmental phenomenon in Ghana since the operation of open pit mines in 1990. In Tarkwa, a key mining community in the western region of Ghana, waste rocks which were later found to be acid producing, were dumped around the resulting pit during mining without any protection. With time, these rocks oxidize and started generating acidic effluent from the toe of the waste dump with high levels of dissolved constituent like manganese, aluminum and other metals, which drain into a natural wetland and eventually flow into streams and forest reserves (Kuma 2003). The livelihood of the local folks, wild life, and farmlands in this area are all at risk, because of their dependency on these streams for domestic and irrigation purposes.

With the realization of AMD occurrence in the area, the need to manage it became necessary. Hence, the objectives of this study are to confirm the AMD potential of the waste rocks (as established by Kuma 2003), to identify the microbial community in the natural wetland and to compare dilution and constructed wetland options for remediation and thus to propose the most sustainable option or method to remediate the AMD in the area.

2 Materials and Methods

Mineralogy of Rock Samples: Fresh rock samples were taken from the working face of a Pit in the study area for mineralogy, Acid Base Accounting (ABA) and Net Acid Generation (NAG) capacity test. Thin sections of samples were prepared and studied under the microscope Lietz Laborlux 11 Pol.

Physico-chemical Properties of Water: Water samples were taken during the month of June and January using properly cleaned polyethylene bottles. Important water quality parameters like temperature, pH, electrical conductivity (EC) and turbidity were measured on site using a portable water analysis kit (Horiba U/10). Whereas other parameters such as concentration of ions were determined in the laboratory according to the Standard Examination of Water and Waste Water of the American Public Health Association (2000) and the HACH methods of water and wastewater analysis. The sample points are as shown in Fig. 1.

Sediment Analysis: Sediment samples were taken from E3, E4, and E5 for bacteria analysis using Standards Nutrients 11 and 16 as recommended by Karaivko et al. (1977).

Dilution Test: A pH dilution test was carried out between E1, E3, E4 and a prepared H_2SO_4 to ascertain whether dilution could be a sustainable method of mitigating the AMD. This test was conducted in the field.

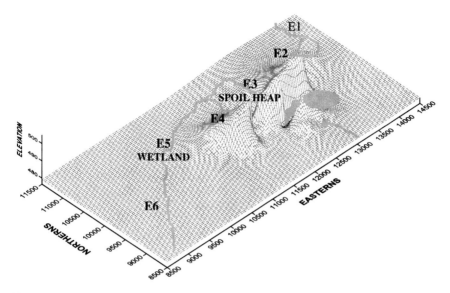

Fig. 1 A surface plot showing sampling points and relief

3 Results and Discussions

Mineralogy of Rock Samples: Anhedral pyrites were the sulphide minerals identified during the mineralogical analysis. The Neutralisation Potential Ratio (NPR) of the waste rock using the ABA concept (Sobeck et al. 1978) was given as;

$$\text{NPR} = \text{NP}/\text{AP} = 1/18.3 \qquad (2)$$

This value implies the waste rocks have the potential of generating AMD. Where, NP is the Neutralisation Potential and AP is the Acid Potential. The NAG capacity in Kg H_2SO_4 was 8; this value however only gives the measure of the rock's ability to generate AMD.

Physico-chemical Properties of Water: The pH of water samples in the natural wetland was observed to have decreased gradually from a value of 7.8 at E1 to 4.4 within the environs of E3 and then increased gradually to pH of 7.4 at E5. Results of the water analysis are presented in Table 1 and Fig. 2. High SO_4^{2-} concentration of 600 mg/L was found at E3 while relatively low concentrations of 13 and 60 mg/L were found at E4 and E5 respectively. Comparing the expected SO_4^{2-} concentration of 248 mg/L downstream (by way of dilution) to the actual value of 60 mg/L, a loss of 188 mg/L was obtained. This loss reflects biogeochemical activities in the natural wetland; meanwhile, corresponding rises in sulphide concentrations from 0.002 to 0.015 mg/L from E3 to E5 presumably suggest the possibility of biochemical activities in the natural wetland.

Table 1 Result of some water quality parameters

Sample point	E1	E2	E3	E4	E5	E6
Temp. (°C)	26.0	26.3	26.2	26.8	25.8	24.6
EC (µS/cm)	228	76	1160	208	332	296
Turb. (NTU)	8	55	7	5	3	265
pH	7.8	6.0	4.4	7.1	7.42	7.2
Fe (mg/L)	0.6	7.26	1.85	1.83	0.413	0.425
SO_4^{2-} (mg/L)	32	8	600	13	60	46
Pb (mg/L)	0.09	0.04	0.36	0.24	0.18	0.1
Ni (mg/L)	0.05	0.014	0.19	0.01	0.03	<0.01
S^{2-} (mg/L)	0.01	0.053	0.002	0.05	0.015	0.013

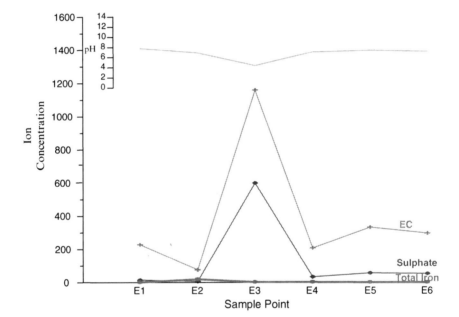

Fig. 2 A graph showing some physico-chemical properties of water

Sediment Analysis: Two main types of bacteria were identified during the bacteria analysis.

➢ *Thiobacillus ferro-oxidans*, responsible for the oxidation of ferrous ions to ferric ions and therefore a catalyst for the AMD at E3.
➢ *Desulfovibrio desulfurican*, responsible for sulphate reduction under anaerobic conditions were present at E4 and E5.

The restoration of pH at E5 was largely due to microbial activity in the wetland among others. Hence, through the enhancement of *Desulfovibrio desulfurican* activity in a constructed anaerobic wetland, the impact of AMD can be sustainably managed at E3.

Dilution Test: From Fig. 3, the dilution ratio of E1 and E3 was 5:1 and that of E1 and a prepared H_2SO_4 was 12:1. In addition, the ratio of dilution between E4 and H_2SO_4 was 12:1. With an average discharge of $0.079\,m^3/s$ at E3 (see Table 2) in the rainy season and $0.005\,m^3/s$ in the dry season, the quantity of E1 required to dilute E3 in an extreme AMD condition (ratio 12:1) is determined to be 0.941 and $0.059\,m^3/s$ respectively.

The restoration of pH by dilution option will not be efficient, particularly during the dry season. This is because the discharge of E1 in the dry season is only $0.008\,m^3/s$ which is much low than the required $0.059\,m^3/s$ in the dry season to dilute E3. The method will also not be reliable as it is dependant on weather conditions. Hence, this option is less attractive compared to anaerobic wetland mitigation.

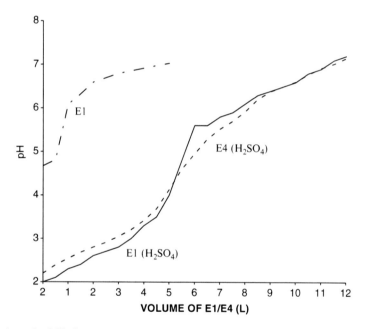

Fig. 3 A graph of dilution test between E3, E1 and E4

Table 2 Discharges of some surface waters

Sample point	Discharge (m^3/s)	
	June (wet period)[a]	January (dry period)[a]
E1	0.154	0.008
E2	0.012	0.002
E3	0.079	0.005
E4	0.114	0.003
E5	0.131	0
E6	0.125	0.003

[a]Classification at the weather condition when discharges were measured.

4 Conclusion

The study identified pyrites and traces of marcasites as the sulphide minerals present in the rock samples and the NPR of 1:18.3 indicates that the study area has the potential to generate AMD. The pH values range from 4.4 at E3 which is at the toe of the waste dump (acidic effluent discharge point) to 7.4 at E5 in the natural wetland. The study confirmed that the presence of AMD was catalysed by the bacteria *Thiobacillus ferro-oxidans*. Restoration of pH by dilution option will not be efficient particularly during the dry season because the discharge of dilutant was less than the amount required to effect the dilution. It will also not be reliable because it is dependant on weather conditions. Hence, the sustainable method of AMD management at E3 can be achieved through the enhancement of activities of *Desulfovibrio desulfurican* in a constructed anaerobic wetland, which would emulate the microbial activities in the natural wetland at E5. If the biochemical reactions of the microorganisms identified can be mimicked, then it will pave the way for a new technology of managing AMD in Ghanaian mines.

References

Karaivko, G. I., Kuznetsov, S. I. and Golonizik, A. I. (1977), "The Bacterial Leaching of Metals from Ores", Technicopy Limited, England, pp. 9–16.

Kuma, J. S. (2003), "Passive Treatment of Acid Mine Drainage – Laboratory Studies on a Soil Heap from the Tarkwa Area, Ghana", *Ghana Mining Journal*, Vol. 7, p. 46.

Sobek, A. A., Schuller, W. A., Freeman J. R. and Smith, R. M. (1978), "Field and Laboratory Methods Applicable to Overburden and Mine Soils", United States Environmental Protection Agency, Cincinnati, Ohio, 600/2-78-054, 203 pp.

Warhurst, A. and Noronha, L. (2000), "Environmental Policy in Mining: Corporate Strategy and Planning for Closure", Lewis Publishers, Washington DC, pp. 118–139.

A Multi-disciplinary Approach to Reclamation Research in the Oil Sands Region of Canada

C.J. Kelln, S.L. Barbour, B. Purdy, and C. Qualizza

Abstract A 7-year research project into the long-term performance of reclaimed landscapes on saline-sodic overburden from oil sands mining in north-central Canada has demonstrated the importance of a multi-disciplinary approach. The land capability assessment tool used by the industry evaluates three key areas: available soil moisture, salt impact, and biological response (including nutrients). Detailed field monitoring and sampling demonstrated the relative performance of three different layered covers (35, 50 and 100 cm) and one monolayer cover (100 cm) through the tracking of water content, suction, stored water volumes, interflow/runoff and water availability for plant growth. Salt ingress into the cover from the underlying waste and salt release through interflow flushing has also been monitored. This long-term monitoring has provided physically based measurements of cover performance that clearly highlight the inability of thin (35 cm) or monolayer covers for providing sufficient moisture to meet all demands throughout a growing season. Interpretation of this data has also provided key insights into the mechanisms governing cover performance. This physically based evaluation was supported by direct measurements of tree development and tree ecophysiology. Vegetation indicators included plant species composition and abundance, tree growth rates, foliar nutrient contents, and plant ecophysiology.

Keywords Mining · reclamation · covers · water balance · salts · vegetation

---

C.J. Kelln (✉) and S.L. Barbour
Department of Civil and Geological Engineering, University of Saskatchewan,
Saskatoon, Canada, S7N 5A9;
e-mails: cjk123@mail.usask.ca; lee.barbour@uask.ca

B. Purdy
Department of Renewable Resources, University of Alberta, Edmonton, Alberta, Canada,
T6G 2H1;
e-mail: brett.purdy@ualberta.ca

C. Qualizza
Syncrude Canada Limited, Fort McMurray, Alberta, Canada,
e-mail: qualizza.clara@syncrude.com

1 Introduction

Land reclamation on saline-sodic shale overburden from oil sands mining in the boreal forests of north-central Canada will occur on an unprecedented scale. Approximately 70 km^2 of saline-sodic shale overburden will have to be reclaimed at the Syncrude Canada Ltd. mine alone, and there are currently three similar open-pit mining operations in the region. Government legislation (Alberta Environmental Protection 1994) requires oil sands operators to return the disturbed areas to an equivalent land capability (ELC). This capability is currently assessed using the Land Capability Classification System (Leskiw 1998, 2006), which is based on a comprehensive suite of soil and landscape parameters, with three of the most heavily weighted areas being Available Water Holding Capacity (AWHC), salt impacts, and nutrient availability.

The research efforts to date have reinforced the idea that an evaluation of the long-term performance of these reclaimed landscapes must: (1) adopt a multi-disciplinary approach that tracks the performance of the hydrological, geochemical, and biological functions of the cover; and, (2) be conducted on a 'watershed' scale in order to address the spatial variability in the cover-performance functions (Qualizza et al. 2004).

The objective of this paper is to demonstrate the value of a multi-disciplinary approach to reclamation cover research at a 'watershed scale'. Various hydrometric and geochemical data are presented that, in isolation, provide limited insight into cover performance. Taken as a whole, the data helps elucidate the mechanisms and processes that govern the spatial controls on the bio-physical response of the cover. A brief synthesis of the cover vegetation research is provided to demonstrate the direct correlation between the ecological and physical responses of the cover.

2 Background

2.1 Study Site

The study was conducted at the SW30 Overburden Research Site (57°2′ N, 111°33′ W), located in north-central Canada at the Syncrude Canada Ltd. mine site. The climate of the region is classified as semi-arid to sub-humid with a mean annual air temperature of 1.5 °C and a mean annual precipitation is 442 mm (1945–1995). Monthly mean air temperatures range from 18 °C in July to −20.7 °C in January. Potential evapotranspiration (PET) as estimated by Penman (1948) typically exceeds 500 mm per year with daily maximums (~7 mm/day) occurring during July and August.

An instrumented watershed was commissioned in 1999 on a 2 km^2 saline-sodic shale overburden dump. The shale overburden is excavated during open-pit mining to gain access to the underlying oil-rich sand. It is then deposited in large waste dumps and re-contoured to create a natural landscape form. The shale is of marine origin (Cretaceous age) and is both saline and sodic. Reclamation is achieved by placing a two layer cover comprised of a thin, organic rich, peat/clay mixture over a thicker clay-rich layer (herein called 'secondary') which is salvaged from natural glacial deposits. This layering creates surrogate 'A' and 'B' horizons similar to that of a natural soil profile. It is expected that this cover will continue to evolve over several decades until capital and circulation rates for moisture and nutrients can be established that are similar to those for natural ecosystems (Qualizza et al. 2004).

Four alternative prototype test covers of varying thickness (35, 50, and 100 cm) were constructed on north-facing slopes and were comprised of: 15 cm of peat-mineral (PM) mix overlying 20 cm of secondary; 20 cm of PM overlying 30 cm of secondary; 20 cm of PM over 80 cm of secondary; and a 100 cm monolayer secondary cover. The three layered covers are each 50 by 200 m in area and were constructed on a 5:1 slope. Each cover drains into a single swale ditch located at the toe of the slope, which connects into the overall drainage system for the hill. The monolayer cover was constructed on a 4:1 slope and was placed 3 years earlier than the other covers (1996).

2.2 Instrumentation and Testing

Details of the soil instrumentation, laboratory testing, and field programs can be found in Barbour et al. (2004). A soil station was installed in the centre of each cover to measure matric suction, ground temperature, and soil moisture throughout the cover profile and into the shale. Two meteorological stations were installed on the site to measure air temperature, net radiation, wind speed and direction, relative humidity, and precipitation. Soil moisture conditions were also measured in neutron and capacitance access tubes at various locations traversing the entire watershed. Surface run-off was monitored at the down-slope edge of each cover using v-notch weirs equipped with automated water level monitoring and data acquisition systems. Subsurface saturated flow (i.e. interflow) was collected over the entire width of each cover using a sub-surface drainage system installed at the cover-shale interface near the toe of the slope. Shallow monitoring wells were installed at approximately 80 locations to monitor the development of a perched water table on the cover-shale interface and provide sampling points for water chemistry.

The in-situ hydraulic conductivity (K_s) of the cover and underlying shale was measured every year using a Guelph permeameter. Material characterization included particle size distribution (PSD), bulk density, and porosity along with a laboratory-based soil water characteristic curve. Major ion chemistry and stable isotope analyses (δD and $\delta^{18}O$) were performed on waters collected from the interflow system and shallow monitoring wells. One-dimensional profiles of *in-situ* pore water chemistry and $\delta^{18}O$ were determined at over 20 locations. Finally, vegetation surveys were conducted to evaluate cover performance including plant species composition and abundance, tree growth rates, and foliar nutrient contents.

3 Presentation and Discussion of Results

3.1 Soil Moisture

Multiple lines of evidence demonstrated that the performance of thin (i.e. 35 cm) multiple layer covers and thicker monolayer covers did not provide sufficient moisture storage to supply all of the evapotranspirative demands from vegetation. This evidence included measured and modelled moisture dynamics and evapotranspiration rates, frequency of wilting point conditions, the frequency and magnitude of water use from the saline shale overburden, the frequency of preferential flow during high intensity infiltration events, and the volumes of stored water (Barbour et al. 2006b). Due to space limitations, only a few examples of this evidence are summarized.

Soil moisture monitoring between 1999 and 2005 (Fig. 1) demonstrated that the volume of water stored in the layered covers approaches field capacity (FC) during the spring, when surface run-off rates are high due to snow melt, and net infiltration is greater due to spring rain and low evapotranspiration rates. These covers also reach FC occasionally throughout the summer and often in the autumn months due to heavy rainfalls. The monolayer cover appears unable to reach FC on an annual basis. This may be due to infiltrating water bypassing the clay matrix, passing directly through the cover along fractures and macropores. The presence of the peat/mineral layer appears to provide sufficient storage for infiltrating water so that it can be absorbed over a longer time period by the underlying clay matrix.

Soil water storage in the 35 cm cover did not reach wilting point (WP) conditions during the driest summer months (Fig. 1); however, the cover approaches the WP during most years due its lower available water holding capacity (AWHC). In contrast, soil water storage in the thicker layered covers remains significantly above WP conditions, even in the driest years, due to the higher AWHC of these covers. The 100 cm mono-layer cover appears to perform no better, and possibly poorer, than the 50 cm layered cover in limiting the onset of WP conditions.

Long-term soil-vegetation-atmosphere (SVA) modelling (Shurniak 2003) suggested that: (1) the peat-mineral layer plays a critical role in maintaining sufficient moisture in the glacial soil of all covers; and, (2) the optimal cover thickness is greater than 60 cm to accommodate extreme dry conditions. In summary, the soil moisture data and SVA modelling indicated that moisture would likely be a limiting factor for vegetation growth in the thinnest layered cover, and even in thicker monolayer covers, during dry climatic years.

3.2 In-Situ Hydraulic Conductivity

The average *in situ* values of saturated hydraulic conductivity (K_s) are shown in Fig. 2. The K_s increased several orders of magnitude in all materials within 3 to 4 years after cover placement in 1999. The increase in K_s over time is consistent with

A MULTI-DISCIPLINARY APPROACH TO RECLAMATION RESEARCH IN THE OIL SANDS REGION OF CANADA

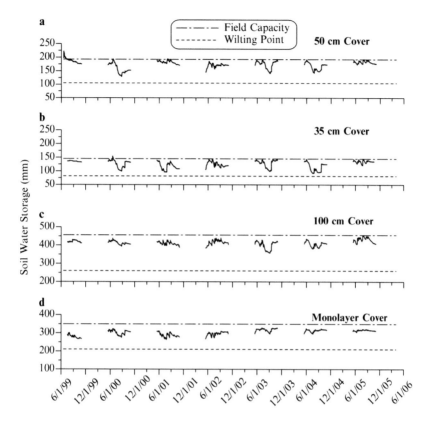

Fig. 3.1 Soil water storage in the 35, 50, 100 cm and monolayer covers

other cover performance studies in which K_s increases are either inferred by increasing drainage rates (e.g., Khire et al. 1997) or measured directly in the field and laboratory (e.g., Albright et al. 2006). In this case, the increase in K_s was attributed to the development of macrostructure caused by freeze-thaw and wet-dry cycling (Meiers et al. 2003). This data demonstrates the importance of tracking K_s over time as bio-physical processes can alter hydraulic properties and possibly the hydrologic response of the cover from the intended design (MEND 2004).

3.3 Hydrometric Data

Figure 3 presents the cumulative volume of water collected in the interflow system for 2005 along with measured precipitation. Only specific hydrometric data are

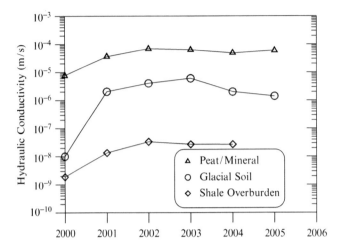

Fig. 3.2 Measured hydraulic conductivity over time (After Meiers et al. 2006)

presented for brevity. Annual interflow volumes ranged from about 1,000 L in 2001 to over 60,000 L in 2005. The shallow monitoring well data indicated that a perched water table, extending approximately 15 m upslope from the toe, develops annually on the cover-shale interface. A comparison between yearly ground temperature data and interflow response revealed the following annual pattern: (1) snow-melt (air temperatures above 0 °C) is complete approximately 1-month prior to the onset of interflow; (2) the majority of snow-melt water reports as surface-runoff; (3) a perched water table develops and interflow begins almost immediately when the ground thaws; and, (4) the cessation of interflow coincides with a recession of the perched water table and increased matric suction in the cover due to elevated evapotranspiration rates in the summer (Kelln et al. 2006).

The most interesting aspect of these observations is the time-lag between snow melt and the initiation of ground thaw. It is well known that surface run-off occurs in cold northern climates during spring due to frozen ground conditions. Infiltration into frozen unsaturated soil is restricted due to pore ice blockage (Newman and Wilson 1997). Consequently, snow melt infiltration must occur via preferential pathways (i.e. macrostructure) when the ground is frozen. Water stored in the macrostructure then migrates into the soil matrix as the ground thaws, creating a perched water table (interflow) on the cover-shale interface.

Finally, it is interesting to note that interflow monitoring is somewhat unconventional for engineers, but more commonplace for hydrologists. In this study, the SVA modelling suggested that saturated conditions would not develop at the base of the cover, particularly in the thickest covers, due to the large amount of AWHC and the semi-arid climate. In contrast, both preferential flow and interflow are reported as near ubiquitous phenomenon in hillslope hydrology studies (Grayson and Western 2001; McDonnell 2003) and form an integral component of watershed monitoring

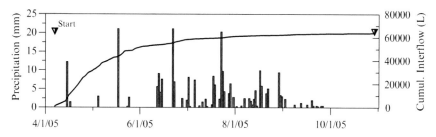

Fig. 3.3 Cumulative interflow and precipitation measured in 2005

programs in all climatic regions. Both interflow and preferential flow have implications for the moisture and salt balance of reclamation covers and should therefore be monitored on a watershed scale (MEND 2004).

3.4 Chemistry and Isotopes

Figure 4 presents the SO_4 concentration in the interflow water (Fig. 4a) and the $\delta^{18}O$ (Fig. 4b) of interflow water, snow, and monitoring wells during the 2005 snow melt. Solute concentrations in the interflow water increase each year from near 0 mg/L at the start of spring melt to a value equivalent to the pore water concentration near the interface. Similarly, the $\delta^{18}O$ interflow water evolves from a fresh snow melt water signature, to a $\delta^{18}O$ value that is consistent with measured *in-situ* $\delta^{18}O$ values of pore water. The *in-situ* $\delta^{18}O$ values were characteristic of summer and fall precipitation, indicating that connate pore water in the cover was derived from precipitation in the following year. An isotope hydrograph separation demonstrated that the interflow waters are eventually comprised of about 80% connate pore water.

The 'evolution' of interflow SO_4 and $\delta^{18}O$ concentrations suggests that the interflow evolves from preferential waters to flow through the soil matrix. The fresh water in the preferential flow paths nearest the interflow pipe drain into the interflow collection system almost immediately, followed by connate water that is transported down slope as the ground thaws and a perched water table develops upslope of the interflow collection system. The time to transition from preferential to predominantly connate water would depend on the upslope continuity and connectivity of preferential flow paths and the volume of water stored in the macrostructure.

These observations are consistent with hillslope hydrology studies, which report that, on average, 75% of 'stormflow' reporting to streams is attributed to connate water stored in the catchment before the episode (Buttle 1994). The geochemistry is also in keeping with the hydrometric data, which indicated that macrostructure is the dominant pathway available for infiltration during snow melt. More importantly, the interflow is responsible for the down-slope translocation of soil moisture and the flushing of salts from the cover in lower-slope positions.

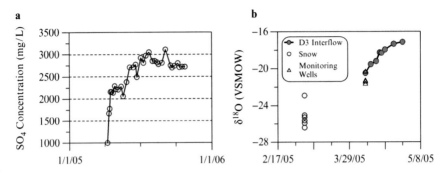

Fig. 3.4 Sulphate concentration and $\delta^{18}O$ signature of interflow waters

3.5 Salt Ingress

In-situ profiles of pore water solute revealed the following (Barbour et al. 2006c): (1) salt ingress is ubiquitous across the watershed; (2) the extent of salt ingress varies from cover to cover and is also a function of location within each cover (i.e. spatial variability); and, (3) interflow and deep percolation are key processes acting to attenuate salt ingress. On average, salt ingress has occurred 15–30 cm above the cover-shale interface in the 35 and 50 cm thick covers, and about 40 cm above the interface in the 100 cm cover. The increased salt ingress in the 100 cm cover is attributed to excess moisture conditions at the base of the cover, which increases the coefficient of diffusion (Lim et al. 1998). Although salt ingress has occurred to a greater extent in the thickest cover, the thinner covers have the greatest potential for 'failure' over time due to the encroachment of salt-rich pore water on the rooting zone.

A comparison of measured and modelled one-dimensional pore water Na^+ profiles demonstrated that salt ingress is greater in lower-slope positions than upper-slope positions due to spatial variability in the soil moisture regime and groundwater flux conditions. The upper-slope regions are drier due to down-slope moisture translocation and a lack of up-slope contributing area for surface run-off during the spring melt. As a result, upward salt transport in upper-slope regions is attenuated by the reduced coefficient of diffusion. The profiles also suggested that at least some deep percolation and/or interflow is occurring in upper-slope regions and is acting to minimize salt ingress, which is consistent with the preferential flow mechanism discussed above.

In lower-slope positions, interflow and deep percolation both appear to counteract the increased diffusive transport rate caused by increased soil moisture. The pore water chemistry demonstrated that current reclamation landscapes are (1) effective in reducing salt ingress, which will ultimately bode well for the development of a healthy eco-system, and (2) the thickest cover is the optimal design for mitigating the effects of salt ingress over the short and long-term. The latter observation is again consistent with the soil moisture monitoring and demonstrates that pore water chemistry is a valuable tool for evaluating long-term cover performance.

3.6 Vegetation Growth

A tree-growth survey was conducted in 2005 to determine if there is spatial variability in vegetation response on the three layered covers. In general, the overall health of vegetation depends on the edaphic factors (moisture and nutrients) and the potential for toxicity caused by salt ingress into the covers. The average measured height and diameter of planted trembling aspen trees on the 50 and 100 cm covers together was about 220 and 28.5 cm, respectively. In contrast, the average height and diameter of trembling aspens on the 35 cm cover was 195 and 26.4 cm, respectively. The datasets for both the 50 and 100 cm covers proved to be statistically different from the 35 cm cover tree data.

The trembling aspen tree data demonstrate that tree growth rates have been greater on the thicker covers (50 and 100 cm). This observation is in keeping with other vegetation studies at the site that have shown a greater number of species have established themselves on the thicker covers, and that the thicker covers have remained more amenable for germination and establishment than the thin cover over time (Purdy, unpublished data). The discrepancy between the two thickest cover prescriptions and the 35 cm cover is likely attributable to the measured differences in soil moisture. Previous studies have shown that the foliar nutrient levels in all three covers were generally within ranges considered acceptable and non-limiting in aspen and white spruce based on comparisons of measured and literature values (Purdy, unpublished data). Finally, the lower AWHC of the thinner cover is consistent with lower transpiration rates that were measured in both white spruce and trembling aspen saplings (Elshorbagy and Barbour 2007).

4 Summary and Conclusions

A multi-disciplinary approach to monitoring and evaluating the long-term performance of a reclamation cover provided important insight into the various mechanisms and processes that will control the hydrological and biological function of the reclaimed landscape. Conventional monitoring of soil moisture in the alternative prototype covers indicated that thin (35 cm) layered covers cannot meet all the moisture demands placed on them and plant communities characteristic of drier sites would likely develop. The saturated hydraulic conductivity of the clay-rich cover material increased by two orders of magnitude within 4 years of cover placement due to freeze-thaw effects. These preferential flow paths play a pivotal role in the formation of interflow as a result of infiltration of fresh snow melt water to the base of the cover during the spring melt. Interflow and deep percolation are important mechanisms for salt flushing, which protect the cover from ongoing salt ingress from the underlying shale. Finally, there is a correlation between the physical and ecological response of the covers. Tree growth and physiologic indicators showed that aspen growth on the shallower covers was lower than that observed on thicker covers. This is attributed to the measured lower volumes of water in the shallower covers.

Acknowledgements The funding provided my Syncrude Canada Limited and the National Science and Engineering Research Council are gratefully acknowledged. In addition, this work would not have been possible without the enthusiastic contribution of numerous graduate students, research engineers, and collaborating engineers and scientists.

References

Alberta Environmental Protection 1994. A Guide for Oil Production Sites, Pursuant to the Environmental Protection and Enhancement Act and Regulations. Queen's Printer.

Albright, W.H., Benson, C., Gee, G.W., Abichou, T., McDonald, E.V., Tyler, S.W., and Rock, S.A. 2006. Field performance of a compacted clay landfill final cover at a humid site. Journal of Geotechnical and Environmental Engineering, **132**(11): 1393–1403.

Barbour, S.L., Chapman, D., Qualizza, C., Kessler, S., Boese, C.D., Shurniak, R., Meiers, G.P., O'Kane, M., Hendry, J., and Wall, S.N. 2004. Tracking the evolution of reclaimed landscapes through the use of instrumented watersheds – a brief history of the Syncrude Southwest 30 overburden reclamation research program. *In* International Instrumented Watershed Symposium, Edmonton, Canada, pp. 622–644.

Barbour, S.L., Chanasyk, D., Hendry, J., Macyk, T., Mendoza, C., Nichol, C., Naeth, A., Leskiw, L.A., Purdy, B., Qualizza, C., Quideau, S., and Welhelm, C. 2006a. Phase 4 Technology Transfer Integration Document (Draft), Syncrude Canada Ltd.

Barbour, S.L., Chanasyk, D., Hendry, J., Macyk, T., Mendoza, C., Nichol, C., Naeth, A., Leskiw, L.A., Purdy, B., Qualizza, C., Quideau, S., and Welhelm, C. 2006b. Phase 1 Technology Transfer: Soil Structure and Moisture Regimes, Syncrude Canada Ltd.

Barbour, S.L., Chanasyk, D., Hendry, J., Macyk, T., Mendoza, C., Nichol, C., Naeth, A., Leskiw, L.A., Purdy, B., Qualizza, C., Quideau, S., and Welhelm, C. 2006c. Phase 2 Technology Transfer: Salt Impacts on Soil Covers in Saline Sodic Overburden Landscapes, Syncrude Canada Ltd.

Buttle, J.M. 1994. Isotope hydrograph separations and rapid delivery of pre-event water from drainage basins. Progress in Physical Geography, **18**(1): 16–41.

Elshorbagy, A. and Barbour, S.L. 2007. A probabilistic approach for design and hydrologic performance assessment of reconstructed watersheds. Journal of Geotechnical and Environmental Engineering.

Grayson, R. and Western, A. 2001. Terrain and the distribution of soil moisture. Hydrological Processes, **15**: 2689–2690.

Kelln, C.J., Barbour, S.L., Elshorbagy, A., and Qualizza, C.V. 2006. Long-term performance of a reclamation cover: The evolution of hydraulic properties and hydrologic response. *In* The Fourth International Conference on Unsaturated Soils. *Edited by* ASCE. Carefree, Arizona. April 2–6.

Khire, M.V., Benson, G.H., and Bosscher, P.J. 1997. Water balance modelling of earthen final covers. Journal of Geotechnical and Environmental Engineering, **123**(8): 744–754.

Leskiw, L.A. 1998. Land Capability Classification for Forest Ecosystems in the Oil Sands Region, Alberta Environment C&R IL.

Leskiw, L.A. 2006. Land Capability Classification System for Forest Ecosystems in the Oilsands Region, Alberta Environment C&R IL.

Lim, P.C., Barbour, S.L., and Fredlund, D.G. 1998. The influence of the degree of saturation on the coefficient of aqueous diffusion. Canadian Geotechnical Journal, **35**: 811–827.

McDonnell, J. 2003. Where does water go when it rains? Moving beyond the variable source area concept of rainfall-runoff response. Hydrological Processes, **17**: 1869–1875.

Meiers, G.P., Barbour, S.L., and Meiers, M.K. 2003. The use of field measurements of hydraulic conductivity to characterize the performance of reclamation soil covers with time. *In* 6th International Conference on Acid Rock Drainage. Cairns, Australia.

Meiers, G.P., Barbour, S.L., and Qualizza, C.V. 2006. The use of *in situ* measurements of hydraulic conductivity to provide an understanding of cover system performance over time. *In* 7th International Conference on Acid Rock Drainage. St. Louis, MO, March 26–30.

MEND 2004. Report 2.21.4: Design, Construction, and Performance Monitoring of Cover Systems for Waste Rock and Tailings, p. 93.

Newman, G.P. and Wilson, G.W. 1997. Heat and mass transfer in unsaturated soils during freezing. Canadian Geotechnical Journal, **34**: 63–70.

Penman, H.C. 1948. Natural evapotranspiration from open water, bare soil and grass. Proceedings of the Royal Society of London, **A193**: 120–145.

Qualizza, C., Chapman, D., Barbour, S.L., and Purdy, B. 2004. Reclamation research at Syncrude Canada's mining operation in Alberta's Athabasca oil sands region. *In* 16th International Conference, Society for Ecological Restoration. Victoria, Canada, August 24–26.

Shurniak, R. 2003. Predictive Modeling of Moisture Movement Within Soil Cover Systems for Saline/Sodic Overburden Piles. M.Sc., University of Saskatchewan, Saskatoon, SK.

Intelligent Machine Monitoring and Sensing for Safe Surface Mining Operations

S. Frimpong, Y. Li, and N. Aouad

Abstract The creation and maintenance of a healthy surface mining environment require advanced research initiatives for developing powered excavation and haulage technologies. The shovel-truck system is widely used in surface mining due to flexibility, economics and maintainability. Advances in technology have resulted in large shovels and trucks for economic, bulk operations. These advances have resulted in high-impact shovel loading operations (HISLO), which cause significant vibrations and affect an operator's health. Dump truck operators also face challenges in interacting with mine layouts, which include limited vision due to extensive "blind" areas and truck stability in difficult conditions. This paper contains a summary on frontier research on truck control, vision and collision avoidance and vibrations and their effects. Dynamic modeling is used to capture truck-road interactions. Intelligent sensing and collision avoidance system is used to develop an integrated system for a 360° vision. Theoretical models are also used to capture the effects of vibrations from HISLO. Virtual simulators are used to simulate the response of the integrated system to HISLO vibrations. This research is significant because it will provide a strong basis for developing technologies to improve shovel-truck haulage safety in surface mines and construction sites.

Keywords Truck vibrations · dynamic control · situational awareness · operating safety

1 Introduction

Dump truck operators face challenges in interacting with mine layouts. These challenges include limited vision due to extensive "blind" areas and truck stability during haulage. This blind area depends on truck size, cabin location, road geometry

S. Frimpong (✉)
University of Missouri-Rolla, Missouri, USA

and visibility. Figure 1 shows the blind areas around a 150-t truck. The ground level can only be seen after 105 ft away from the cabin (right side) and 16 ft (left side). A 6-ft person can be seen by an operator 70 ft away (right side) and 10 ft away (left side) (Miller 1975). Haul road terrains may also be uneven, curvy with high grades, which are augmented by ice or snow, mud, pot holes and tire wear. Depending on prevailing conditions, vehicular control can result in steering control losses and fatal accidents, with human injuries and deaths (Adlinger and Keran 1994). Out of the 250 fatalities (1998–) for US surface mines, 40% was due to powered haulage (MSHA 2004). An average of 675 accidents and 21 fatalities involving powered haulage equipment occur each year in surface mining and 20% of these accidents involve dump trucks (Ruff 2002). Large equipment, such as P&H 4100XPB, 4100 BOSS, CAT 797 and HE 930, is being used for economic, bulk production operations. These large machines have improved efficiency and economics. However, these successes have been achieved with negative side effects. Figure 2 illustrates a typical HISLO.

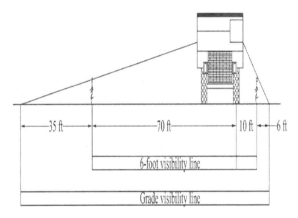

Fig. 1 Blind areas around 150-t truck

Fig. 2 High-impact shovel loading (HISLO)

Research has improved truck-road interactions and reduced collisions (Changirwa et al. 1999; Frimpong et al. 2003; Harpster et al. 1996; Schiffbauer 2001). However, tire-road interactions, and collision problems persist in surface mining (Ruff 2004). Frimpong et al. (2003) developed a simulation algorithm to test an automated dump truck. Vibration research has focused on dynamics of light trucks, cars and trains and their responses to stimuli. Active and semi-active control methods have been used to reduce vibrations with better results than passive technologies (Leeming and Hartly 1981; Margolis and Nobles 1991; Wilkinson 1994).

2 Global Intelligent Truck Dynamic Control and Sensing

The challenge is to develop intelligent sensing technologies to assist operators in mitigating extreme dynamic deviations on haul roads, and reduce blind areas around trucks. These technologies will predict impending collisions, reduce false alarms and help operators in making intelligent decisions. To achieve this objective, current research focuses on: (i) truck control dynamics; (ii) truck-road interactions; (iii) intelligent sensing and collision avoidance system; (iv) integrated system for situational awareness; and (v) tolerance of developed technologies.

2.1 Dump Truck Kinematics for Effective Machine Control

Kinematics modeling is used for displacement, velocity and acceleration profiling and steering control associated with truck motions (Zorriassatine et al. 2003). Truck kinematics research has revealed its potential for surface mining (Frimpong et al. 2003; Changirwa et al. 1999). Current research focuses on vehicle speed, braking, rolling, grade and acceleration and effects on steering system dynamics. Research also focuses on lateral steering system flexibility using combined steering and vehicular equations of motion. Given specific geometric planes, the forward motion, the pitch, roll angle and the yaw rate interact with the truck to perform its kinematics functions (in Fig. 3).

2.2 Dump Truck-Road Dynamics

Truck-road interactions allow haul road profiling and assist in screening out intrusive objects. Figures 3 and 4 illustrate the tuck dynamic behavior, the truck-road interactions and road deformation (Frimpong and Li 2006; Frimpong et al. 2005). Current research has advanced virtual truck-road simulators, which provide a basis for dynamic sensory system. The truck-road dynamics consist of: (i) dynamic truck model; (ii) load deformation model; and (iii) truck-soil contact model. During

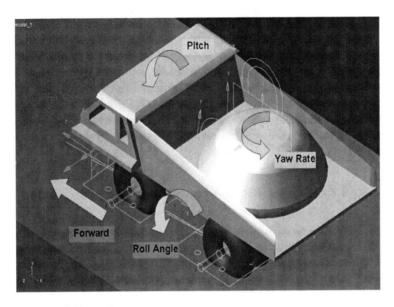

Fig. 3 Dump truck kinematics

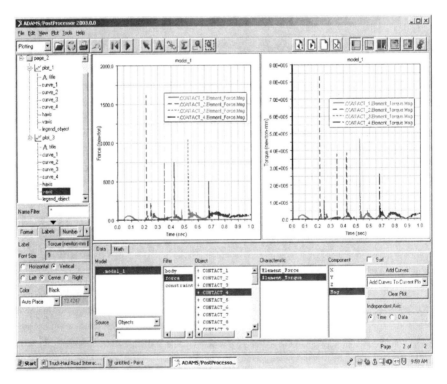

Fig. 4 Tire contact force and torque

Intelligent Machine Monitoring and Sensing

haulage, the road bed is subjected to a contact force, which includes the normal reactive, lateral resistant and longitudinal reactive forces. Kinematics and dynamic results are used to develop optimum truck-road performance parameters, which serve as threshold parameters for dynamic sensing.

2.3 Truck Vision and Collision Avoidance Research

Collision sensing augments situational awareness. Situational awareness is critical in two areas: (a) loading and dumping; and (b) empty and loaded haulage. The possible sources of collision include: (i) veering off course; (ii) human, animal or other obstacle; (iii) lateral collision with vehicle in opposite lane. Figure 5 shows a schematic of the sensor suite for enhancing an operator's situation awareness. It consists of: (i) front-side stereo-camera system; (ii) front-side Doppler radar; (iii) close range sonar proximity sensors; and (iv) rear-side camera system. For effective interface, only the relevant information must be displayed (Wallace and Donald 1992; Jansson et al. 2002).

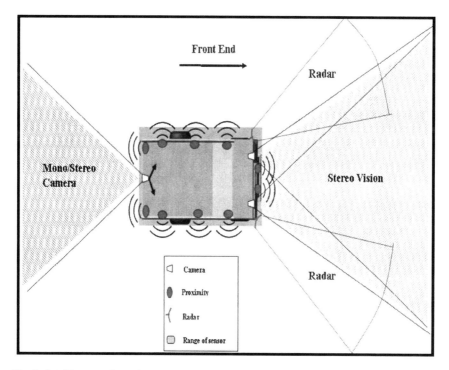

Fig. 5 Intelligent truck sensing

3 Dump Truck and Operator Vibrations in HISLO

The primary objective is to provide a basis for developing appropriate technologies to prevent long-term operator health problems. The elements of this objective include developing: (i) HISLO and shock wave generation and propagation (SWGP) models; (ii) models to capture an operator body movements due to SWGP; (iii) a virtual simulator of (i) and (ii) in ADAMS; and (iv) carrying out laboratory bench-scale studies.

3.1 Loading Impact Force and Generated Shock Waves

The impact force is a function of material property, truck weight, truck-road interactions, stiffness and damping coefficients of the truck and its suspension. Other environmental and operating factors include: (i) physical and mechanical properties of material; (ii) shovel operating conditions; (iii) prevailing environmental conditions; and (iv) dumping process. The increasing truck mass, from the net vehicle weight (NVW) to the gross vehicle weight (GVW), affects the properties of vehicle vibration. Frimpong et al. (2004) showed that significant vibrations are produced from HISLO.

3.2 Shock Waves and Vibrations Propagation

Actual suspension system is used to develop vibration models using Newton-Euler-Lagrange theory. The resultant vibration vector is a function of truck vibrations. A time-dependent shock wave response is obtained using finite element modeling (FEM) techniques. Modal and parameter sensitivity analyses are used to model the shock waves and the suspension system response. Control theory is used to analyze transmissibility functions between transmission segments. The models are combined to develop an integrated truck vibration model to capture the 3D multi-body HISLO vibrations.

3.3 Virtual HISLO Simulators

Virtual simulators are used to simulate the behavior of the integrated 3D truck-operator vibration model within ADAMS and ALGOR. These packages have excellent post-processor tools for 3D solid models, FEA tools, simulation graphics, animation, contouring, curve analysis, and continuous modifications. Figures 6 and 7 show a 3D simulator and a solid model for simulating truck vibrations (Frimpong et al. 2004). Figure 8 shows the vibration acceleration response. This model incorporates varying loading impact forces and the tire-road interactions. The suspension mechanisms of the rear of the dump truck will is expanded to account for the dual-tire geometry. The simulator will be used to test various operating conditions to understand the human body responses to truck vibrations.

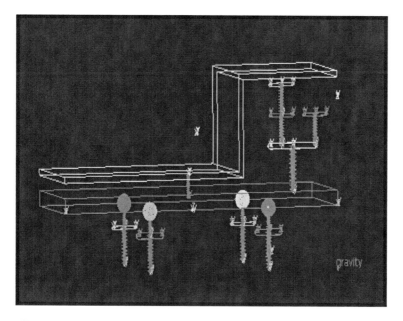

Fig. 6 3D Truck vibration simulator

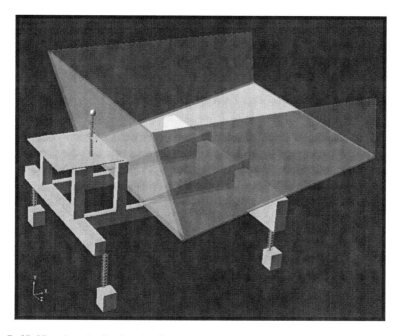

Fig. 7 3D Virtual truck vibration simulator

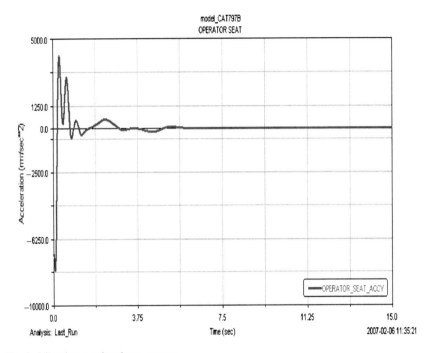

Fig. 8 Vibration acceleration response

4 Research and Technology Challenges

4.1 Sensor Selection and Layout Design

One design challenge is the front-side stereo camera and radar systems. The stereo system's accuracy depends on the separation between the cameras. The camera's range and field view depend on the focal length of the lenses. In order to allow maximum area of stereo overlap between the two views of the camera, the camera's CCD sensors are placed off-center from the optical axis (Bergendahl et al. 1996).

4.2 Sensor Integration and Calibration

Once the camera systems and other sensors are mounted on the truck, they are registered and calibrated with respect to each other and with respect to the truck so that all the sensors are situated in common coordinate systems. An automated procedure and appropriate hardware will be developed to calibrate all the sensors.

4.3 Collision Detection and Sensor Control Algorithms

Most of the sensors are commercial with their own processing modules. However, an in-house vision algorithm is required for "recognition" or characterization of the targets using 2-1/2 D information from the stereo cameras. The main collision detection challenge is the information fusion and the sensor control algorithms. In particular, the rear view camera direction should be appropriately controlled to provide the correct view of the spotting area. The sweep area of the radar should be similarly controlled depending on the steering commands to capture data from the resulting blind spots of the stereo camera. The multi-sensor data from different sensors around the truck need to be fused into a single operating view of the environment (Takano et al. 2004; Terrien et al. 2000).

4.4 False Alarm Mitigation

False alarms are a compromise between conflicting requirement of maximizing detection of actual impending collisions and minimizing false detections. A metric of false alarms that has operational significance for the dump truck operators is the false alarm rate. If the system make a few false alarms per hour, it may be useful in keeping the operator alert. However, frequent false alarms may cause an operator to shut it off. While overlapping information from multiple sensors is one mechanism to reduce false alarm, integration of information over time and processing can help reduce false alarms (Wallace and Donald 1992).

4.5 Operator-Interface Information Overload

Effective operator response to audio-visual alarm system must preclude operator-interface information overload. The strategy is to ensure that data acquisition, processing, and all interchanges are carried out in the background. Operators are exposed to audio-visual alarm systems for mitigating hazards/dangers.

4.6 Filtering Vibration Noises

To understand HISLO vibration effects, all other residual noises from external stimuli, such as shovel banging must be filtered from recorded vibration signals.

5 Research Significance to the Surface Mining Industry

This research covers a broad spectrum including truck kinematics and dynamics, truck-haul road dynamics, soil load deformation, intelligent target sensing and vision engineering and truck and operator vibration control in HISLO. These areas constitute a scientific envelope for a healthy surface mining operation. This initiative is significant because it will develop technologies to improve shovel-truck loading and truck haulage safety in surface mines and construction sites. These improvements will save many lives, promote efficient operating practices and equipment longevity. Intelligent sensing will improve dump truck operator's control and visibility in difficult environmental conditions. This research program will also advance intelligent sensing through fundamental and applied research. Collision avoidance systems will improve operator and equipment safety. This research will further create understanding into the long-term effects of HISLO on an operator's general health conditions.

The study will also provide measures to reduce and/or eliminate the sources of operator health problems in HISLO operations. The effective control and/or elimination of HISLO problems will result in safe working environments, healthy work force, production efficiency and economic operations. It will also develop new computational intelligent algorithms, simulation platforms, and truck operating standards for addressing fundamental problems in surface mining.

6 Conclusions

Dump truck operators face enormous challenges in interacting with mine layouts and haul roads during truck haulage. Two major challenges are an operator's limited vision due to the extensive "blind" areas around the truck and its stability during an operator's response to intrusive dangers. The extent of this blind area is a function of truck size, operator's cabin location, haul road geometry and environmental challenges to visibility. These problems may be augmented by the presence of ice or snow, mud, pot holes and tire tread wear. Depending on prevailing road conditions and the proximity of intrusive objects, vehicular control in response to perimeter sensing can be very challenging. The challenges can result in steering control losses and fatal accidents with corresponding human injuries and deaths, equipment and production losses and irreversible capital investment losses. Other problems include the operator low back and general health conditions in HISLO conditions over an extended time period. Using the current body of scientific knowledge, advanced research initiatives are being carried out by to address health problems associated with truck haulage and HISLO.

References

Adlinger, J.A. and C.M. Keran (1994), A Review of Accidents During Surface Mine Mobile Equipment Operation, *Proc. of 25th AIMHSR*, Blacksburg, VA, pp. 99–108.

Bergendahl, J., I.Masaki and B.K.P. Horn (1996), Three-Camera Stereo Vision for Intelligent Transportation Systems, *Transp. Sensors Controls Proc., Vol. 2902.* © SPIE, Boston, MA.

Changirwa, R., S. Frimpong and J. Szymanski (1999), Intelligent Modeling of Multi-phase Slurry Freight & Vehicle for Traffic Chaos Control, *Proc. of ISAS/SCI'99 Conference*, Orlando, FL.

Frimpong, S. and Y. Li (2006), Current Research on Intelligent Truck Sensing for 360° Perimeter Sensing, *Mining and Nuclear Engineering*, Missouri S&T, Rolla, MO.

Frimpong, S., R. Changirwa and J. Szymanski (2003), Simulation of Automated Dump Trucks for Large-Scale Surface Mining, *IJSM, Vol. 17, No. 1.* © Swets & Zeitlinger, The Netherlands.

Frimpong, S., Z. Chang and V. Kecojevic (2004), Biomechanics of Operator Vibrations Control in High-Impact Shovel Loading Operations. *SCI Conference*, Orlando, FL (July).

Frimpong, S., Y. Hu and K. Awuah-Offei (2005), Mechanics of Cable Shovel-Formation Interactions in Surface Mining Excavations, *Journal of TerraMechanics, Vol. 42.* © International Society for Terrain Vehicle Systems, Moline, IL, pp. 15–33.

Harpster, J.L., R.W. Huey, N.D. Lerner and G.V. Steinberg (1996), Backup Warning Signals – Driver Perception and Response, *Technical Report No. DOT HS 808 536.* © US DOT.

Jansson J., J. Johansson and F. Gustafsson (2002), Decision Making for Collision Avoidance Systems, *SAE 2002 World Congress & Exhibition* (Intelligent Vehicles: Advances in Crash Avoidance), Doc. 2002-01-0403, March 2002, Detroit, MI.

Leeming, D.J. and R. Hartly (1981), Heavy Vehicle Technology (2nd ed.), Hutchinson: London.

Margolis, D. and C.M. Nobles (1991), Semi-active Heave and Roll Control for Large Off-road Vehicle, *Com. Vehicle Suspensions, Steering Systems & Traction.* © SAE, Warrendale, PA.

Miller, W.K. (1975), Analysis of Haul Truck Visibility Hazards at Metal and Non-metal Surface Mines, *MESA Information Report 1038*: 19.

MSHA (2004), Distribution of Total Days Lost by Accident Class: Surface Mining Operations, © MSHA; http://www.cdc.gov/niosh/mining/topics/data/images-pdfs/acds.gif.

Ruff, T. (2002), Hazard Detection and Warning Devices: Safety Enhancement for Off-Highway Dump Trucks. *Compendium of NIOSH Research.* © NIOSH, Washington DC.

Ruff, T.M. (2004), Advances in Proximity Detection Technologies for Surface Mining Equipment, *Proc. of 34th AIMHSR*, Salt Lake City, UT.

Schiffbauer, W.H. (2001), An Active Proximity Warning System for Surface and Underground Mining Applications, *SME Annual Meeting; Pre-print No. 01-117*, Denver, CO.

Takano, K., T.Monji, H. Kondo and Y. Otsuka (2004), Environment Recognition Technologies for supporting Safe Driving, *Hitachi Review*, Vol. 53, No. 4 (November).

Terrien G., T. Fong, C. Thorpe and C. Baur (2000), Remote Driving with a Multisensor User Interface, International Conference on Environmental Systems (Tolerobotics), Doc. 2000-01-2358, July 2000, Toulouse, France.

Wallace, J.S. and Donald, L.F. (1992), Interface Between the Driver and Collision Warning Systems: Lessons in Complexity, *Proc. of SPIE*, Vol. 2902, pp. 299–305.

Wilkinson, P.A. (1994), Advance Hydraulic suspension Ride Simulation Using a 3-D Vehicle Model; Vehicle Suspension System Advancements, *SAE*, Warrendale, PA.

Zorriassatine, F., C. Wykes, R. Parkin and N. Gindy (2003), A Survey of Virtual Prototyping Techniques for Mechanical Product Development, *Inst. Mech. Eng. Part B: J. Eng. Manuf.* Vol. 217, pp. 513–530.

Neutralization Potential of Reclaimed Limestone Residual (RLR)

H. Keith Moo-Young

Abstract Mining activities that lead to the exposure of iron pyrite and sulfite minerals associated with coal deposits to air and water result in the problem of acid mine drainage (AMD). In the U.S., AMD and other toxins from abandoned mines have polluted 180,000 acres of reservoirs and lakes, and 12,000 miles of streams and rivers. Acidity is a characteristic of AMD, and due to low pH conditions metals such as iron, aluminum, copper, zinc, manganese, magnesium, and calcium are leached from soil and rock contaminating streams.

Reclaimed Limestone Residual (RLR) is a co-product of the steel making process, and is developed during the refining of crude iron products to steel. It has been shown to have oxidation-reduction capabilities that facilitate metals reduction, and also has significant acid neutralizing potential. In this study, the neutralization potential of RLR was studied.

Keywords RLR · acid mine drainage · neutralization · slag · x-ray diffraction

1 Introduction

Mining activities that lead to the exposure of iron pyrite and sulfite minerals associated with coal deposits to air and water result in the problem of acid mine drainage (AMD). In the U.S. AMD and other toxins from abandoned mines have polluted 180,000 acres of reservoirs and lakes, and 12,000 miles of streams and rivers. It has been estimated that cleaning up these polluted waterways will cost US taxpayers between $32 billion and $72 billion (Speart 1995). AMD is the product formed by the atmospheric (i.e. by water, oxygen and carbon dioxide) oxidation of the

---

H. Keith Moo-Young (✉)
California State University Los Angeles, 5151 State University Dr., Los Angeles, CA 91105, USA
e-mail: kmooyou@calstatela.edu

relatively common iron-sulfur minerals, pyrite (FeS_2) and pyrrhotite (FeS), in the presence of (catalyzed by) bacteria (Thiobacillus ferrooxidans), and any other products generated as a consequence of these oxidation reactions. Two key features of AMD should be understood to appreciate why it is a big problem:

1. Rainwater flowing over unearthed sulfide minerals may become acidic. When water becomes acidic, it dissolves toxic heavy metals, which are abundant in abandoned coalmines.
2. As yet, we cannot predict whether or not water will become acidic, when acid-generation will start, or whether it will continue for months or decades.

Due to these factors, AMD may be a liability for mines long after they cease to operate, and the environmental impact it causes can be severe.

The equation for AMD formation is:

$$FeS_2 + 3.75\,O_2 + 3.5\,H_2O \rightarrow Fe(OH)_3 + 2\,H_2SO_4 \qquad (1)$$

Pyrite (FeS_2) is commonly present in coal seams and in the rock layers overlying coal seams and mining operations that involve the breaking of the rocks to get at the coal expose these minerals to the atmosphere resulting in AMD.

Acidity is a characteristic of AMD, and due to low pH conditions metals such as iron, aluminum, copper, zinc, manganese, magnesium, and calcium are leached from soil and rock contaminating streams (Nadakavukaren 1995). In regions such as the western part of the country where there is abundant iron, water resources can be devastated and aquatic life virtually disappears due to the low pH values that result in the formation of iron hydroxide that coats stream bottoms, forming the familiar orange colored "yellow boy" common in areas with abandoned mine drainage. Typical compositions of sulfide and coal mine operations are shown in Table 1.

Legislatively mandated regrading and revegetation of abandoned mine strips have been helpful in reducing runoff of sediment from abandoned mine sites, but have had minimal impact on acid drainage (Nadakavukaren 1995). Sealing openings to abandoned underground mines, thereby cutting off the supply of oxygen and water that permit acid formation has been the chief means of combating acid mine drainage. Although the practice significantly reduces acid loadings it

Table 1 Typical mine drainage compositions (USEPA 2003)

Composition metals (mg/L)	Coal mines	Cu-Pb-Zn Sulfide mines	Industrial effluent limits
Ph	2.6–6.3	2.0–7.9	6–9
Fe	1–473	8.5–3,200	3.5
Zn		130–350	0.38
Al	1–58		8.2
Mn	1–130	0.4	2
Cu		0.005–76	0.05
Pb		0.02–90	0.2

does not eliminate them, and therefore there is the need to find better more cost effective ways of dealing with the problem. In recent years the use of engineered wetlands has yielded some promising results especially when coupled with alkaline recharge zones to neutralize the acidity (Nadakavukaren 1995). To demonstrate the feasibility of utilizing RLR to remediate Acid Mine Drainage (AMD), the acid neutralization potential of RLR was determined and subsequent neutralization tests utilizing synthetic AMD groundwater were also conducted.

2 Acid Neutralization Potential (ANP) Results

The results of the neutralization tests by the four different methods are shown in Table 2. From these results it is determined that this particular RLR has an average neutralization potential of approximately 83% as calcium carbonate ($CaCO_3$). A comparison of the determined neutralization potential of this particular RLR with other similar material is shown in Table 3.

3 Neutralization Testing Results

Testing to determine the dissolved heavy metal removing capability of RLR from AMD were conducted by utilizing neutralization tests between RLR and synthetic AMD comprised of various concentrations of heavy metals at different pH values.

Table 2 ANP of RLR from the Waylite Corporation in Bethlehem, PA

Digestion method	Neutralization potential	
	%	Tons $CaCO_3$/ w1,000 t RLR
Sobek	81.6	816
Boil	84.1	841
H_2O_2	84.4	844
Sobper	80.5	805

Table 3 ANP of various steel slags

Steel slag type	Neutralization potential	
	%	Tons $CaCO_3$/1,000 t steel slag
RLR: Waylite, Bethlehem PA	83	827
Slag fines: Weirton, WV	76	760
Recmix: Washington, PA	69	690
Slag fines: 1/8 in., Mingo Jct, OH	66	660
EAF: Waylite, Johnstown PA	59	590
Slag fines, 1/8 in., USX, Fairfield, AL	53	530

The results from these tests are summarized in Table 4, and show that for the most part RLR was effective in removing over 99% of the dissolved metals from solution. Of interest to note is the removal of hexavalent chromium Cr (VI) since the commonly employed technique of precipitation does not remove it from solution. Based on these finding additional tests were conducted to identify the mechanism responsible for this added benefit of RLR, and are discussed later on.

4 RLR Analysis

Testing was conducted on RLR to determine both its physical and chemical properties. Water content, specific gravity, organic content, sieve analysis, and hydrometer analysis, were conducted according ASTM procedures. For a chemical characterization of RLR samples were finely ground and sieved to create a proper sample for both energy dispersive X-ray spectroscopy (EDX) and X-ray Diffraction (XRD). EDX is a micro analytical technique that is based on the characteristic X-ray peaks that are generated when the high-energy beam of the scanning electron microscope (SEM) interacts with the specimen. Each element yields a characteristic spectral fingerprint that may be used to identify the presence of that element within the sample. The relative intensities of the spectral peaks were used to determine the relative concentrations of each element in the specimen. XRD is an analytical technique that is used to study the atomic and molecular structure of substances by directing X-rays at them and causing a slight spreading of the waves (diffraction of the rays) around the atoms. By using measurements of the position and intensity of the diffracted waves, it was possible to calculate the shape and size of the atoms in the samples and subsequently determine what compounds RLR is composed of.

The physical characteristics of RLR used in this study has the consistency of coarse-grained sand, and Fig. 1 is an image of what a 5 g sample looks like. The results of the major physical properties tested showed that RLR has a water content of 0.01%, an organic content of 1.5%, a porosity of 30%, and a specific gravity G_s of 3.46. Grain size analysis of the material from which RLR is obtained showed that it was well graded, and the results of the analysis are shown in Table 5 and Fig. 2.

Table 4 Heavy metal neutralization test results

Metal	Initial concentration range (mg/L)	Final concentration range (mg/L)	Initial pH range	Final pH range	Removal efficiency (%)
Fe	0.5–200	0.001–0.009	2.7–4.8	10.3–11.3	99.50–99.99
Zn	0.5–200	0.000–0.250	3.0–7.0	10.6–11.3	99.86–100.00
Al	0.5–200	0.002–0.087	2.1–5.9	10.0–11.3	83.80–99.99
Cu	0.5–200	0.010–0.060	2.1–6.1	10.7–11.4	96.00–99.99
Pb	0.5–200	0.001–0.003	3.2–7.0	10.9–11.5	99.50–99.99
Cr (VI)	0.5–100	0.010–29.98	1.4–7.0	10.8–11.4	70.02–96.00

Fig. 1 RLR (approximately 5 g)

Table 5 Grain size distribution results

Sieve No.	Mass retained (g)	Percent finer (%)	Opening (mm)
4	0	100.00	4.75
10	18,800	19.56	2
40	3,505.95	4.53	0.425
60	437.08	2.66	0.25
80	176.14	1.91	0.18
100	68.6	1.61	0.15
200	190.04	0.80	0.075
pan	186.62	0	0

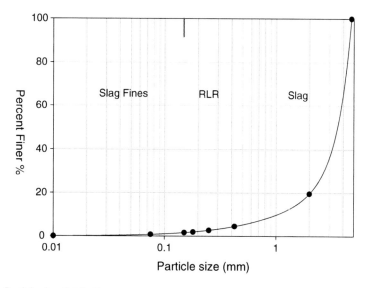

Fig. 2 Particle size distribution curve

Table 6 Elemental composition of RLR

Element	Mg	Al	Si	S	K	Ca	Ti	Mn	Fe
wt %	6.2	3.9	17.9	0.7	nd	58.6	nd	2.0	11.3

Table 7 Compound comprising RLR listed in order of abundance

Compound	MgO	Ca_2SiO_4	$Ca_5MgSi_3O_{12}$	MnAlO	Fe_2SiO_4	Mn	$Ca_2Al_2O_4$	MnO
Order	1	2	3	4	5	6	7	8

5 Chemical Characterization Results

The EDX analyses of RLR are shown in Table 6 by weight percents, and are averages of analyses from three areas of the SEM preparations. Results of the X-ray diffraction showed that RLR is approximately 92% crystalline and is composed of several compounds including Magnesium oxide, Calcium silicates, Calcium magnesium silicates, Manganese aluminum oxide, Iron silicate, Manganese, Calcium aluminum oxide, and Manganese oxide but not exclusive to these compounds. Table 7 lists these compounds in approximate order of abundance in the XRD pattern.

6 Conclusion

The viability of utilizing Reclaimed Limestone Residual (RLR) to remediate Acid Mine Drainage (AMD) was investigated. Physical and chemical characterization of RLR showed that it is composed of various minerals that contain significant quantities of limestone or calcium bearing compounds that can be exploited for acid neutralization. Acid Neutralization Potential (ANP) test results showed that RLR has a neutralization potential of approximately 83% as calcium carbonate ($CaCO_3$). Neutralization tests with most of the heavy metals associated with AMD showed removal efficiencies of over 99%.

Acknowledgement The authors would like to thank the U.S. Department of Energy for funding this work under grant number DE-PS26-03NT41634-14. The views and opinions of authors expressed herein do not necessarily state or reflect those of the United States Government or any agency thereof.

References

Nadakavukaren, A. (1995). *Our Global Environment*. Waveland Press, Prospect Heights. IL.
Speart, J. (1995). A Lust for Gold, *Mother Jones*, p. 60.
USEPA. (2003). Development Document for the Proposed Effluent Limitations Guidelines and Standards for the Metal Products & Machinery Point Source Category (EPA #: 821-B-00-005).

Quantification of the Impact of Irrigating with Coalmine Waters on the Underlying Aquifers

D. Vermeulen and B. Usher

Abstract It is predicted that vast volumes of impacted mine water will be produced by mining activities in the Mpumalanga coalfields of South Africa. The potential environmental impact of this excess water is of great concern in a water-scarce country like South Africa. Research over a period of more than 10 years has shown that this water can be used successfully for the irrigation of a range of crops (Annandale et al. 2002). There is however continuing concern from the local regulators regarding the long-term impact that large scale mine water irrigation may have on groundwater quality and quantity. Detailed research has been undertaken over the last 3 years to supplement the groundwater monitoring program at five different pilot sites, on both virgin soils (greenfields) and in coal mining spoils. These sites range from sandy soils to very clayey soils. The research has included soil moisture measurements, collection of in situ soil moisture over time, long-term laboratory studies of the leaching and attenuation properties of different soils and the impact of irrigation on acid rock drainage processes, and in depth determination of the hydraulic properties of the subsurface at each of these sites, including falling head tests, pumping tests and point dilution tests.

This has been supported by geochemical modelling of these processes to quantify the impacts. The results indicate that many of the soils have considerable attenuation capacities and that in the period of irrigation, a large proportion of the salts have been contained in the upper portions of the unsaturated zones below each irrigation pivot. The volumes and quality of water leaching through to the aquifers have been quantified at each site. From this mixing ratios have been calculated in order to determine the effect of the irrigation water on the underlying aquifers.

Keywords Coal mining · irrigation · gypsum-saturated · waters · attenuation capabilities

---

D. Vermeulen (✉) and B. Usher
Institute for Groundwater Studies, University of the Free State, P.O. Box 339, Bloemfontein, South Africa, 9300
e-mail: VermeulenD.SCI@.ufs.ac.za

1 Introduction

South Africa is a water-poor country. With increased industrialisation and population growth, the demands on this resource are increasing. South Africa is the fourth largest producer of coal in the world, and the 224 million metric tons of coal produced per year directly supports employment for approximately 50,000 employees. Unfortunately, several water-related problems, largely associated with water quality deterioration due to pyrite oxidation, occur, as a result of mining.

Huge volumes of mine water, impacted on by the phenomenon of acid mine drainage, are presently being produced as a result of mining activities in the Mpumalanga coalfields. When released into water environments, the high salinities of this water are responsible for unacceptable water quality degradation.

1.1 Objectives

In a water-stressed country like South Africa, all water must be regarded as a potential resource, and there is potentially a tremendous resource that can be utilised by activities such as irrigation, provided the environmental impact is not excessive. Irrigation provides for a novel approach to the utilisation and disposal of mine water, under the correct conditions. The significance of these findings lies in the versatility of this irrigation. Communities which often have very few other resources can utilise mine water to generate livelihoods. When one considers social aspects like job creation, especially after mine closure, it is clear that irrigation with certain mine waters, on carefully selected and managed sites, could form a sustainable, economically feasible and socially uplifting strategy in the developing world. This research investigated the impact of these activities on groundwater resources at five irrigation pivots at collieries across the coalfields of South Africa, where mine water irrigation has been done for periods ranging from several months to more than 7 years.

1.2 The Mpumalanga Coalfields and Associated Mine Water

Coal extraction has been ongoing at the Mpumalanga Coalfields for more than 100 years. Coal is generally mined by opencast- or underground methods in South Africa. (Grobbelaar 2001). The depth of mining ranges from less than 10 m below surface to more than 100 m. The coal seams generally increase in depth to the south. Mining methods are bord-and-pillar, stooping and opencast. Opencast mining has been introduced during the late 1970s. Underground mining on the 2-Seam comprises in excess of 100,000 ha, while opencast mining is expected to eventually exceed 40,000 ha (Grobbelaar et al. 2002). Several sources of water influx are expected in South African collieries. In opencast areas, much of the influx is dependent on the state of post-mining rehabilitation, while in underground mining, factors such as the mining type, depth and degree of collapse and interconnectivity is important.

After the closure of mines, water in the mined-out areas will flow along the coal seam floor and accumulate in the lower-lying areas. These voids will fill up with water, and hydraulic gradients will be exerted onto peripheral areas (barriers) or compartments within mines. This results in water flow between mines, or onto the surface eventually. This flow is referred to as intermine flow (Grobbelaar 2001). Projections for future volumes of water to decant from the mines have been made by Grobbelaar et al. (2002). In total, about 360 ml/day will decant from all the mines in combination.

1.3 Water Quality Impacts

Associated with coal mining in South Africa, the phenomenon of acid mine drainage (AMD) occurs. Acid mine drainage occurs when sulphide minerals in rock are oxidised, usually as a result of exposure to moisture and oxygen. This results in the generation of sulphates, metals and acidity. Pyrite (FeS_2) is the most important sulphide found in South African coalmines. When exposed to water and oxygen, it can react to form sulphuric acid (H_2SO_4). The following oxidation and reduction reactions give the pyrite oxidation that leads to acid mine drainage.

$$FeS_2 + 7/2\ O_2 + H_2O \Rightarrow Fe^{2+} + 2SO_4^{2-} + 2H^+ \quad (1)$$

$$Fe^{2+} + 1/4O_2 + H+ \Rightarrow Fe^{3+} + 1/2\ H_2O \quad (2)$$

$$Fe^{3+} + 3H_2O \Rightarrow Fe(OH)_3 + 3H^+ \quad (3)$$

$$FeS_2 + 14Fe^{3+} + 8H_2O \Rightarrow 15Fe^{2+} + 2SO_4^{2-} + 16H^+ \text{ (Stumm and Morgan 1996).} \quad (4)$$

In the South African coalfields there are co-existing carbonates such as calcite and dolomite, which can neutralise the acidity generated (Usher 2003) Alternatively the acidity can be neutralised by lime addition, as occurs with acidic water pumped from the Kleinkopje Colliery workings. From the overall reaction of calcite as buffering mineral, it is evident that calcium and sulphate will increase in concentration:

$$FeS_2 + 2CaCO_3 + 3,75O_2 + 1,5H_2O \Leftrightarrow Fe(OH)_3 + 2SO_4^{2-} + 2Ca^{2+} + 2CO_2 \quad (5)$$

This increase in Ca^{2+} and SO_4^{2-} can only occur up to a point, where the aqueous solubility of these ions becomes limited by the solubility of gypsum ($CaSO_4.2H_2O$). Using the PHREEQC geochemical model (Parkhurst and Appello 1999), the saturation state of the neutralised mine water used to irrigate at Kleinkopje's Pivot No1's was determined. The results show that the gypsum approached saturation (SI = 0) for most of the values. The implication of this is that when irrigation takes place, some evaporation, together with the selective uptake of essential nutrients, will result in gypsum precipitation. Gypsum is a partially soluble salt. Concentrating the gypsiferous soil solution through crop evapotranspiration precipitates gypsum in the soil profile and therefore removes it from the water system (see Table 1 for irrigation water quality), reducing potential pollution (Annandale et al. 2002).

Table 1 Average water quality of the irrigation water at Kleinkopje 1

pH	EC mS/m	Ca mg/L	Mg mg/L	Na mg/L	K mg/L	Alkalinity As mg/L CaCO3	Cl mg/L	SO4 mg/L	Fe mg/L	Mn mg/L	Al mg/L
6.21	344	578	242	52	12.9	34	12	2,550	3.1	10.3	0.01

1.4 Sustainability of Irrigation with Gypsiferous Mine Water

Annandale et al. (2002) did the initial work regarding irrigation with gypsiferous water in South Africa. Their research was to ascertain:

- Whether gypsiferous mine water can be used on a sustainable basis for irrigation of crops and/or amelioration of acidic soils
- The effects of gypsum precipitation and salt accumulation on soil characteristics and predict depths of salinisation of soil over time

The commercial production of several crops irrigated with gypsiferous mine water was tested in a field trial at Kleinkopjé Colliery from 1997 to 2000, and also at the other pivots at Syferfontein, Optimum and New Vaal Collieries since 2000. From these trials, it was observed that no foliar injury was observed due to sprinkle irrigation with gypsiferous mine water, and that possible nutritional problems, such as deficiencies in K, Mg and NO_3, occurring due to Ca and SO_4 dominating the system, can be solved through fertilization. The soil salinity at Kleinkopje increased compared to the beginning of the trial, but the values of soil saturated electrical conductivity fluctuated around 200 mS/m, which is typical for a saturated gypsum solution.

Crops like sugarbeans, wheat and maize were found to be commercially viable. The finding from this research was that gypsiferous mine water for irrigation proved to be sustainable for crop production in the short term (3 years) with negligible impact on the soil salinity. Groundwater monitoring has been undertaken at these sites by Grobbelaar and Hodgson (1997–2001), Usher and Ellington (2002) and by Usher and Vermeulen (2003–2006). Observation of limited water quality impacts in the groundwater over time has prompted much of the research currently underway in the vadose zone below the root zone of each irrigated area.

2 Discussion and Results

2.1 Kleinkopje

Five irrigation pivots have been established as field tests, but this paper concentrates mainly on the observations from Pivot 1 at Kleinkopje Colliery, where the longest trials have been running (since 1997). Data of Syferfontein is also included.

Methodology

- Seven boreholes, of which farming activities later destroyed two, were drilled to monitor the influence of irrigation on the aquifer. These were constructed in such a way that no run-off from the irrigation enters the boreholes. Water levels are shallow, as expected from an irrigation area (Fig. 1).
- These boreholes have been sampled quarterly since 1997 at the same depth with a Solinst specific-depth sampler. On each of these occasions, full chemical analyses of major ionic species and selected trace elements were done.
- In the more recent investigations, trenches were dug with an excavator until the water level of the water table aquifer at 3.5–4 m was reached. Samples were taken at various depths for soil characteristic analysis. Mineralogical determinations (XRD and XRF analyses using Norish and Hutton methods) were also done at the Geology Dept of the Free State University.
- Three core boreholes were drilled within the pivot area; two inside the cultivated and irrigated portion and one outside for background values. The core samples were analysed at specific intervals for water-soluble constituents to determine the major cations and anions.
- Porous cups were installed in the core boreholes at meter intervals, and bedded in silica flower. The holes were backfilled with material originating from the hole. Each layer between the porous cups was sealed with a thin layer of cement.

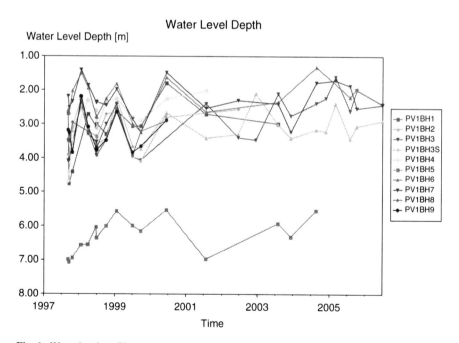

Fig. 1 Water levels at Pivot 1

- Tensiometers were installed next to the porous cups to determine the moisture content of the unsaturated zone at each depth. From this, the migration of salt between the root zone and aquifer was monitored over time.

2.1.1 Results of Groundwater Monitoring

Groundwater monitoring results since 1997, when irrigation started, have shown that the impact on the underlying aquifer is not significant. Water levels have not risen to any degree, although some seasonal variation, consistent with the Karoo sediments in the area, occurs (Fig. 1). Of greater importance is the lack of increase in salinity, and more particularly sulphate (Fig. 2). It is clear from this data series that, as yet, irrigation activities have very little discernible influence on the underlying aquifer's water quality.

The data from the porous cups monitoring provide some insight into the reasons for this apparent lack of salinity increase in the aquifer. The data (Fig. 3) show a steady decrease in sulphate concentrations with depth. This suggests that the majority of salts are contained within the uppermost portions of the soil profile, above and in a clay layer present between 1 and 2.5 m. The percentage clay in this layer is 20% compared to about 5% in the soil above and below it. To verify this observation, leaching tests were done on representative samples obtained in the soil profiles. Background values outside the irrigation area are also compared with the values obtained inside the irrigation area (Fig. 4). In these tests, an excess of deionised

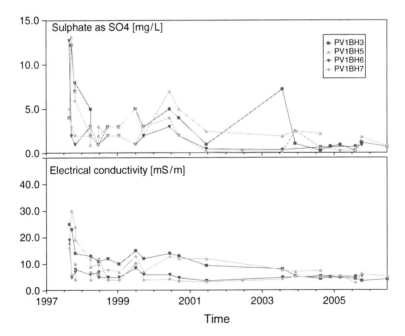

Fig. 2 EC and sulphate concentrations in monitoring boreholes at Pivot 1

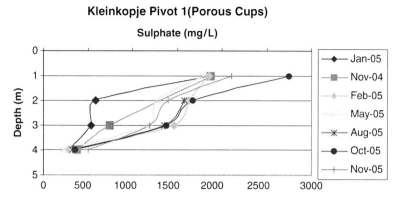

Fig. 3 Sulphate concentrations with depth in the porous cups

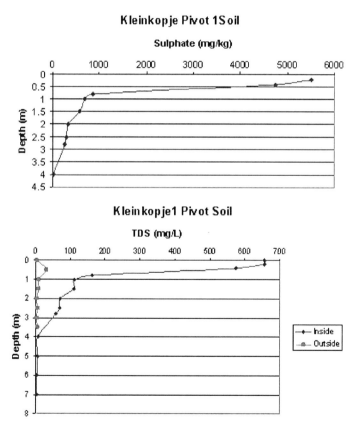

Fig. 4 Sulphate concentrations liberated from the soil with depth (above) and TDS-values of the soil profile inside and outside the pivot area (below)

water was added to each soil sample and the liberated ions in the supernatant analysed. The results of the soil tests suggest that the porous cup results are consistent with the trapped pore water, and adsorbed/precipitated ions at each of these levels. Figure 4 also indicates that most of the salts are present in the first metre of soil, above the layer with increased clay content.

One of the implications of this would be that there are hydraulic and attenuation factors preventing the salts in the mine water used for irrigation from being mobilised through the soil profile and into the aquifer. The soil composition and associated sorption and hydraulic properties may be informative. The typical composition determined by standard soil composition tests shows a marked increase in clay content below the depth of 1 m. This is confirmed by XRD, which indicates >40% quartz, 2–10% clay, and 2–10% hematite. No gypsum could be detected in the soils.

2.2 Syferfontein

In contrast, the results of another pivot site can be considered. The Syferfontein Colliery lies further south, and the irrigation water is less gypsiferous and has a stronger Na-SO$_4$ character. This would be expected to be more "mobile" irrigation water. However, this pivot is underlain by a heavy clay soil. For the first meter in depth the clay percentage is 63%, and for 1–4.5 m it varies between 40% and 31%. The results again indicate that only the very shallow soils and soil water show elevated salinity or sulphate (Figure 5). The result is that groundwater-monitoring systems around this pivot have also not shown any significant changes in water quality in the period 1999–2004.

3 Quantification of the Salt

In order to determine the hydraulic behaviour, salt balances and attenuation, and the movement of the salts at the various irrigation sites, tensiometer experiments have been performed on site. Moisture potentials were calculated from the tensiometer data using a method determined by Hutson (1983). An example of Kleinkopje is illustrated in Fig. 6 and the results of the different sites are summarised in Table 2.

To calculate the salts retained in the soil, the following calculations was made:

Bulk density (obtained from Lorentz et al. 2001). * *thickness of the layer * sulphate concentration in the soil*

To calculate the percentage salts retained in the soil:

Sulphate retained/total sulphate applied over the period of irrigation

To calculate the sulphate retained in the soil water:

*Sulphate concentration in soil water * thickness of layer * moisture content at that depth*

From these calculations the following results were obtained per hectare of irrigation (Table 3):

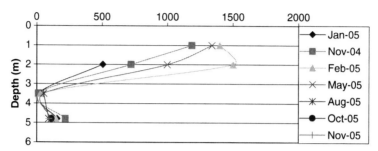

Fig. 5 Sulphate concentrations with depth in the porous cups

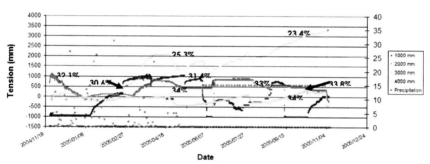

Fig. 6 Tensiometer data for Kleinkopje with estimated volumetric water content

Table 2 Tensiometer data at Irrigation sites

Depth	1,000 mm	2,000 mm	3,000 mm	4,000 mm
Kleinkopje 1	0.32	0.373	0.34	0.34
Kleinkopje 2	0.358	0.339	0.34	0.28
Syferfontein	0.384	0.35	0.274	0.262

Table 3 Salt balance calculation at the irrigation sites.

Site	% Total sulphate applied through irrigation retained in vadoze zone	% Total sulphate applied through irrigation in soil water	% of Salts retained that are in Soil water	Sulphate in soil (ton/ha)	Sulphate in soil water (ton/ha)
Kleinkopje 1	67%	25-38	37-56	56	21-38
Kleinkopje 2	77%	49-57	63-74	48	32-35
Syferfontein	82%	21-36	25-44	42	13
New Vaal	85%	30-38	35-45	18	7

Because the calculations are so sensitive to the input parameters, they cannot be regarded as exact figures, but rather in a range of ±10% accuracy.

4 Conclusion

The results to date point to several potentially significant findings for the wider application of mine water irrigation. Where the soils are richer in clay content, there is a significant attenuation of salts in the shallower zones. While the attenuation capacity of clays is a well-established concept, the long-term viability of irrigating with water influenced by coal mining is not widely accepted by regulators in South Africa. The groundwater monitoring results indicate that this attenuation makes mine water irrigation a viable option in the short to medium term where gypsum-saturated waters are used, as analysis of the water in the aquifers below show limited increase in degradation in quality, except in sandy soils.

Analysis of the tensiometer data over time, continued groundwater and soil water monitoring and detailed analysis of the soil characteristics as far as hydraulic and mass transport properties at each site, allowed the development of accurate conceptual models of the interaction between irrigation and the underlying soils and aquifers. A general model for irrigation sites indicates the following:

1. Tensiometer data indicates that the soil throughout the profile is high in moisture content, with the exception of the top 0.5–1 m. On average the moisture content is above 30%. The tensiometer data also indicates that the deeper layers dry out during winter.
2. The clay rich layers play an important role in the moisture content, with a build-up of moisture above these layers. This indicates that clay does play a role in the vertical flux.
3. Data from soil analysis with depth through the profile indicates that most of the salt is contained in the top 2 m of the profile. Chemical modelling of the soil water indicates saturation of the water with respect to gypsum above 1 m, implicating gypsum precipitation. Deeper down the soil water is unsaturated with regard to gypsum. Approximately 80% of the salts applied over the years of irrigation are retained. Data from soil water analysis obtained of the porous cup sampling indicates that most of these salts occur in the soil water (about 60% of the total salts applied), and that the balance precipitates in the soil or gets adsorbed. This implies that over the short to medium term the irrigation with coal mine water does not influence the aquifers to a great degree. Dissolved salts leach to the aquifers at a very low rate and are diluted at such a fast rate because of lateral groundwater flux. As a result low concentrations are detected through borehole sampling.

Acknowledgements Funding from Water Research Commission and Coaltech 2020.

References

Annandale, J.G., Jovanovic, N.Z., Claassens, A.S., Benade, N., Lorentz, S.A., Tanner, P.D., Aken, M.E. and Hodgson, F.D.I. (2002) The influence of irrigation with gypsiferous mine water on soil properties and drainage. WRC Report No. 858/1/02 ISBN No: 1-86845-952-7.

Grobbelaar, R. (2001) The long-term impact of intermine flow from collieries in the Mpumalanga Coalfields. Unpublished M.Sc. thesis, University of the Free State, Bloemfontein, South Africa.

Grobbelaar, R. and Hodgson, F.D.I. (1997–2001) Groundwater monitoring at irrigation pivots. Confidential Report for University of Pretoria, South Africa.

Grobbelaar, R, Usher, B.H., Cruywagen, L.-M., de Necker, E. and Hodgson F.D.I. (2002) The long-term impact of intermine flow from collieries in the Mpumalanga coalfields, Water Research Commission Report.

Hutson, J.L. (1983) Estimation of hydrological properties of South African soils. Unpublished Ph.D. thesis, University of Natal, Pietermaritzburg, South Africa.

Lorentz, S., Goba, P. and Pretorius, J. (2001) Hydrological process research: Experiments and measurements of soil hydraulic properties. WRC Report No. 744/1/01.

Parkhurst, D.L. and Appelo C.A.J. (1999) User's guide to PHREEQC (version 2) – a computer program for speciation, batch-reaction, one-dimensional transport, and inverse geochemical calculations. Water-Resources Investigations, Report No. 99-4259, U.S. Geological Survey, U.S. Department of the Interior.

Stumm, W. and Morgan, J. J. (1996) Aquatic Chemistry, 2nd Ed. Wiley, New York.

Usher, B.H. (2003) Development and evaluation of hydrogeochemical prediction techniques for South African coalmines. Unpublished Ph.D. thesis, University of the Free State, Bloemfontein, South Africa.

Usher, B.H. and Ellington, R. (2002) Groundwater monitoring at irrigation pivots. Confidential Report for University of Pretoria, South Africa.

Usher, B.H. and Vermeulen, P.D. (2003–2006) Monitoring reports for irrigation project: Syferfontein Colliery, Kleinkopje Colliery and New Vaal Colliery. Confidential Report for University of Pretoria, South Africa.

Application of Coal Fly Ash to Replace Lime in the Management of Reactive Mine Tailings

H. Wang, J. Shang, Y. Xu, M. Yeheyis, and E. Yanful

Abstract Acid mine drainage (AMD), the acidic water discharged from active, inactive, or abandoned mine sites, is probably the single largest environmental liability facing the mining industry. Mining companies use large quantities of lime for AMD control in daily operation. The manufacture of lime contributes significant emission of greenhouse gas (GHG). This paper presents an innovative approach of using coal fly ash, a by-product from coal-fired generation stations, to replace lime in AMD control.

The potential market and scientific principles of the approach are discussed first, followed by discussion on its effectiveness, benefits and limitations. In a site-specific case study, the effects of using coal fly ashes from Atikokan Generation Station to control reactive mine tailings from the Musselwhite Mine in Northern Ontario, Canada, are investigated. The laboratory results of different mixtures demonstrated that coal fly ash is an effective and cost-efficient approach with minimum environmental impacts when calcium-rich coal fly ash is accessible to the mining site. A preliminary procedure is recommended to assess the feasibility of this approach, including the compatibility of reactive mine tailings and coal fly ash, treatment effectiveness, cost and environment impacts.

Keywords Coal fly ash · reactive mine tailings · acid mine drainage · lime, greenhouse gas

1 Introduction

Mine tailings are the solid waste generated from mineral processing. Exposed to the atmosphere, sulphide-bearing minerals contained in reactive tailings oxidize easily, and so generate acid mine drainage (AMD) in tailings impoundments. The AMD

H. Wang (✉), J. Shang, Y. Xu, M. Yeheyis, and E. Yanful
Department of Civil and Environmental Engineering, University of Western Ontario,
London, Canada

often has pH of less than 2.0 with total acidity exceeding total alkalinity. Discharge of AMD with high concentrations of heavy metals cause degradation of both groundwater and surface water quality. The common technologies of minimizing oxidation of reactive tailings include preventing oxygen ingress, preventing water infiltration, encapsulating or blending sulphide minerals, adding neutralization agents to treat tailings, and using bactericides to slow down the rate of oxidation.

Lime products are primary neutralization chemicals used by the mining industry to control AMD generation. The quantity of lime required for AMD control at a mine site depends on the characteristics of the waste. Guidelines are often provided based on empirical data and engineering practice. For example, Karam and Guay (1994) suggested that the application of lime for mine tailings is 118 t per hectare to a depth of 15 cm, which shows that an enormous quantity of lime is required for the management of reactive mine tailings. In average, 0.8 t of CO_2 (an important greenhouse gas, GHG) is generated for each tonne of lime produced. The use of lime in AMD control contributes significantly to GHG emission. The shipment of large quantities of lime to mine sites plays an important role in GHG emission as well, especially in cases of long distances between mine sites and lime providers.

Industrialized countries contributed more than two-thirds of the annual CO_2 emission. However, it is anticipated that emissions by developing countries would exceed those of industrialized countries by 2002 (Reid and Goldemberg 1998). China, India, Mexico, South Africa, Iran, Indonesia, and Brazil are currently on the top 20 list of CO_2 emitting countries and are facing the challenge of taking prompt actions to reduce GHG emissions. Common measures to reduce GHG emissions include reducing the use of gas or fossil-fuel generated electricity, developing more energy efficient technologies, adopting renewable energy sources and specific clean energy technologies (World Business Council for Sustainable Development and World Resources Institute 2004).

Recent studies show that fresh fly ash (FFA) can provide effective neutralization capacity to control tailings oxidation (Shang and Wang 2005; Shang et al. 2006; Wang et al. 2006). The studies also show that landfill fly ash (LFA) may be used as a low cost alternative to FFA with comparable effectiveness. The fly ash approach may provide a significant potential for mining industries to reduce or eliminate the use of lime, which would contributes to the global effort in reducing GHG emission.

Coal fly ash is a fine and non-cohesive combustion residue from coal-fired utilities. In Ontario, Canada, Ontario Power Generation (OPG) owns five coal-fired power generation stations (GS), and generates approximately 1.29 million tonnes of ash each year (Ontario Clean Air Alliance 2007). About 49% of this material is re-used while the rest is landfilled on site. Even with the planned closure of most coal-fired generation stations by 2009, there will be abundant coal fly ash available from landfill sites. The physical and chemical properties of coal fly ash depend on the composition, combustion temperature, mixing ratio of air and coal, pulverized coal particle size and rate of combustion (Lohtia and Joshi 1995). The dominant constituents of coal fly ash are amorphous oxides (70–90%, Qafoku et al. 1999), notable silicon, aluminium, iron, magnesium and calcium oxides. Among these oxides, silica, alumina, and free lime are chemically reactive, which contribute to the cementitious properties of coal fly ash.

2 Principles of Coal Fly Ash Applications in Reactive Mine Tailings

Coal fly ash is a by-product from coal-fired utilities and using it to replace lime can directly reduce CO_2 emission from lime manufacture. Coal fly ash can serve two major functions in the management of reactive mine tailings, which are outlined as follows:

1. Neutralization and precipitation. The neutralization capacity of coal fly ash is mainly determined by its CaO and MgO contents. Upon contact with low pH AMD, the dissolved Ca^{2+} and Mg^{2+} ions in coal fly ash react with SO^{2-} ions present in AMD to form calcium and magnesium sulphates. Thus, dissolved metals are precipitated; the pH is raised as well. Experimental data have shown that 10% mass ratio of coal fly ash addition to tailings could maintain pH over 11. The effect, in terms of pH, conductivity and element concentration, is equivalent to approximately 5% lime addition, a typical mass ratio in practice. The results also showed that the neutralization capacity of coal fly ash lasts over a long period. The pH remained stable after more than ten pore volumes of AMD permeation through compacted coal fly ash (Shang and Wang 2005; Shang et al. 2006; Wang et al. 2006).
2. Cementation. Coal fly ash contains significant SiO_2 (~40%) and Al_2O_3 (~20%) contents in total oxides. These two reactive silica and alumina are major contributors of cementation. The reactive silica and alumina, as well as lime, react with water and result in cementitious process. The reactions become more vigorous when AMD is involved, as it contains high concentrations of dissolved solids. When coal fly ash is applied to reactive mine tailings, dissolved metals are precipitated due to pH increase. Furthermore, secondary minerals are formed from the cementitious reactions. The products clog voids and decrease air and water permeability in ash and ash-tailings mixture. It has been proven that the hydraulic conductivity of compacted coal fly ash was reduced more than three orders of magnitude when the coal fly ash was permeated with AMD, along with a ten-fold increase in break-through hydraulic gradient (Shang and Wang 2005; Shang et al. 2006; Wang et al. 2006). In other words, the coal fly ash formed a hydraulic barrier when in contact with AMD. Such a barrier can separate oxygen and water from reactive mine tailings.

3 A Case Study of Coal Fly Ash Application in Reactive Mine Tailings

At the Musselwhite Mine in Northern Ontario, Canada, about 22 million tonnes of tailings containing ~2% sulphur will be generated by 2012. Six million tonnes of total Musselwhite mine tailings (MMT) will be disposed on land because of limited underwater disposal capacity at the site. The measured net neutralization potential from acid-base accounting tests is -14.5 kg $CaCO_3$/t, indicating the acid generating nature of the tailings. Signs of oxidation can be seen in the in-situ tailings, after 1–2 years of disposal. Therefore, actions must be taken to prevent further tailings oxidation and hence AMD generation. If lime is applied in the tailings treatment, at least 300,000 tonnes (~5% mass ratio) would be required, which would be responsible for approximately 240,000 tonnes of CO_2 emission. In addition, the distance to the nearest lime production plant at Faulkner, Manitoba is more than 1,000 km from the mine site.

Atikokan GS is located approximately 500 km south of Musselwhite Mine. Between 40,000 and 60,000 t of lignite fly ash are generated every year. The coal fly ash has high CaO content (15–16%) with 38–44% SiO_2 and 20–21% Al_2O_3, which makes it an ideal candidate for lime replacement. Currently, 80% of Atikokan fly ash is utilized in concrete cement manufacturing and other industries, whereas the remaining 20% is landfilled. In the landfill at the site, more than 350,000 t of coal fly ash is available for potential applications, and this could be a source of coal fly ash application in mine tailings management.

3.1 Methods and Materials

A site-specific study was carried out to assess the suitability of utilizing Atikokan coal fly ash in the management of reactive Musselwhite mine tailings. The physical, chemical and mineralogical properties of the FFA and MMT were summarized in the previous published paper (Wang et al. 2006). The kinetic column leaching tests were carried out to monitor the leaching properties of the coal fly ash, and the ash-tailings mixtures. The results of column leaching tests indicate that the pH of the leachate was increased from acidic (pH 4) to alkaline (above pH 8) values. The hydraulic conductivities of the coal fly ash and ash-tailings mixtures were reduced significantly (~3 orders of magnitude) upon contact with AMD. Chemical analyses further indicate that the concentrations of regulated elements in the leachate from the ash-tailings mixtures were well below the guidelines set by Ontario environmental authority under accelerated flow conditions. To facilitate field trial, further laboratory studies were carried out to evaluate the neutralization effect of FFA and LFA from Atikokan GS.

3.2 Results and Discussion

It was noted that the LFA have basically identical SiO_2 and Al_2O_3 contents and slightly lower (15%) CaO content than the FFA (16.4%), as shown in Table 1. The grain size distributions of the FFA and LFA are noticeably different, as indicated in Fig. 1. The portion of fine solids (<2 µm) in the LFA is only 4% as against 22% in the FFA, which is mainly attributed to the hydration of CaO that prompted finer

Table 1 Major oxides analyses of Atikokan FFA and LFA

Major oxides	FFA %	LFA %
SiO_2	37.45	37.95
TiO_2	0.71	0.67
Al_2O_3	18.91	17.61
Fe_2O_3	6.31	4.99
MnO	0.03	0.02
CaO	16.41	15.03
K2O	0.85	0.83
Na_2O	9.12	7.46
P_2O_5	0.41	0.45
Cr_2O_3	0.003	0.004
LOI[a]	0.4	7.5
Total	94.26	95.80

[a]Loss on ignition.

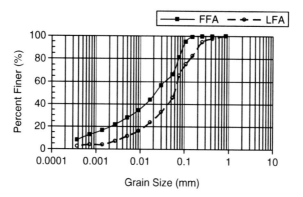

Fig. 1 Grain size distribution of Atikokan coal fly ashes

particles to form lumps (Fig. 2). Another significant difference between FFA and LFA is that the latter has prominently higher maximum dry density and lower optimum moisture content, as determined by compaction tests (Fig. 3). Therefore the LFA may be a better candidate as a solid cover for land disposal of reactive mine tailings.

The comparison of the neutralization effect of FFA, LFA and lime at various mass ratios is shown in Table 2. It is noted that, at the same mass ratios, the mixtures of lime-MMT and FFA-MMT mixtures have similar pH (within one unit) at equilibrium. The LFA-MMT mixtures, on the other hand, are characterized by lower pH and lower electrical conductivities at equilibrium. This is in general agreement with the slightly lower CaO content in the LFA (Table 1). Acid-base accounting tests showed that the net neutralization potential of FFA and LFA were 306 and 300 kg $CaCO_3$/t, respectively. Therefore, both ashes possess strong neutralization potential. The regulated concentrations of elements in the supernatants of both ashes are well below Ontario Regulation 558 (Leachate Criteria).

Fig. 2 SEM image of Atikokan coal fly ashes: (**a**) FFA (x500), (**b**) FFA (x2000), (**c**) LFA (x500), (**d**) LFA (x2000)

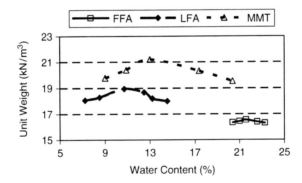

Fig. 3 Compaction test results on FFA, LFA and MTT

The results of this study indicate that, in general, the FFA and LFA from Atikokan GS have favourable physical and chemical properties for AMD control at the Musselwhite Mine tailings disposal site. Compared to the FFA, the LFA is slightly cemented due to some degree of hydration. Both FFA and LFA demonstrate noticeable neutralization capacity. In addition, the LFA is especially suitable for applications as solid covers because of its compaction characteristics. There will be about 400,000 t of LFA available at the time of closure of Atikokan GS in 2007. Therefore, the fly ash may be a cost-efficient and effective alternative to lime, especially considering the distance of shipping lime to the Musselwhite Mine site

Table 2 Water leaching test (ASTM D 3987-85) pH and electrical conductivity of Atikokan FFA-MMT, Atikokan LFA-MMT and lime-MMT

Material mass ratio	Leach method	Initial pH	Final pH	Conductivity (mS/cm)
FFA:MMT (10:90)	Water	10.21	11.03	1,0415
FFA:MMT (20:80)	Water	10.44	11.07	1,442.5
FFA:MMT (30:70)	Water	10.70	11.09	1,760.5
FFA:MMT (100:0)	Water	11.18	11.53	3,280.0
LFA:MMT (10:90)	Water	9.86	9.94	510.5
LFA:MMT (20:80)	Water	9.96	10.36	647.0
LFA:MMT (30:70)	Water	9.88	10.06	744.0
LFA:MMT (100:0)	Water	10.06	10.11	1,324.5
LIME:MM (10:90)	Water	11.69	11.73	9,235
LIME:MM (20:80)	Water	11.78	11.76	9,295
LIME:MM (30:70)	Water	11.80	11.74	9,085
MMT	Water	9.35	9.27	156.0

(more than 1,000 km). A field trial is underway at the Musselwhite Mine site to evaluate long-term performance of coal fly ash approach on the effectiveness and environmental impacts.

4 Feasibility Assessment

The factors affecting the feasibility of using coal fly ash in reactive mine tailings management include: characteristics of coal fly ash and reactive mine tailings; the required quantity of coal fly ash for effective AMD prevention and control; the comparison of coal fly ash and lime shipping to the mine site; and environmental concern. These factors are site-specific and vary considerably depending on the characteristics of coal fly ash, lime and mine tailings. A feasibility study is suggested prior to an implementation.

The following is a recommended procedure for the feasibility assessment:

1. *Assessment of compatibility and effectiveness.* The effectiveness of using coal fly ash at a mine site can be estimated from characterization of the coal fly ash and mine tailings for a number of parameters including: chemical and mineralogical composition, acid-base accounting, sulphur content, regulated trace elements concentrations, grain size distribution of mine tailings; neutralization potential, major oxides, grain size distribution and compaction characteristics of coal fly ash; compatibility study on ash-tailings in kinetic column tests.

2. *Assessment of economic feasibility.* The information, including available quantity, type and cost of neutralization agents, distance and transportation method from the mine site to the neutralization agent supplier, should be collected beforehand to carry out total cost comparison between coal fly ash option and lime option.

3. *Assessment of environmental feasibility*. GHG reduction from lime option to coal fly ash option, long term acid neutralization capacity and regulated trace elements release should be evaluated.

5 Summary and Conclusion

This study demonstrated that, with appropriate design, coal fly ash can replace lime in AMD control for reactive mine tailings. This approach can significantly reduce GHG emission from lime manufacturing. The major benefits of this approach include:

1. Coal fly ash, upon characterization, may replace lime to serve as a neutralization and stabilization material to prevent or mitigate AMD generation in reactive mine tailings. Thus, this approach will contribute to GHG emission reduction.
2. Coal fly ash may be beneficial for tailings management because of its cementitious nature. When applied in tailings as additive or cover, it can form a low permeability barrier to inhibit oxygen ingress and water infiltration.
3. Landfill fly ash has additional benefits, such as low cost, superior compaction characteristics and neutralization capacity comparable to that of fresh fly ash.

References

Karam, A. and Guay, R. 1994. Inondation artificielle du parc à résidus miniers Solbec-Cupra: études microbiologique et chimique. Université Laval.

Lohtia, R.P. and Joshi, R.C. 1995. Concrete admixtures handbook – properties, science, and technology, 2nd Edition. In: Ramachandran V.S. (Ed.), Noyes, Park Ridge, NJ, pp. 657–739.

Ontario Clean Air Alliance 2007. OPG: Ontario's Pollution Giant. Ontario Clean Air Alliance, Toronto.

Qafoku, N.P., Kukier, U., Sumner, M.E., Miller, W.P. and Radcliffe, D.E. 1999. Arsenate displacement from fly ash in amended soils. Water, Air, and Soil Pollution, 114: 185–198.

Reid, W.V. and Goldemberg, J. 1998. Developing countries are combating climate change. Energy Policy, 26(3): 233–237.

Shang, J.Q. and Wang, H.L. 2005. Coal fly ash as contaminant barrier for reactive mine tailings. Proceedings of World of Coal Ash Conference, Lexington, KY, April 11–15.

Shang, J.Q., Wang, H.L., Kovac, V. and Fyfe, J. 2006. Site-specific study on stabilization of acid generating mine tailings using coal fly ash. ASCE Journal of Materials in Civil Engineering, 18(2): 140–151.

Wang, H.L., Shang, J.Q., Kovac, V. and Ho, K.S. 2006. Utilization of Atikokan coal fly ash in Musselwhite mine tailings management. Canadian Geotechnical Journal, 43(3): 229–243.

World Business Council for Sustainable Development and World Resources Institute 2004. The greenhouse gas protocol: a corporate accounting and reporting standard (Revised edition). World Business Pub.

Part V
Soil Stabilization

Consolidation and Strength Characteristics of Biofilm Amended Barrier Soils

J.L. Daniels, R. Cherukuri, and V.O. Ogunro

Abstract Experimental work was conducted to investigate the influence of biofilm on the consolidation and strength characteristics of two barrier soils. Biofilm has potential as a low-cost additive for soil stabilization, and it may be formed naturally in landfills throughout the developing world. The EPS-producing bacterium *Beijerinckia indica* was used to prepare solutions of varying concentration of exopolymeric substances (EPS). These solutions were then used as the molding moisture for compacted specimens of locally available clay ("red bull tallow," RBT) as well as a mix of 65% sand and 35% bentonite (65:35 mix). As compared to tap water, the influence of the nutrient solution or biofilm on RBT is to increase the compression index (C_c), although this trend is variable for increasing EPS concentration. While the effect of biofilm on the 65:35 mix is less uniform, the largest increase in C_c was observed for the highest level of biofilm amendment (EPS-5, 300 mg/L). Amendment with biofilm results in both increases and decreases in the rate of consolidation (c_v). The c_v values ranged from 0.4 to 13.6 m²/year and from 0.2 to 19.3 m²/year for RBT and 65:35 mix, respectively. In general, EPS has a decreasing effect on observed strength. For example, the peak unconfined compressive strengths for unmodified RBT and 65:35 mix were found to be 667.0 and 395.3 kPa, respectively. Many of these values decreased with increasing biofilm amendment, and for the highest level of amendment, the observed peak strengths were 159.1 and 98.8 kPa. To the extent that naturally-occurring methanotrophic activity in landfill cover systems results in biofilm production, the results suggest potential concerns with cover stability.

Keywords Polymers · biofilm · waste management · barriers · strength

---

J.L. Daniels (✉) and V.O. Ogunro
Associate Professor, Department of Civil and Environmental Engineering and Global Institute for Energy and Environmental Systems, University of North Carolina, 9201 University City Blvd, Charlotte, NC 28223, USA

R. Cherukuri
Graduate Research Assistant, Department of Civil and Environmental Engineering and Global Institute for Energy and Environmental Systems, University of North Carolina, 9201 University City Blvd, Charlotte, NC 28223, USA

1 Introduction

Despite the recent emphasis on source reduction as well as the growth of recycling and incineration, land disposal continues to be the dominant form of municipal solid waste (MSW) disposal, particularly in developing countries. In addition to MSW, industrial development and military legacies lead to the production of hazardous and radioactive waste. The primary objective of landfill facilities is to isolate waste from the natural environment. While the geotechnical performance of landfill components has been tested extensively in response to physical and chemical stress, little is known about microbiological effects. At issue is the recently documented presence of biofilm-producing methanotrophic bacteria in landfill cover soils (Hilger et al. 1999, Hilger and Barlaz 2000; Wise et al. 1999; Borjesson et al. 1998a, b; Kightley et al. 1995). Biofilm is composed largely of exopolymeric substances (EPS) and polysaccharides in particular. The presence of biofilm is expected to influence both the friction angles and adhesion/cohesion properties of various landfill components, yet little effort has been made to investigate its significance. Indeed, work performed by Daniels et al. (2002), Yen et al. (1996) and Yang et al. (1993) as well as literature from soil science suggests that the presence of biofilm-type substances can result in either decreases or increases in shear strength. Any change in shear strength is of direct interest for the stability of landfill covers with respect to sliding. Given the multi-component nature of virtually all landfill covers and liners, sliding failure is a particularly common concern. Sliding failure may occur in covers or liners, the most notable of which occurred at the Kettleman Hills Landfill as described by Seed et al. (1990) and Mitchell et al. (1990). The nature of biofilm is also quite similar to aqueous polymers used to mitigate freeze-thaw and desiccation induced stresses in landfill cover systems, as proposed by Daniels et al. (2003) and Daniels and Inyang (2004). The polymers considered included guar gum polysaccharide and polyacrylamide, which were dissolved in solution and used as the molding water at compaction. Results from this work indicate that these macromolecules can bind soil particles, ostensibly increasing the resistance to crack and fissure formation. Daniels et al. (2005) observed that biofilm amendment appears to reduce strength, while having a modest effect on compression and desiccation characteristics. However, while a range of biofilm concentrations may be relevant in field situations, the previous investigation included only one level of biofilm concentration.

Considering the potential presence of biofilm in landfill cover systems and the previous use of similar polymeric substances in barrier material improvement, the objectives of this paper are to (1) evaluate a range of biofilm concentrations on the consolidation and strength characteristics of two soils that could be used in waste containment applications and (2) comment on the significance of biofilm production in landfill cover soils as related to geotechnical performance.

2 Background

Biofilm research is broad in scope, and includes efforts related to problems connected to artificial recharge and/or sewerage infiltration (Mitchell and Nevo 1964; Wood and Bassett 1975; Okubo and Matsumoto 1979, 1983), efforts to enhance oil recovery (Hart et al. 1960; Shaw et al. 1985); biofouling or clogging of well systems for water supply or groundwater remediation (Clement et al. 1996) and drainage systems in landfill leachate collection systems (Brune et al. 1991; Rowe et al. 1998). In essence, the production of EPS and the formation of biofilm is the mechanism by which microorganisms can control their local environment. In porous media, a network of EPS forms a complex geometry on particle surfaces in response to spatial and temporal variations in nutrients, temperature, contaminants and predators (Costerton et al. 1978). Microbiological colonies use EPS to develop localized channels and pathways as a circulatory system to control such processes as nutrient delivery and waste rejection (Wingender et al. 1999). The result of this microbiological activity is a reduction in hydraulic conductivity as well as a change in the frictional and cohesive properties of a soil mass. One of the first reports regarding the influence of biofilm on hydraulic conductivity was given by Allison (1947). Extensive work has been conducted since that time and several investigators have assembled comprehensive summaries, including Taylor and Jaffe (1990), Wu et al. (1997), Dennis and Turner (1998) and Daniels and Cherukuri (2005).

While it is generally agreed that biofilm formation results in a decrease in hydraulic conductivity, the effect on strength and compressibility is less clear. Much depends on the extent to which the biofilm matrix bonds particles together versus simply coating the particles. Depending on the intensity of biofilm production, it may only consist of dispersed microcolonies with vast separation in between, or with continued growth, these may coalesce to form a contiguous film that coats all available surface area (Clement et al. 1996). Yang et al. (1993) investigated the influence of different commercial biopolymers (polyhydroxybutyrate, xanthan gum and sodium alginate) as well as a "slime-forming" bacterial suspension on the shear strength of sand and clay. The authors reported increases in strength for all cases, with improvements ranging from 2% to nearly 200%. Yen et al. (1996) continued the research by evaluating the effect of xanthan gum and a bacterial suspension composed of *Alcaligenes eutrophus* on the shear strength of silt as measured through triaxial compression tests. The authors noted that either of the additives increased the maximum deviatoric stress by 50% after an "aging" period of 10 to 15 days. This strength was maintained for the duration of the 45 day testing period, while concomitant permeability measurements revealed sustained reductions (implying continued presence) for 6 months. The authors concluded that biopolymers may represent a possible soil modifier with applications in waste containment. Though not discussed by the authors, it appears that polymeric bridging among particles may have contributed to the observed strength increases. The influence of a dissolved polysaccharide (from guar gum) on the unconfined compressive strength was investigated by Daniels et al. (2002). The soil tested contained a large amount of illitic clay, and is commonly referred to as Boston Blue Clay (BBC). Results showed that increasing the concentration of polysaccharide, analogous to increasing the level of EPS in the matrix, led to lower strength and increased ductility. Thus, for this particular case, it appears the polymer coated the particles, reducing the internal angle of friction.

3 Materials and Methods

The materials and sample preparation are largely as reported by Daniels et al. (2005), with the exception that for this contribution, solutions of varying biofilm concentration were prepared. The EPS-producing bacterium *Beijerinckia indica* was used in this research and obtained from the American Type Culture Collection (ATCC). Certain strains of *Beijerinckia* have since been reclassified as *Sphingomonas* (Gibson 1999). Dennis and Turner (1998) observe that *B. indica* has several characteristics that make it favorable from a research perspective. It is a free-living, non-pathogenic species that produces significant EPS. Although *B. indica* is classified as a strict aerobe, it tends to tolerate relatively wide fluctuations in oxygen partial pressure, pH, and functions optimally at 26 °C.

Cultures of *B. indica* were grown in Azotobacter #13 broth nutrient solution (ATCC 2002). The nutrient solution was prepared by adding the following constituents to distilled water, raised to 1,000 mL: 20 g glucose, 1 g NaCl, 1 g yeast extract, 5 mL of 10% $MgSO_4$, 8 mL of 10% K_2HPO_4, 2 mL 10% KH_2PO_4, 150 µL 5% $FeCl_2$, 2% glucose solution, and 100 mL of soil extract. The soil extract was prepared by adding 77.0 g of African violet soil, 0.2 g of Na_2CO_3 and 200 mL of distilled water. The pH of the suspension was adjusted to 6.0 and then autoclaved for 1 h, after which 100 mL of the extract was obtained by filtering the suspension through a sterilized Whatman filter paper #1. In certain cases, 10% sucrose (i.e. 100 mL of total 1,000 mL nutrient solution) was added to enhance EPS production. Optical density (OD) measurements at 520 nm were made as a function of time to assess growth of bacteria and EPS. A calibration curve was developed to relate OD data to a known concentration of hyaluronic acid, which is used as an indicator of EPS (Blumenkrantz and Asboe-Hansen 1973). Solutions of varying EPS concentration were obtained by preparing and growing solutions for different lengths of time, up to 12 days. The EPS and nutrient solution were then used as the molding moisture content for compacted samples, discussed as follows.

Earthen barrier materials generally fall into two categories, locally-available fine-grained soils which have a low hydraulic conductivity, or coarser sands and silts which have to be amended with an expansive clay such as bentonite. With this in mind, two types of material were used in this research, a locally available expansive soil known as "Red Bull Tallow" (RBT) and a mixture of sand and bentonite. Samples of RBT were obtained from a construction site northeast of Charlotte, NC. Intermittent deposits of RBT exist throughout the region where its undesirable shrink/swell and strength properties are well-known locally. While unsuitable as a foundation soil, the low-permeability aspect of RBT makes it an ideal barrier material. The RBT had a natural in situ moisture content of 20%, and it was subsequently air dried and ground to obtain a homogeneous mixture for laboratory evaluation. The RBT classifies as a CL (lean clay) according to the Unified Soil Classification System. The other barrier material selected was a mixture of 65% sand and 35% bentonite, by weight (65-35 mix). The sand was obtained from Humboldt, Inc., given as density sand with the gradation defined between the #20 (0.85 mm) and #30 (0.60 mm) sieve sizes while the bentonite was obtained from the Texas Sodium Bentonite, Inc. The liquid limit, plasticity index, specific surface area (BET-N_2 Adsorption), optimum moisture content, specific gravity, coefficient of concavity and coefficient of uniformity are provided in Table 1.

The optimum moisture content was determined with both the Harvard Miniature (HM) device (Bowles 1992) and by Standard Proctor Effort, ASTM D698 (ASTM 2000a) as the former was to prepare samples for unconfined compression testing while the latter was used to prepare samples used in consolidation. The HM device has a spring loaded tamper, which was applied ten times for each of three layers. The resulting specimen dimension was 3.3 cm in diameter and 7.2 cm in length. Samples for consolidation testing were extracted from that which was compacted in the Proctor mold with a cutting ring and cut so that the specimen dimensions were 6.4 cm in diameter and 2.3 cm in height. In both cases, the optimum moisture content was determined using water alone, and applied to mixes which used the nutrient and/or nutrient+EPS solution. While there are slight differences in the moisture density relationship obtained from using water alone as compared to nutrient/EPS solution, this was neglected herein. In soil specimens mixed with similar nutrient and EPS solutions, Dennis and Turner (1998) note that there is essentially no difference as compared to water alone. Moreover, in the work performed herein, no obvious differences in workability during sample mixing and preparation were observed. In general, samples were prepared by mixing the appropriate amount of soil and molding liquid (water, nutrient solution, or nutrient solution+EPS), allowing hydration for 24 h and then compacting at a moisture content of approximately 1% post-optimum. The solutions used, expressed in terms of initial aqueous concentrations and subsequent solid phase concentrations upon compaction, are provided in Table 2 for RBT and 65-35. Solid phase concentrations vary somewhat between the RBT and 65-35 mixes because of the difference in molding moisture content.

Table 1 Barrier materials evaluated

Material	LL (%)	PI (%)	Specific surface area (m²/g)	ω_{opt} (%)		$\gamma_{d(max)}$ (g/cm³)		G_s	Coefficient of concavity	Coefficient of uniformity
				SP	HM	SP	HM			
RBT	30	14	2.61	15.5	12.0	1.74	2.33	2.71	0.18	62.5
65-35	40	24	10.10	16	12.3	1.73	2.36	2.60	0.27	13.3

Note: SP – standard proctor, HM – harvard miniature.

Table 2 Aqueous and solid phase concentrations of EPS tested

Solution ID	Initial solution concentration (mg/L)	Solid phase concentration (mg/kg)			
		Consolidation tests		Unconfined compression	
		RBT	65:35 Mix	RBT	65:35 Mix
TW	0	0	0	0	0
NS	0	0	0	0	0
EPS-1	11.6	1.9	2.0	1.5	1.6
EPS-2	51.7	8.5	8.8	6.7	7.0
EPS-3	79.9	13.2	13.6	10.4	10.7
EPS-4	112.0	18.5	19.0	14.6	15.1
EPS-5	300.0	49.6	50.9	39.0	40.5

One dimensional consolidation tests were performed using lever-arm type consolidometers in accordance with ASTM D2435 (ASTM 2000b). Specimens for the consolidation test were obtained by pressing the consolidation ring into a previously compacted mass of soil. The consolidation ring had a diameter of 6.37 cm and a height of 2.27 cm. The loading sequence consisted of pressures of 30.8, 61.6, 123.1, 246.2 and 492.5 kPa while rebound was measured with two load decrements, from 492.5 to 246.2 kPa and 246.2 to 123.1 kPa. Each load increment and decrement was completed in 24 h. Unconfined compression tests were conducted following ASTM D2166 (ASTM 2000c). Three tests were conducted for each material (RBT or 65-35 Mix) and molding moisture source (distilled water or nutrient solution + EPS). The shear rate for conducting all the experiments was kept constant at 0.3175 mm/min.

4 Results and Discussion

4.1 Consolidation

Compression and swelling indices are tabulated in Table 3, while Tables 4 and 5 summarize the coefficients of consolidation. The relationship between void ratio and effective stress is given in Figs. 1 and 2. Note that while all specimens were prepared with the same level of energy (Standard Proctor Effort), the initial void ratio was different for each sample. This difference in initial void ratio is attributed to the different molding moisture composition – i.e., tap water, nutrient solution and EPS levels 1–5 as shown in Table 2. The overall compressibility of these soils is very low, with the compression index (C_c) not exceeding 0.295 for all samples tested. These low values reflect the remolded and compacted nature of the material, in contrast with naturally occurring soils that would tend to have higher compression indices. For example, use of correlations based on natural soils that relate liquid limit (Terzaghi and Peck 1967) or plasticity index (Wroth and Wood

Consolidation and Strength Characteristics of Biofilm Amended Barrier Soils

Table 3 Summary of consolidation test data

Solution ID	Compression index, C_c (31–123 kPa)		Compression index, C_c (123–493 kPa)		Swelling index, C_s (493–123 kPa)	
	RBT	65:35 Mix	RBT	65:35 Mix	RBT	65:35 Mix
TW	0.063	0.104	0.128	0.165	0.018	0.019
NS	0.157	0.142	0.252	0.222	0.040	0.029
EPS-1	0.101	0.087	0.171	0.151	0.012	0.023
EPS-2	0.122	0.091	0.169	0.160	0.018	0.026
EPS-3	0.071	0.155	0.145	0.195	0.016	0.021
EPS-4	0.149	0.062	0.214	0.178	0.024	0.024
EPS-5	0.131	0.095	0.207	0.295	0.015	0.006

Table 4 Coefficient of consolidation data – RBT

Solution ID	Coefficient of consolidation (c_v) (m²/year)				
	Load 1 30.8 kPa	Load 2 61.6 kPa	Load 3 123 kPa	Load 4 246 kPa	Load 5 493 kPa
TW	8.2	6.1	5.8	3.8	3.0
NS	7.2	13.6	6.5	2.5	2.2
EPS-1	5.9	2.6	6.2	2.7	9.6
EPS-2	8.5	10.9	2.9	1.4	0.8
EPS-3	7.8	11.8	5.4	12.4	4.2
EPS-4	8.6	7.0	7.9	7.8	2.9
EPS-5	4.0	9.4	1.8	0.4	11.0

Table 5 Coefficient of consolidation data – 65:35 Mix

Solution ID	Coefficient of consolidation (c_v) (m²/year)				
	Load 1 30.8 kPa	Load 2 61.6 kPa	Load 3 123 kPa	Load 4 246 kPa	Load 5 493 kPa
TW	5.1	6.0	4.9	4.9	5.5
NS	8.4	10.3	3.1	1.6	2.3
EPS-1	9.5	7.4	6.7	11.2	7.2
EPS-2	6.6	19.3	6.1	0.5	1.6
EPS-3	8.5	10.9	5.7	5.1	6.9
EPS-4	13.4	16.2	3.1	3.2	5.7
EPS-5	8.9	8.4	5.2	0.2	8.7

1978; Kulhawy and Mayne 1990) to C_c would predict values several times larger than measured for the unmodified RBT and 65:35 mix. While remolded samples such as those tested do not exhibit clearly defined preconsolidation pressures, there is a change in C_c as a function of applied load. In particular, C_c for RBT ranged from 0.063 to 0.157 when loaded from 31 to 123 kPa. This range increased to 0.128 to 0.252 for the load increment between 123 and 495 kPa. Similarly, C_c for the 65:35 mix ranged from 0.062 to 0.155 when loaded from 31 to 123 kPa, while the range increased to 0.151 to 0.295 for the load increment between 123

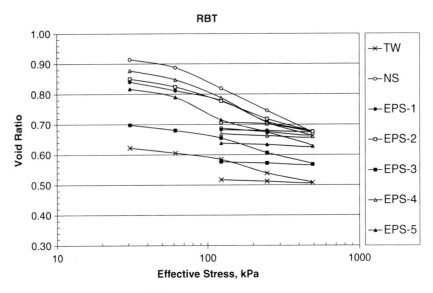

Fig. 1 Consolidation curves – RBT

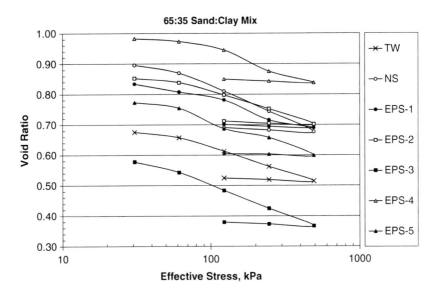

Fig. 2 Consolidation curves – 65:35 mix

and 495 kPa. As compared to tap water, the influence of the nutrient solution or biofilm on RBT is to increase the C_c, although this trend is variable for increasing EPS concentration. While the effect of biofilm on the 65:35 mix is less uniform, the largest increase in C_c was observed for the highest level of biofilm amendment

(EPS-5). In terms of the swelling index, observed values were generally less than the compression index by a factor of 10, which is typical for a variety of soils (Winterkorn and Fang 1975). The influence of biofilm on swelling is modest but consistent for both soils tested. While the extent to which changes in indices reflect natural variability instead of biofilm amendment is not certain, the overall tendency is toward increasing compressibility.

Amendment with biofilm results in both increases and decreases in the rate of consolidation. As shown in Table 4, the c_v values ranged from 0.4 to 13.6 m²/year and from 0.2 to 19.3 m²/year for RBT and 65:35 mix, respectively. By way comparison to natural clays, a particular study found that the range for montmorillonite, illite and kaolinite was 0.019 to 0.095, 0.095 to 0.757 and 3.78 to 28.38 m²/year, respectively (Cornell University 1950; Mitchell and Soga 2005). As such, the values obtained herein are comparable to low plasticity clays. Based on correlations between the liquid limit and c_v, as published by the U.S. Navy (NAVFAC 1982), the unmodified RBT approaches the upper limit for completely remolded samples while the 65:35 mix exceeds this limit, suggesting behavior closer to an undisturbed sample. For both RBT and the 65:35 mix, the lowest c_v values were observed for the highest level of biofilm amendment (EPS-5). Since c_v values are directly proportional to hydraulic conductivity, this would imply that high values of biofilm amendment result in lower permeability. In fact, a negative exponential relationship was observed between hydraulic conductivity and biofilm concentration for the same combinations of soil, nutrient solution and bacteria when tested in flexible wall permeameters, as reported in Daniels and Cherukuri (2005). A complicating factor affecting all consolidation data is the extent to which bacteria continued to synthesize EPS during the several weeks over which the incremental loading and unloading was applied. Given that the soil specimens remained saturated with the same nutrient solution in which the bacteria were grown, it is possible that actual concentrations of EPS were greater than reported in Table 2. The kinetics of this growth is not sufficiently characterized to allow extrapolation, nor was any attempt made to determine actual concentrations at the conclusion of the consolidation test. It is also possible that EPS concentrations decreased during the course of the consolidation testing, as EPS is known to be subject to a variety of biotic and abiotic degradation processes (Wingender et al. 1999), and no attempts were made to limit bacterial activity to the source *B. indica*. While it is not clear whether EPS concentrations increased, decreased or remained constant during individual tests, the presumption is that the trend is similar for the different EPS solutions – such that the specimens remain comparable on the basis of initial EPS concentration. Continued EPS production in a compacted matrix of soil is such that intermittent micro-zones of EPS are possible. For example, Vandevivere and Baveye (1992) observed clusters of bacteria and EPS, separated by vast swaths of inactivity in columns of sand. The net effect of these micro-zones would be to introduce greater heterogeneity into the soil specimens.

4.2 Unconfined Compressive Strength

The maximum unconfined compressive strength, initial tangent modulus and secant modulus at 50% of the peak strength are provided in Table 6. Figure 3 illustrates the general decreasing effect of EPS concentration on observed strength. For example, the peak unconfined compressive strengths for unmodified RBT and 65:35 mix were found to be 667.0 and 395.3 kPa, respectively. Notwithstanding significant standard deviations, many of these values generally decreased with increasing biofilm amendment, and for EPS-5, the observed peak strengths were 159.1 and

Table 6 Summary comparison of strength and moduli

Solution ID	Peak unconfined compressive strength (kPa)		Initial tangent modulus (MPa)		Secant modulus at 50% maximum load (MPa)	
	RBT	65:35 Mix	RBT	65:35 Mix	RBT	65:35 Mix
TW	667.0 ± 9.9	395.3 ± 53.7	24.3 ± 0.2	15.2 ± 3.2	17.6 ± 0.8	11.0 ± 3.1
NS	615.9 ± 59.8	255.7 ± 68.4	3.8 ± 1.3	4.8 ± 0.7	7.0 ± 1.7	4.2 ± 0.9
EPS-1	615.4 ± 53.4	386.2 ± 35.4	3.8 ± 1.7	8.2 ± 1.5	7.4 ± 0.2	6.6 ± 0.5
EPS-2	585.4 ± 157.5	158.1 ± 67.3	5.5 ± 3.1	9.4 ± 8.7	9.6 ± 3.4	13.8 ± 9.9
EPS-3	561.6 ± 75.8	368.2 ± 13.3	7.8 ± 2.6	15.9 ± 2.6	13.1 ± 4.9	17.9 ± 5.6
EPS-4	663.8 ± 47.1	264.9 ± 28.9	37.2 ± 4.1	6.7 ± 2.0	59.0 ± 21.3	5.7 ± 3.0
EPS-5	159.1 ± 78.3	98.8 ± 27.4	4.4 ± 1.4	4.2 ± 1.6	3.6 ± 0.8	3.1 ± 0.6

Note: Values are reported as the average, plus or minus one standard deviation.

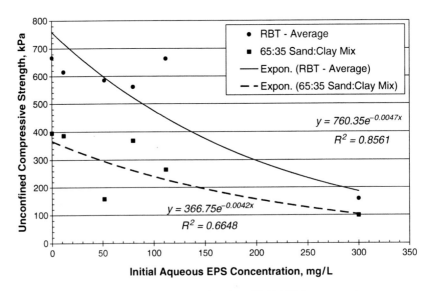

Fig. 3 Unconfined compressive strength as a function of initial EPS concentration

98.8 kPa. Likewise, the material becomes less stiff as observed from the initial tangent and secant modulus. The initial tangent moduli for RBT and 65:35 mix were found to be 24.3 and 15.2 MPa, respectively. For EPS-5, the corresponding moduli were 4.4 and 4.2 MPa, respectively. However, a closer inspection of the data suggests a more complicated relationship between biofilm amendment and strength or stiffness. In the case of RBT, increasing modification with EPS decreased the measured strength until EPS-4 (112 mg/L), at which point the strength was essentially the same as the unmodified soil (663.8 vs. 667.0 kPa) while the stiffness was even greater (37.2 vs. 24.3 MPa). Interestingly, this spike in strength is preceded by a modest but steady increase in initial tangent and secant modulus. Considering that the initial tangent modulus dropped from 24.3 to 3.8 MPa when mixed with the nutrient solution alone (which contains no bacteria or EPS), it might be argued that subsequent increases in stiffness (i.e., 3.8, 5.5, 7.8 and 37.2 MPa for EPS-1, EPS-2, EPS-3 and EPS-4, respectively) reflect the presence of EPS. These increases continue until EPS-5, where the initial tangent modulus is reduced to 4.4 MPa. One explanation is that low dosages of EPS results in particle binding up to some threshold value. Particle binding manifests when the polymers attach to multiple particles and prevent their relative movement, as discussed in Daniels and Inyang (2004). According to this model, strength is derived from both the interparticle friction and the apparent cohesion created by the polymeric bridging. As EPS concentration increases (i.e., to 300 mg/L as in EPS-5), it may well be that polymers cover entire particle surfaces. At this point, there is less resistance to particle movement as EPS has an inherently lower coefficient of friction as compared to soil. Similar behavior is observed with the 65:35 mix, although the noticeable peak in moduli occurred at the EPS-3 level, where the material was more stiff than the unmodified soil.

5 Significance of Biofilm in Waste Containment Systems

The foregoing results suggest that, at the least, biofilm production influences barrier material characteristics in general and strength in particular. Most of the effort, as reflected through both literature and regulations, given to the design and analysis of landfills has been directed toward minimizing impacts on groundwater quality through infiltration and leachate escape. However, system stability is also critical to landfill performance and yet has received relatively less attention, despite several significant failures (Qian et al. 2002). Landfills are susceptible to a number of different rotational, translational and sliding failure modes, including base failure of native soils as well as failure through the waste matrix itself. Within a multilayer cover system, there are several surfaces across which gravitational shear stresses must be transferred. While the critical interface in any given landfill system may vary considering the number of geosynthetic products on the market, different soil types and acceptable configurations, common interfaces of concern are those between a geosynthetic (i.e., geomembrane, geotextile, or geonet) and clay as well as between geosynthetics (e.g., between a geomembrane and a geosynthetic clay liner).

For materials such as compacted clay and geosynthetic clay liners used in cover systems, consideration of both internal and interface shear resistance is important. Because the mechanisms of strength in cover systems are unique and sensitive to slight changes in product specification and loading, the use of literature values for shear and interface strength is completely inappropriate (Qian et al. 2002; Koerner 1997). As such, current design practice involves laboratory testing of the proposed materials under anticipated loads to determine the requisite design parameters. Likewise, it is suggested herein that the initial design strength at construction may change after methane exposure and/or biofilm production. Site-specific field measurements of methane concentrations may help to identify the extent to which these changes are significant.

6 Conclusions

The results of this study suggest that biofilm amendment can influence the consolidation and strength characteristics of barrier materials. Specifically, modification with a mixture of nutrient solution, *B. indica* and EPS results in both increases and decreases in compression index and coefficient of consolidation for a local soil (RBT) and a 65:35 mixture of sand and bentonite clay. As compared to tap water, the influence of the nutrient solution or biofilm on RBT is to increase C_c, although this trend is variable for increasing EPS concentration. While the effect of biofilm on the 65:35 mix is less uniform, the largest increase in C_c was observed for the highest level of biofilm amendment (EPS-5, 300 mg/L). Amendment with biofilm results in both increases and decreases in c_v. The c_v values ranged from 0.4 to 13.6 m^2/year and from 0.2 to 19.3 m^2/year for RBT and 65:35 mix, respectively.

In general, EPS has a decreasing effect on observed strength. For example, the peak unconfined compressive strengths for unmodified RBT and 65:35 mix were found to be 667.0 and 395.3 kPa, respectively. Many of these values decreased with increasing biofilm amendment, and for the highest level of amendment, the observed peak strengths were 159.1 and 98.8 kPa. The relationship between EPS concentration and consolidation characteristics may be influenced by bacterial growth or decay which can occur with the time of testing. In terms of strength, the overall influence of EPS is one of weakening the peak strength, while intermediate increases in initial tangent and secant modulus were observed. Subsequent investigation of EPS on soil properties should involve efforts to distinguish between the nutrient solution, bacteria and EPS. To the extent that naturally-occurring methanotrophic activity in landfill cover systems results in biofilm production, the results suggest potential concerns with cover stability, given the observed strength reductions. As such, while biofilm has potential as a low-cost additive for soil stabilization, more work remains before it can be reliably deployed in that capacity.

Acknowledgements This research was supported by University of North Carolina at Charlotte and Duke Energy. Special thanks are given to Dr. James Oliver, Cone Distinguished Professor of Biology and his students Maya Dougler and Tonya Bates for their microbiological work.

References

Allison, L.E. (1947) Effect of microorganisms on permeability of soil under prolonged submergence. *Soil Science*, 63, 439–450.
ASTM (2000a) Standard test methods for laboratory compaction characteristics of soil using standard effort (12,400 t-lbf/ft^3) American Society for Testing of Materials, D698, Philadelphia, PA.
ASTM (2000b) D2435 Standard test method for one-dimensional consolidation properties of soils. American Society for Testing of Materials, Philadelphia, PA.
ASTM (2000c) D2166 Standard test method for unconfined compressive strength of cohesive soil. American Society for Testing of Materials, Philadelphia, PA.
ATCC (2002) American Type Culture Collection, growth instructions accompanying ATCC strain #9038.
Blumenkrantz, N. and Asboe-Hansen, G. (1973) New method for quantitative determination of uronic acids. *Analytical Biochemistry*, 54, 484–489.
Borjesson, G., Sundh, I., Tunlid, A., Frostegard, A., and Svensson, B.H. (1998a) Microbial oxidation of CH_4 at high partial pressures in an organic landfill cover soil under different moisture regimes. *FEMS Microbiology Ecology* 26, 207–217.
Borjesson, G., Sundh, I., Tunlid, A., and Svensson, B.H. (1998b) Methane oxidation in landfill cover soils, as revealed by potential oxidation measurements and phospholipid fatty acid analyses. *Soil Biology and Biochemistry*, 30(10/11), 1423–1433.
Bowles, J.E. (1992) *Engineering Properties of Soils and Their Measurement*, 4th edition, McGraw-Hill, Boston, MA.
Brune, M., Ramke, H.G., Collins, H.J., and Hanert, H.H. (1991) Incrustation processes in drainage systems of sanitary landfills. *Proceedings, Sardinia'91, 3rd International Landfill Symposium*, Cagliari, Italy, pp. 999–1035.
Clement, T.P., Hooker, B.S., and Skeen, R.S. (1996) Macroscopic models for predicting changes in saturated porous media properties caused by microbial growth. *Ground Water*, September–October, 934–942.
Cornell University (1950) Final Report, Soil Solidification Research Cornell University, Ithaca, NY.
Costerton, J.W., Geesey, G.G., and Cheng, K-J. (1978) How bacteria stick. *Scientific American* 238, 86–95.
Daniels, J.L. and Cherukuri, R. (2005) Influence of biofilm on barrier material performance. *ASCE Practice Periodical of Hazardous, Toxic and Radioactive Waste Management*, 9(4), 245–252.
Daniels, J.L. and Inyang, H.I. (2004) Contaminant barrier material textural response to interaction with aqueous polymers. *ASCE Journal of Materials in Civil Engineering*, 16(3), 265–275.
Daniels, J.L., Inyang, H.I., and Kurup, P. (2002) The influence of dissolved polymers on the properties of earthen barriers used in waste containment applications. In W.-P. Hong (Ed.), *Proceedings of the 6th International Symposium on Environmental Geotechnology and Global Sustainable Development*. Seoul, Korea, July 2–5, pp. 363–370. *Environmental Geotechnology*, Nanjing, PR China, pp. 326–333.
Daniels, J.L., Inyang, H.I., and Iskandar, I. (2003) Durability of Boston blue clay in waste containment applications. *ASCE Journal of Materials in Civil Engineering*, 15(2), 144–152.
Daniels, J.L., Cherukuri, R., Hilger, H.A., Oliver, J.D., and Bin, S. (2005) Engineering behavior of biofilm amended earthen barriers used in waste containment. *Management of Environmental Quality, An International Journal*, 16(6), 691–704.
Dennis, M.L. and Turner, J.P. (1998) Hydraulic conductivity of compacted soil treated with biofilm. *ASCE Journal of Geotechnical and Geoenvironmental Engineering*, 124, 120–127.
Gibson, D.T. (1999) Beijerinckia sp strain B1: A strain by any other name… *Journal of Industrial Microbiology and Biotechnology*, 23, 284–293.
Hart, R.T., Fekete, T., and Flock, D.L. (1960) The plugging effect of bacteria in sandstone systems. *Canadian Mining and Metallurgical Bulletin*, 53, 495–501.
Hilger, H.A. and Barlaz, M.A. (2000) Methane oxidation and microbial exopolymer production in landfill cover soil. *Soil Biology and Biochemistry*, 32, 457–467.

Hilger, H.A., Liehr, S.K., and Barlaz, M.A. (1999) Exopolysaccharide control of methane oxidation in landfill cover soil. *ASCE Journal of Environmental Engineering*, 125, 1113–1123.

Kightley, D., Nedwell, D.B., and Cooper, M. (1995) Capacity for methane oxidation in landfill cover soils measured in laboratory-scale soil microcosms. *Applied and Environmental Microbiology*, 61(2), 592–601.

Koerner, R.M. (1997) *Designing with Geosynthetics*, 4th edition, Prentice Hall, Upper Saddle River, NJ, 761 p.

Kulhawy, F.H. and Mayne, P.W. (1990) Manual on Estimating Soil Properties for Foundation Design, Final Report, Project 1493–6, EL-6800, Electric Power Research Institute, Palo Alto, CA.

Mitchell, R. and Nevo, Z. (1964) Effect of bacterial polysaccharide accumulation on infiltration of water through sand. *Applied Microbiology*, 12(3), 219–223.

Mitchell, J.K. and Soga, K. (2005) *Fundamentals of Soil Behavior*, 3rd edition, Wiley, New York.

Mitchell, J.K., Seed, R.B., and Seed, H.B. (1990) Kettleman Hills waste landfill slope failure. I. Liner-system properties. *ASCE Journal of Geotechnical Engineering*, 116(4), 647–668.

NAVFAC (1982) Soil Mechanics, DM 7.1, Naval Facilities Engineering Command, Alexandria, VA.

Okubo, T. and Matsumoto, J. (1979) Effect of infiltration rate on biological clogging and water quality changes during artificial recharge. *Water Resources Research*, 15, 1536–1542.

Okubo, T. and Matsumoto, J. (1983) Biological clogging of sand and changes of organic constituents during artificial recharge. *Water Resources Research*, 17, 813–821.

Qian, X., Koerner, R.M., and Gray, D.H. (2002) *Geotechnical Aspects of Landfill Design and Construction*, Prentice Hall, Upper Saddle River, NJ, 717 p.

Rowe, R.K., Armstrong, M.D., and Cullimore, D.R. (1998) Effect of particle size on the rate of clogging of a granular media using municipal solid waste leachate. Research Report, Geotechnical Research Center, University of Western Ontario, Ontario, Canada.

Seed, R.B., Mitchell, J.K., and Seed, H.B. (1990) Kettleman Hills waste landfill slope failure. II. Stability analysis. *ASCE Journal of Geotechnical Engineering*, 116(4), 669–690.

Shaw, J.C., Bramhill, B., Wardlaw, N.N., and Costerton, J.W. (1985) Bacterial fouling in a model core system. *Applied Environmental Microbiology*, 49, 693–701.

Taylor, S.W. and Jaffe, P.R. (1990) Biofilm growth and the related changes in the physical properties of a prorous medium. 2. Permeability. *Water Resources Research*, 26(9), 2161–2169.

Terzaghi, K. and Peck, R.B. 1967. *Soil Mechanics in Engineering Practice*, 2nd edition, Wiley, New York, 729 pp.

Vandevivere, P. and Baveye, P. (1992) Relationship between transport of bacteria and their clogging efficiency in sand columns. *Applied Environmental Microbiology*, 58, 2523–2530.

Wingender, J., Neu, T.R., and Flemming, H-C. (1999) What are bacterial extracellular polymeric substances? In J. Wingender et al. (Eds.), *Microbial Extracellular Polymeric Substances: Characterization, Structure, and Function*. Springer, Berlin/Heidelberg, Germany.

Winterkorn, H.F. and Fang, H.Y. 1975. Soil technology and engineering properties of soils, In H.F. Winterkorn and H.Y. Fang (Eds.), *Foundation Engineering Handbook (Chapter 2)*. Van Nostrand Reinhold, New York, pp. 67–120.

Wise, M.G., McArthur, J.V., and Shimkets, L.J. (1999) Methanotroph diversity in landfill soil: Isolation of novel type I and type II methanotrophs whose presence was suggested by culture-independent 16S ribosomal DNA analysis. *Applied and Environmental Microbiology*, 65(11), 4887–4897.

Wood, W.W. and Bassett, R.L. (1975) Water quality changes related to the development of anaerobic conditions during artificial recharge. *Water Resources Research*, 11, 553–558.

Wroth, C.P. and Wood, D.M. 1978. The correlation of index properties with some basic engineering properties of soils, *Canadian Geotechnical Journal*, 15(2), 137–145.

Wu, J., Gui, S., Stahl, P., and Zhang, R. (1997) Experimental study on the reduction of soil hydraulic conductivity by enhanced biomass growth. *Soil Science*, 162(10), 741–748.

Yang, I.C.-Y., Li, Y., Park, J.K., and Yen, T.F. (1993) The use of slime-forming bacteria to enhance the strength of the soil matrix. In E.T. Premuzic and A. Woodhead (Eds.) *Microbial Enhanced Oil Recovery – Recent Advances*. Elsevier, Amsterdam, pp. 89–96.

Yen, T.F., Yang, I.C.-Y., Karimi, S., and Martin, G.R. (1996) Biopolymers for geotechnical applications. Proceedings of the North American Water and Environment Congress, ASCE, 6 p.

Bagasse Ash Stabilization of Lateritic Soil

K.J. Osinubi, V. Bafyau, and A.O. Eberemu

Abstract A lateritic soil was treated with an agro-industrial waste product of sugar mills – Bagasse Ash. The study focused on the effect of up to 12% bagasse ash by weight of dry soil on the geotechnical properties of the deficient lateritic soil. Test specimens were subjected to particle size analysis, compaction, unconfined compressive strength (UCS), California bearing ratio (CBR) and durability tests. The compactions were carried out at the energy of the British Standard Light (BSL).

The study showed changes in moisture – density relationships resulting in lower maximum dry densities (MDD), higher optimum moisture contents (OMC), reduction in fine fractions with higher bagasse ash content in the soil – stabilizer mixtures. A 2% bagasse ash treatment of lateritic soil yielded peak 7 days UCS and CBR values of 836 kN/m^2 and 16%, respectively. Since these values are below 1,700 kN/m^2 and 180% for UCS and CBR, respectively, recommended for adequate cement stabilization, it implies that bagasse ash cannot be used as a 'stand alone' stabilizer but should be employed in admixture stabilization.

Keywords Bagasse ash · California bearing ratio · compaction · durability · lateritic soil · unconfined compressive strength

---

K.J. Osinubi (✉)
Professor, Department of Civil Engineering, Ahmadu Bello University Zaria
KadunaState Nigeria
e-mail: kosinubi@yahoo.com

V. Bafyau and A.O. Eberemu
Post Graduate Students, Ahmadu Bello University, Zaria, Nigeria

1 Introduction

Lateritic soils abound in most parts of the tropics. Such soils have over the years found wide applications in such areas as pavements, embankments, low-cost houses, etc. In some cases the properties of the soils in the immediate vicinity of the construction works may not meet the required specifications. The need thus arises to improve the properties of the available materials. Soil stabilization is a method used to improve soil strength, bearing capacity and durability under adverse moisture and stress conditions (Gidigasu 1976). It refers particularly to the mixing of the parent soil with other soil, cement, lime, bituminous products, silicates and various other chemicals and natural or synthetic, organic and inorganic materials.

The recent trend in soil stabilization is to consider the beneficial reuse of waste products from industries. Ingles and Metcalf (1972) reported the use of fly ash for soil stabilization, while Osinubi (1998a, 2000a, 2006) showed that phosphatic waste, a by–product from the production of superphosphate fertilizer, pulverized coal bottom ash and blast furnace slag can be effectively used to improve deficient lateritic soil and tropical black clay.

The by-product or residue of milling sugarcane is bagasse (the fibre of the cane) in which the residual juice and the moisture from the extraction process remain. Also, bagasse is produced locally by individuals, through the chewing of sugarcane. Most of the bagasse produced, amounting to about one-third of all the cane crushed in some cases supplies the fuel for the generation of steam (Bilba et al. 2003), which eventually results in bagasse ash. As a result of electrification and availability of other sources of fuel, sugar factories have an excess of bagasse during the regular production season. The locally generated bagasse and those from sugar factories present a problem, of handling due to the bulk of the material. When left in the open, it ferments and decays, thus necessitating the safe disposal of the pollutant. Also, when the pollutant is inhaled in large doses it can cause a respiratory disease known as *bagassiosis* (Laurianne 2004). The treatment of soil with bagasse ash could be a safe way of reducing the menace. Therefore, the aim of the study was to determine the suitability of bagasse ash as a stabilizer in the improvement of the engineering properties of lateritic soils in place of the traditional additives such as cement, lime and bitumen.

2 Justification for the Study

In Nigeria sugarcane is one of the numerous agricultural crops commonly cultivated to enhance the economy of the individual farmers. The estimated land under sugarcane cultivation is 23–30,000 ha. Large scale cultivation is done at Bacita and Numan with an estimated annual output of 96,000 t (Misari et al. 1998). The use of bagasse to produce heat from boilers leaves behind ash, commonly known as bagasse ash. Bagasse ash can also be generated by burning bagasse in purpose made furnace.

Generally, pozzolanas are siliceous materials which, while having no cementitious value themselves, will in finely divided form and in the presence of moisture, chemically react with calcium hydroxide (lime) at ordinary temperatures to form compounds possessing cementitious property. According to Ahmad and Shaikh (1992), the physical and chemical properties of sugarcane bagasse ash are found to be satisfactory and conform to the requirements for class N pozzolanas (ASTM C618-78). The major oxides commonly associated with bagasse ash are SiO_2, Al_2O_3, Fe_2O_3, CaO, MgO, Na_2O, K_2O, P_2O_5 and MnO. The loss on ignition, which is indicative of total carbon content, is often below the specified value of 10% for class N pozzolanas. Although researchers (Ola 1983; Osinubi 1998a, 2000a, b, 2006) reported the stabilization of lateritic soils using such additives as Portland cement, lime, bitumen, phosphatic waste and pulverized coal bottom ash to improve the strength properties of these soils in the past; but stabilization using bagasse ash has received little attention.

Most lateritic soils in their natural conditions are at best suitable mainly for sub-base but not for standard pavement base construction material (Ola 1983; Osinubi 2000b). However, the constraint in the availability of standard base course materials may be overcome by designing the road to fit the sub-standard site materials or replacing the sub-standard material with a standard fill material. Alternatively, the unsuitable material may be improved by stabilization. The use of cement, lime and bitumen as stabilizers has enhanced the engineering properties of lateritic soils. However, the cost of such additives is on the increase and bagasse ash is relatively inexpensive and readily available locally through the burning of bagasse.

The major part of Nigeria is underlain by basement complex rocks, the weathering of which had produced lateritic materials spread over most of the area (Osinubi 1998b). It is virtually impossible to execute any construction work in Nigeria without the use of lateritic soils. Thus, the use of bagasse ash for the treatment of lateritic soil would go a long way in providing an overwhelming quantity of the construction material.

3 Materials and Methods

3.1 *Materials*

Soil: The soil samples used are natural reddish-brown lateritic soils, obtained by method of disturbed sampling from a borrow pit in Zaria (latitude 11°15′N and longitude 7°45′E), Nigeria. The dominant clay mineral present in the natural soil of material passing the BS No. 200 sieve with 75 ∞m aperture was qualitatively and quantitatively assessed by differential thermal analysis (DTA) and X-ray diffraction (XRD), respectively, and found to be kaolinite (Ola 1983; Osinubi 1998b).

Bagasse Ash: The bagasse ash used, with a specific gravity of 1.92, was obtained locally from a bagasse dump in the outskirts of Jimeta in Yola North Local Government Area, Adamawa State, Nigeria. The air dried bagasse was burnt in a

Table 1 Detectable oxide composition of bagasse ash

Property	Concentration (% by weight)
CaO	3.23
SiO_2	41.17
Al_2O_3	6.98
Fe_2O_3	2.75
MgO	0.11
K_2O	8.72
SO_3	0.02
TiO_2	1.10
Loss on ignition	17.57

locally constructed incinerator and the bagasse ash obtained was passed through BS No. 200 sieve. The detectable oxide composition of the bagasse ash obtained using X-ray fluorescence (XRF) analysis is summarized in Table 1.

3.2 Methods of Testing

Compaction: Tests involving moisture – density relationship, unconfined compression and California bearing ratio (CBR) were carried out using samples air dried for 1 day. The British Standard Light (BSL) compactive effort was utilized because it is easily achieved in the field.

Strength: Samples of natural soil and soil stabilized with bagasse ash were prepared and tested in accordance with BS 1377 (1990) and BS 1924 (1990), respectively. However, the UCS tests were performed on cubic specimens (70 × 70 × 70 mm). In the preparation of all specimens, the required amounts of bagasse ash by dry weight of soil and soil were measured and mixed in the dry state, before addition of water at the respective OMCs. The soil – bagasse ash mixture was compacted into a detachable mould, specimens removed and wrapped with polyethylene sheets to prevent moisture loss. The specimens were air-cured for periods of 7, 14 and 28 days in the case of UCS, whereas CBR specimens were cured for 6 days and soaked in water for 1 day before testing in accordance with the provisions of Nigerian General Specifications (1997).

Durability: The durability assessment of the soil-bagasse ash mixture was done by immersing specimens in water for measurement of resistance to loss in strength rather than the wet-dry and freeze-thaw tests as highlighted in ASTM (1992) that are not effective under tropical conditions. The resistance to loss in strength was determined as ratio of the UCS of specimens wax-cured for 7 days, de-waxed top and bottom and later immersed in water for another 7 days to the UCS of specimens wax-cured for 14 days. Three specimens were used for each UCS, CBR and durability experiment conducted. The specimens were waxed to prevent loss of moisture by evaporation and allowed to cure under controlled conditions.

4 Discussion of Results

4.1 Index Properties of Natural Soil

The index properties of the natural soil are summarized in Table 2, while the particle size distribution curve is shown in Fig. 1. The soil is classified as A-7-5 based on the AASHTO (1986) Soil Classification System and CL on the Unified Soil Classification System (USCS) (ASTM 1992). It has a clay content of not more than 2%.

Table 2 Properties of natural soil

Properties	Quantity
Natural moisture content (%)	5.13
Liquid limit (%)	41
Plastic limit (%)	24
Plasticity index (%)	17
Linear shrinkage (%)	6.4
Percent passing BS No. 200 sieve	54.9
Group index	4
AASHTO classification	A-7-5
USCS classification	CL
Maximum dry density (Mg/m^3)	1.93
Optimum moisture content (%)	14.7
Specific gravity	2.62
Colour	Reddish-brown

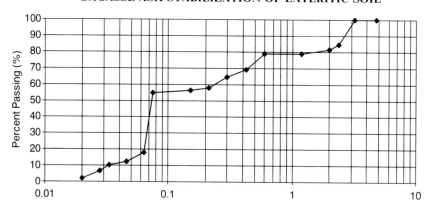

Fig. 1 Particle size distribution curve of natural lateritic soil

4.2 Compaction Characteristics

The effect of bagasse ash content on the maximum dry density (MDD) and optimum moisture content (OMC) of lateritic soil are shown in Fig. 2. The MDD shows a general trend of decrease in value with higher bagasse ash content.

The general drop in MDD could be a result of the flocculated and agglomerated fine particles (caused by cation exchange) occupying larger spaces leading to corresponding decrease in dry density (Ola 1983; Osinubi 1998b). Also, the drop in density with higher stabilizer content may be attributed to the replacement of soil particles with specific gravity of 2.62 in a given volume by particles of the stabilizer which has a comparatively lower specific gravity of 1.92.

The variation of OMC with bagasse ash content (see Fig. 2) shows a general trend of increase with higher bagasse ash content. When the lime in bagasse ash dissociated in the presence of water into calcium and hydroxyl ions either of two situations arose. The calcium ion either replaced cations of other elements present at the exchange sites in the soil: or the calcium ions were absorbed by the soil if there were other unattached anions apart from hydroxyl ions on the soil surface. The increase in OMC was mostly due to pozzolanic reaction of lime in bagasse ash with clay fraction of the soil in conformity with Ola (1983).

The advantage of the increase in OMC with higher stabilizer content and corresponding decrease in MDD of the soil is that it allowed compaction to be easily achieved with wet soil. Thus, there is less of a need for the soil to be dried to lower moisture content prior to compaction in the field.

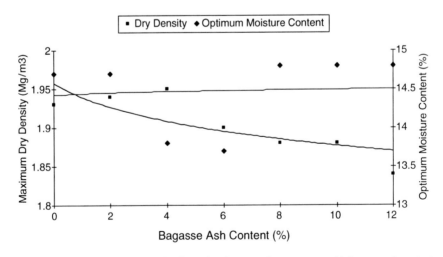

Fig. 2 Variation of maximum dry density and optimum moisture content with bagasse ash content

4.3 Strength Characteristics Unconfined Compressive Strength

Figure 3 shows the effect of bagasse ash content on the UCS of lateritic soil specimens compacted at the BSL energy level and cured for 7, 14 and 28 days, respectively. The improvement in strength of bagasse ash-treated soil has been attributed to soil-bagasse ash reactions, which result in the formation of cementitious compounds that bind soil aggregates. Peak UCS values of 836, 842 and 973 kN/m^2 were recorded at 2% bagasse ash content for 7, 14 and 28 days curing periods, respectively. As expected strength increased with age but decreased with higher bagasse content. It is pertinent to note that the peak 7 day UCS value of 836 kN/m^2 is about half the value of 1,700 kN/m^2 recommended for cement stabilized soils by TRRL (1977).

4.4 California Bearing Ratio

The variation of CBR with bagasse ash content is shown in Fig. 4. The results recorded show that a peak value of 16% was obtained at 2% bagasse ash. This was followed by a decrease in CBR value with bagasse content up to 6% and thereafter increased slightly between 8–12% stabilizer content. However, the increases in CBR values beyond the peak values did not match the maximum values recorded earlier. The results obtained show that if soaked CBR values of 10%, 20–30% and 60–80% (standard Proctor or BSL) for sub-grade, sub-base and base course materials,

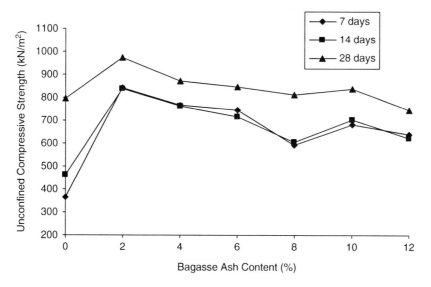

Fig. 3 Variation of unconfined compressive strength with bagasse ash content

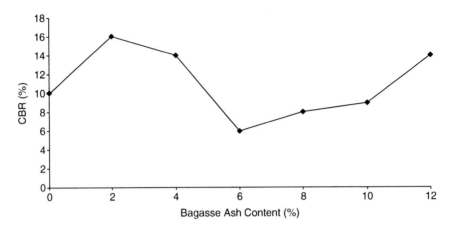

Fig. 4 Variation of California bearing ratio with bagasse ash content

respectively, are considered then only the sub grade requirement of a minimum CBR value of 10% (BSL) is met at 2% bagasse ash treatment. However, if the pozzolanic nature of soil – bagasse ash reaction is taken into consideration, 2% bagasse ash can be said to be adequate for sub-base course construction.

4.5 Durability

The durability of all the UCS specimens were assessed by evaluating the resistance to strength loss of specimens wax-cured for 7 days, dewaxed top and bottom and later immersed in water for an additional 7 days. The resistance to loss in strength for specimens treated with 2% bagasse ash is 9% (i.e., 91% loss in strength). The loss in strength recorded is more than the maximum 20% allowable loss in strength for a 7 day curing dry and 4 day soaking period prescribed by conventional specifications. Regardless of the difference in soaking periods, it is glaring that the specimens did not withstand the durability test. It will therefore be necessary to incorporate other potent additives such as cement or lime in the soil to improve its durability.

5 Conclusion and Recommendations

An experimental programme was undertaken to investigate the effect of bagasse ash on the geotechnical characteristics of a lateritic soil. The following conclusions can be drawn from the study.

The natural soil used in the work is an A-7-5(4) soil according to the AASHTO soil classification system or CL in the USCS. The MDD and OMC of the treated generally showed trends of decrease and increase, respectively, with higher bagasse ash content. The trend can be of advantage in construction involving wet soils, since there is less need for the soil to be dry prior to compaction.

The UCS increased from $366\,kN/m^2$ for the natural soil to 836, 842 and $973\,kN/m^2$ for specimens treated with 2% bagasse ash content and cured for 7, 14 and 28 days, respectively. The soil samples stabilized with bagasse ash, though recorded some gain in UCS did not meet the 7 day $1,700\,kN/m^2$ strength criterion recommended by TRRL (1977) to be met by base course materials. The material, however, can be used for low cost roads with light traffic.

The maximum CBR value of 16% was recorded for soil treated with 2% bagasse ash. This value does not meet the requirement of the Nigerian General Specifications (1997) that a value of 180% be attained in the laboratory for cement-stabilized material to be constructed by the mix-in place method. The resistance to loss in strength achieved at 2% bagasse content is 9% (i.e., 91% loss in strength). It implies that bagasse ash cannot be used as a 'stand alone' stabilizer.

Using the Nigerian General Specifications (1997) and TRRL (1977), strength gains of the soil-bagasse ash mixtures are inadequate using the laterite from the Zaria borrow pit. It is uneconomical to stabilize the material using bagasse ash alone as a stabilizer. On the other hand, if the UCS criterion is used as against the recommended CBR criterion of the Nigerian General Specifications (1997), the 2% bagasse ash content could be recommended for sub base course material for lightly trafficked roads.

References

AASHTO (1986). *Standard Specifications for Transportation Materials and Methods of Sampling and Testing*. 14th edition, American Association of State Highway and Transportation Officials, Washington, DC.

Ahmad, S.F. and Shaikh, Z. (1992). Portland – pozzolana cement from sugar cane bagasse ash. In: Hill, N., Holmes, S. and Mather, D. (Eds), *Lime and Other Alternative Cements*. Intermediate Technology Publlications, London, pp.172–179.

ASTM C618-78. *Specification for Fly Ash and Raw or Calcined Natural Pozzolanas for Use as a Mineral Admixture in Portland Cement Concrete*. American Society for Testing and Materials, Philadelphia, PA.

ASTM (1992). *Annual Book of Standards, Vol. 04.08*. American Society for Testing and Materials, Philadelphia, PA.

Bilba, K., Arsene, M.A. and Ouensanga, O. (2003). Sugar cane bagasse fibre reinforced cement composites part I, Influence of the botanical components of bagasse on the setting of bagasse/cement composite. *Cement and Concrete Composites*, Vol. 25, No. 1, pp. 91–96.

BS 1377 (1990). *Methods of Test for Soil for Civil Engineering Purposes*. British Standard Institute, London.

BS 1924 (1990). *Methods of Test for Stabilized Soils*. British Standard Institute, London.

Gidigasu, M.D. (1976). *Laterite Soil Engineering*. Elsevier, Amsterdam.

Ingles, O.G. and Metcalf, J.B. (1972). *Soil Stabilization Principles and Practice*. Butterworth, Sydney.

Laurianne, S.A. (2004). *Farmers Lungs*. www.emedicine.com/med/topic 771.htm/ 15.11.2005.

Misari, S.M., Busari, L.D. and Agboire, S. (1998). *Current Status of Sugar Cane Research and Development in Nigeria*. Proceedings of National Co-ordinated Research Programme on Sugar Cane, NCRI, Badeggi, pp. 2–12.

Nigerian General Specification (1997). *Roads and Bridges*, Federal Ministry of Works, Abuja.

Ola, S.A. (1983). Geotechnical properties and behaviour of some Nigerian lateritic soils, In: Ola, S.A. and Balkema, A.A. (Eds.), *Tropical Soils of Nigeria in Engineering Practice*. Balkema, Rotterdam.

Osinubi, K.J. (2006). Influence of compactive efforts on lime – slag treated tropical black clay. *Journal of Materials in Civil Engineering, ASCE*, Vol. 18, No. 2, *pp*. 175–145.

Osinubi, K.J. (1998a), Laboratory investigation of engineering use of phosphatic waste. *Journal of Engineering Research*, Vol. 6, No. 2, pp. 47–60.

Osinubi, K.J. (1998b). Permeability of lime-treated lateritic soil. *Journal of Transportation Engineering*, Vol. 124, No.5, pp. 465–469.

Osinubi, K.J. (2000a). Laboratory trial of soil stabilization using pulverized coal bottom ash. *NSE Technical Transactions*, Vol. 35, No. 4, pp. 1–13.

Osinubi, K.J. (2000b). Treatment of laterite with anionic bitumen emulsion and cement: A comparative study. *Ife Journal of Technology*, Vol. 9, No. 1, pp. 139–145.

Transport and Road Research Laboratory (TRRL) (1977). A guide to the structural design of bitumen-surfaced roads in tropical and sub-tropical countries, Road Note 31, H.M.S.O., London.

Strength and Leaching Patterns of Heavy Metals from Ash-Amended Flowable Fill Monoliths

R. Gaddam, H.I. Inyang, V.O. Ogunro, R. Janardhanam, and F.F. Udoeyo

Abstract Solidified flowable fill comprising of Type I portland cement, Class F fly ash, fine sand and water, is a porous monolith. In the case of excavatable fill, material mix proportions in the ash must be such that adequate but inexcessive strength is developed. For non-excavatable fill, maximization of fill strength is the primary objective. Furthermore, being that heavy metals are typically present in fly ash, physico-chemical interactions among mix components must mitigate against leaching out of metals. Herein, flowable fill monoliths containing class F fly ash in weight fractions of 0, 5, 10, 15, and 20, were subjected to unconfined compressive strength (UCS) tests and the American Nuclear Society's ANSI 16.1 leaching test, using de-ionized (DI) water and acidified water (pH = 5.5) as leachants. The results show that comprehensive strength is directly proportional to ash content, reaching $834\,kN/m^2$ for excavatable fill and $3,753\,kN/m^2$ for non-excavatable fill. The diffusion coefficients of arsenic (As) and selenium (Se) from samples decrease sharply with increase in ash content from 5% to 10% and stay relatively low at higher ash content. The leachability indices which are inversely proportional to the quantity of material leached, indicate that the effects of reduction in monolith internal permeability exceed the effects of increasing As and Se content introduced by higher ash content in the monoliths.

Keywords Leaching · flowable fill · fly ash

1 Introduction

Ash-amended flowable fill typically comprises sand aggregate, Portland cement, water, coal ash and possibly, some chemical additives in proportions that are selected for optimal strength, flowability, and permeability. These parameters are

---

R. Gaddam (✉)

important with respect to the use of flowable fill which is frequently called controlled low strength material (CLSM) for utility and pipeline trenches and in pavement repair. In many circumstances, coal ash produced by electric power generating plants, is available in large quantities and at reasonable costs. When the source of ash is close to construction sites, there is usually interest in maximizing its use to reduce the cost of flowable fill projects. Several investigators (Inyang 2003; Sorvari 2003; Reijnders 2005) have devised decision support systems and reviewed methodologies on the use of waste materials such as ash in construction.

A critical constraint to the maximization of ash content in flowable fill is the possibility of an increase in the leachability of contaminants, especially heavy metals, in excessive concentrations, from ash concrete that contains a high weight fraction of ash. When solidified into ash concrete, flowable fill is a porous monolith in which the aggregate particles (sand) are held together by the cementing action of portland cement, coal ash and their reaction products. Generally, fly ash is used in flowable fills because of its high specific surface area relative to that of bottom ash, the larger-grained coal combustion product. This high specific surface area is beneficial in solid-state chemical reactions. Compositionally, fly ash is known to be mostly glass (about 65–70%) formed as spherical particles with surficially attached heavy metals that are condensation products during coal combustion. Some heavy metals are also distributed more uniformly in the matrix of particles as described by Inyang (1992). Coal that is mined at fields in the eastern United States produces class F fly ash which has pozzolanic characteristics. The implication is that in a flowable fill mixture with portland cement, the fly ash reacts with lime that is produced during portland cement hydration reactions, to form cementitious compounds that account for strength development in the flowable fill monolith. In concrete, fly ash can be introduced as a partial replacement for either portland cement or fine aggregate. Usually, the fine aggregate is sand. If fly ash is introduced as partial replacement for portland cement, a net loss in strength may result. If it is replaced partially by fly ash, it essentially amounts to the replacement of a relatively inert material by a reactive one. Thus, monolith strength should be expected to increase as fly ash replaces sand as realized in experiments performed by Sahu and Piyo (2002). In that experiment, the confined compressive strength of sand slurry was found to increase as sand was replaced by fly ash. That observation is consistent with those of Siddique (2003) in which the compressive strength, splitting strength, and modulus of elasticity of concrete increased when fine aggregate was replaced by fly ash at the level of 50% by weight. For a particular ash concrete, the strength achieved depends on the type of fly ash, type of cement, water content and mix proportions of the components (ACI Committee 226 1987; Hurley et al. 1998). Several investigators, including Janardhanam et al. (1992), Gabr and Bowders (2000) and Subramaniam et al. (2005) have analyzed mix designs to assess the effects of fly ash on flowable fill concrete strength.

For facility maintenance considerations, it is most desirable to use flowable fill that is not excessively strong in backfilling of trenches for utilities and communication lines. This is because of the need to excavate the material to gain access to conduits, cables and equipment for repair. At the same time, the flowable fill must be strong enough to carry external (surface) loads and resist the borrowing activities of rodents. Low permeability which often correlates with high strength, is also desirable. ACI Committee 229 (1994), classifies flowable fill that has an unconfined compressive strength below 2,607 kN/m^2 (300 psi) as excavatable fill. Those that have unconfined compressive strength above 2,067 kN/m^2 (300 psi) are classified as non-excavatable fill. Nevertheless, with respect to excavation strength, levels below 689 kN/m^2 (100 psi) are desirable.

Advances in solidification/stabilization (S/S) of waste during the past 20 years have provided some insights to the relationship among strength development, porosity evolution and other transport parameters in cemented systems. A review of fundamental processes of cement solidification and stabilization has been published by Glasser (1997). Therein, there is an acknowledgement that the introduction of fly ash into cemented systems can reduce the connectivity among pores to the extent that permeabilities lower than 10^{-12} m/s can be achieved. However, the generation of cracks during hydration processes can enhance the transport of chemical species in such systems. As noted for cement-stabilized wastes (Gougar et al. 1996; Rosetti and Medici 1995; Park 2000; Asavapisit et al. 1997), heavy metals from fly ash can also enter the structure of evolving chemical phases such as ettringite, during the hydration of cement. Apart from structural inclusions, heavy metals can conceivably precipitate out of internal pore fluids of solidified concrete, chemisorb on other particles out of internal pore fluids of solidified concrete, and other particles in the concrete matrix, react with other constituents to form new compounds, or diffuse out of the concrete matrix under suitable hydraulic and physico-chemical conditions.

Heavy metal diffusion from monolithic systems is a critical phenomenon in the assessment of contaminant leachability and has been analyzed by many researchers, exemplified by Batchelor (1990, 1992, 1998), Batchelor and Park (1998), Van der Sloot (2002), Kim and Batchelor (2001), Inyang et al. (2003), Ogunro and Inyang (2004), Bai et al. (1996), and Poon and Chen (1999). The migration of contaminants through wastes is hindered by intergranular cementation (Reddi and Inyang 2000; Naik et al. 2001). Where portland cement is used, alkaline conditions that can produce the precipitation of some metal from pore fluid can be generated (Cote et al. 1986).

2 Experimentation

In order to evaluate the extent to which the incorporation of Class F fly ash affects both the unconfined compressive strength and contaminant leachability from monolithic samples of flowable fill, experiments were performed using fly ash contents of 0%, 5%, 10%, 15%, and 20% by weight. The other components that were used in the flowable fill mixtures are portland cement (Type 1), fine sand, water and admixture (Darafill Dry). Increase in fly ash content in the mixes was made through reductions in fine aggregate content in the mixes without decrease in cement content.

3 Sources of Ash and Concrete Materials

During the preliminary stages of this research, the materials needed for the mix design of the fill were procured from various companies. A 55-gallon capacity drum of dry fly ash was supplied by Duke Energy (Charlotte, NC) from its Marshall Station. The results of the XRF analysis of the ash used in this research are presented in Table 1 for the elemental composition and Table 2 for the normalized oxide composition. The ash conforms to ASTM C 618 specifications of coal combustible by-products. Lafarge Type 1 cement was supplied by Concrete Supply Co. (Charlotte, NC). This cement conforms to ASTM C 150 specifications. This cement is used commonly in ready mix concrete, and has a specific gravity of 3.15. Hanson Sand that meets ASTM C-33 specifications was supplied by Concrete Supply Co. (Charlotte, NC). Darafill Dry is an additive used in controlled low strength concrete per ASTM C 494. It was supplied by Grace Materials of Charlotte, NC. The moisture content of the sand was determined using the Marshall test. In this test, 400 g of wet sand is subjected to electric drying for nearly 10 min. Once the sand is dry, it is weighed. Then, from the initial and final weights of the sand, the moisture content is determined.

4 Mix Design and Flowable Fill

The mix design of the flowable fill was based on South Carolina Department of Transportation Specifications. These specifications are commonly used in projects in North Carolina from where the ash samples were taken. The air content of the mixes was assumed to be about 15%, and the initial water content was 0.85. Water content was gradually increased beyond 0.85 as ash content was increased to maintain good workability. During initial testing, mixes were designed and tested for

Table 1 XRF analysis on the ash used in the research

Trace metal	Concentration (wt %)	Trace metal	Concentration (wt %)
Si	52.90	Cu	0.03
Al	15.40	Zn	0.02
Fe	3.90	Rb	0.02
K	2.95	Y	0.02
Ca	1.13	Mn	0.02
Ti	1.04	Co	0.01
Mg	0.54	Ga	0.01
S	0.40	Mo	ND
Na	0.20	Hg	ND
Sr	0.09	Cd	0.00
Cr	0.08	Se	0.00
P	0.08	As	0.00
Zr	0.07	Pb	0.01
V	0.04	–	–

Table 2 Normalized metal oxide composition of the ash used in the research

Metal oxide	Concentration (wt %)
SiO_2	53.20
Al_2O_3	25.50
Fe_2O_3	7.74
K_2O	4.26
SO_3	2.50
CaO	2.34
TiO_2	2.25
MgO	0.75
Na_2O	0.26
P_2O_5	0.26
BaO	0.19
SrO	0.17
Cr_2O_3	0.15
V_2O_5	0.09
ZrO_2	0.07
CuO	0.07
ZnO	0.04
Y_2O_3	0.03
Rb_2O	0.03
MnO	0.03
CoO	0.02
Ga_2O_3	0.02

Table 3 Mix design of ash flowable fill for excavatable fill mix

Material content	Percent of fly ash				
	0%	5%	10%	15%	20%
Cement (kg/m³)	61	61	61	61	61
Fly ash (kg/m³)	0	93	186	340	463
Sand (kg/m³)	1,699	1,580	1,418	1,337	1,319
Water (kg/m³)	216	247	279	370	341

strength and suitability for use in subsequent leachability tests. The trial-strength of fill selected was 248 kN/m² which conforms to the ACI 229-R94 excavatable flowable fill strength criteria. However, the mix was developed by increasing the cement content significantly to satisfy not only the field strength criterion but also monolith strength for leaching. Mix designs for both excavatable and non-excavatable fill are shown in Tables 3 and 4 respectively.

5 Strength Tests and their Protocols

The standard proctor test which is a compaction test, was used to determine the dry density and optimum moisture content of the ash. The test meets ASTM D 698 criteria. To perform this test, 2 kg of ash was put in a tray and water was added at the level of 5% by mass. The fly ash absorbed all the water, and after mixing, it still remained as dry powder. Then, 5% more water was added and mixed. The prepared sample was then paced in layers into the compaction mold and compacted 25 blows of the hammer per layer. The sample was then trimmed above the mold, and the weight of the sample with mold was recorded. A portion of compacted ash was removed from the mold and then the procedure was repeated with various moisture contents up to 40%. A plot was drawn to relate moisture content of ash and the dry density calculated for each moisture content. This enabled the maximum dry density and optimum moisture to be established. The maximum dry density of the Duke ash is 1,212 kg/m^3 (75 pcf), and its optimum optimum moisture content is 28%. Based on the grain size distribution curve (Fig. 1), the ash is classified as a silty material, however the result of the compaction test revealed that the maximum dry density of the ash is lower than most natural soils.

The comprehensive strength test was performed in consistence with ASTM D 4832 standards. The sample was mounted on the rubber pad which in turn, was mounted on the load pad. Then, another rubber pad was placed on the top of the sample and the set-up was joggled to be in contact with the load cell. An initial load of 68.9 kN/m^2 (10 psi) was applied. Then, gradually, load was increased at the constant strain rate of 4.3×10^{-3} m/s (0.02 in./min) until each sample crumbled and broke in compression.

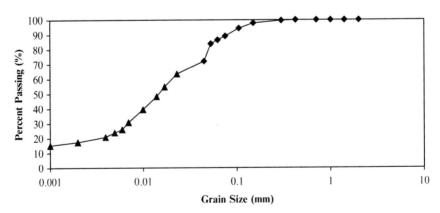

Fig. 1 Combined grain size distribution of the ash tested using both the sieving method and hydrometer method (square shaped points represent sieve analysis and triangle shaped points represent hydrometer analysis)

6 Leaching Tests and Protocols

Monolith leaching tests were performed largely in accordance with American Nuclear Society's ANSI 16.1 test method. Two leachants: de-ionized water and pH5.5 solutions were used. The de-ionized water simulated a leachant of neutral pH while the pH 5.5 leachant simulated mildly aggressive acid rain. The leachant was prepared using concentrated HNO_3 and de-ionized water. Specifically, 3.16×10^{-6} M HNO_3 was transferred into a 15 gallon tank filled with de-ionized water. The leachant volume required to perform the leaching was calculated to be ten times the surface area of the sample. So, for a 2-in. (0.05 m) cubic sample, the required leachant volume was 1,500 ml. The leaching vessel was filled with the estimated leachant and the sample was placed in the vessel to undergo the leaching process. The starting time was noted and after 2 h the leachate was emptied, and a small portion was collected in test tubes for further analyses. Then, the same amount of fresh leachant was transferred into the leaching vessel. The leachant was renewed at 7, 24, 48, 72, 96, 124, 168, and 336 h, and at these times, the leachate was also sampled in the earlier fashion (Fig. 2). The sampled leachate was used in the estimation of the concentrations of the trace elements: arsenic and selenium. Although these two metals are typically found in many ash samples, they were not targeted in this research for regulatory compliance assessment.

Each concentration was determined using the Atomic Absorption Spectrometer (AAS). The process includes standard solution preparation, calibration, and analysis. Based on the absorbance of target elements (As, Se) from the AAS cook book (user's guide), three standards were prepared. The standard tubes were mounted on the standard platform of the AAS. Then, 10 ml samples of leachate were transferred into AAS tubes and placed in trays. The lamps of target elements were mounted onto the lamp chamber and the respective element lamp was turned on in the calibration settings. Calibration was repeated until a smooth calibration curve was

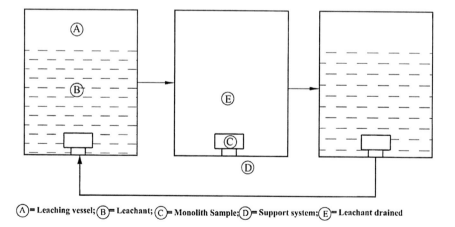

Ⓐ = Leaching vessel; Ⓑ = Leachant; Ⓒ = Monolith Sample; Ⓓ = Support system; Ⓔ = Leachant drained

Fig. 2 Leachant renewal cycle of monolith leaching test

observed. After obtaining the calibration, the analysis button was pressed. The concentration of the target element was then observed. From, the initial observations, it was concluded that the concentration levels of the target elements were too low and undetectable at the sensitivity of the AAS used. Hence, the samples were than spiked with a known higher concentration of each target element and tested in the fashion described earlier. The metals were then detectable. To confirm the latter approach, the samples were tested using a highly sensitive graphite furnace which detects even parts per billion (ppb) levels. The results from both types of tests indicate that the difference was insignificant.

7 Analysis and Discussion of Results

Effects of ash content on monolith strength: The strength of the excavatable fill mixes are 184, 461, 517, 754, and 834 kN/m^2 for 0%, 5%, 10%, 15% and 20% mixes, respectively (Fig. 3). In the case of non-excavatable fill, the compressive strengths of 10%, 15%, and 20% mixes are 2,172, 2,952, and 3,753 kN/m^2 respectively (Fig. 4).

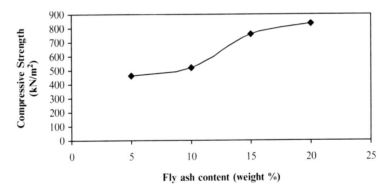

Fig. 3 Compressive strength of excavatable flowable fill

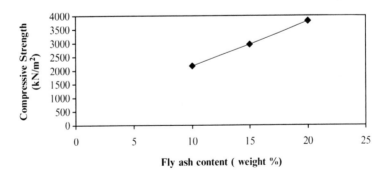

Fig. 4 Compressive strength of non-excavatable flowable fill

Compressive strength increases with increase in ash content. The primary reason for the observed increase in strength is the presence of pozzolanic materials such as silica (53.2%), alumina (25.5%) and calcium compounds (2.34%) in the ash. These materials react with the tricalcium-alumino silicates present in the cement, thereby increasing the strength of the monoliths.

The water content required is directly proportional to the ash content in the mix. The water content demand increases for excavatable fill as ash contents increase from 0% to 5%, 5% to 10%, and 10% to 15%. However, from 15% to 20% ash content, the water demand decreases. A similar trend of increase in water content with increase in ash content was observed in the case of non-excavatable fill. Also, flowability increases with increase in water content. The strength of each mix was found to increase with ash content when the cement content is kept constant. This is due to the fact that pozzolanic activity is enhanced with increase in ash content, resulting in increased strength.

8 Effects of Ash Substitution on Contaminant Leachability Index

The leaching/diffusion test results for arsenic with both water and pH5.5 leachants for a given percentage of ash in excavatable fill mix are shown in Fig. 5. The sample diffusion coefficient (D_{es}) for arsenic for 5% excavatable fill mix is higher than those of other mixes for both pH5.5 and de-ionized water leaching. The highest recorded D_{es} is 5.00×10^{-4} m²/s (pH5.5 leaching). As the ash content in the mix is increased, the diffusion coefficient decreases. The mass leached from the sample

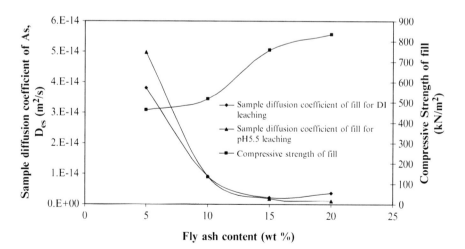

Fig. 5 Test result showing the relationship among fly ash content, arsenic diffusion coefficient and compressive strength of excavatable fill material

Table 5 Mass of arsenic leached from the sample and its leachability index with respect to ash percent and leachant quality

Fly ash percent (%)	Leachant pH	Mass of arsenic leached from sample(mg)	Sample leachability index
5% excavatable fill	DI	0.01	13.4
	pH5.5	0.01	13.3
10% excavatable fill	DI	0.01	14.0
	pH5.5	0.01	14.0
15% excavatable fill	DI	0.01	14.7
	pH5.5	0.01	14.8
20% excavatable fill	DI	0.02	14.4
	pH5.5	0.01	15.0
10% Non-excavatable fill	DI	0.01	14.5
	pH5.5	0.01	14.5
15% Non-excavatable fill	DI	0.01	14.8
	pH5.5	0.01	14.9
20% Non-Excavatable Fill	DI	0.01	14.9
	pH5.5	0.01	15.2

and the sample leachability index of arsenic are given in Table 5. In the case of excavatable fill with de-ionized water leachant, the mass of arsenic released, and leachability index for 5%, 10%, 15%, and 20% mixes are 0.0121, 0.0132, 0.0111, 0.0194 mg; and 13.2, 14.04, 14.66, 14.43, respectively (Tables shows rounded values). In the case of excavatable fill, for pH5.5 leachant, the mass of arsenic released and the leachability index with 5%, 10%, 15% and 20% mixes are 0.0138, 0.0131, 0.010, 0.010 mg; and 13.3, 14.04, 14.75, 14.99 respectively (Tables shows rounded values). The mass released from the sample is greater with de-ionised water leaching than with pH5.5 leaching except for 5 percent excavatable mix. This may be due to temporal changes in leachant chemistry and structural properties of the fill material. Arsenic may have leached more in the alkaline pH range than in the acidic range. In the case of non-excavatable fill, the diffusion coefficient (D_{es}) of arsenic is observed at 10% ash content with pH5.5 leaching and de-ionised water leaching. It decreases with increase in the ash content up to 20% ash content (Fig. 6).

A similar trend concerning the relationship between selenium diffusion coefficient and ash content is shown in Fig. 7. The sample diffusion coefficient (D_{es}) for selenium of 5% excavatable mix is higher than those of other mixes in the cases of both pH5.5 leaching and de-ionised water leaching. The highest recorded D_{es} is 1.4×10^{-14} m²/s (pH5.5 leaching). As the ash content mix was increases, the diffusion coefficient decreases. A similar trend is also observed in the case of selenium leaching from non-excavatable fill mix, but the diffusion coefficients of non-excavatable fill are low in comparison to those of excavatable fill (Fig. 8). The range of selenium diffusion coefficients for excavatable fill is 2.59×10^{-14} m²/s – 1.42×10^{-12} m²/s. The D_{es} values of pH5.5 leaching for 10%, 15%, and 20% mixes are 3.0×10^{-15} m²/s, 1.0×10^{-15} m²/s, and 6.9×10^{-19} m²/s respectively. The mass leached from each sample and the sample leachability index of selenium are given in Table 6.

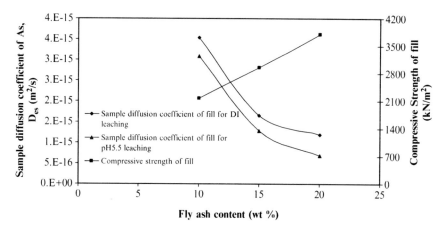

Fig. 6 Test result showing the relationship among fly ash content, arsenic diffusion coefficient and compressive strength of non-excavatable fill material

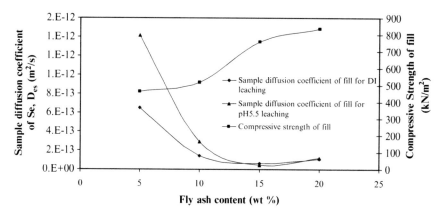

Fig. 7 Test result showing the relationship among fly ash content, selenium diffusion coefficient and compressive strength of excavatable fill material

The mass of selenium leached form the mixes vary from 0.0321 to 0.08 mg. The leachability index range is 11.85–13.49. Except for an ash content of 15%, pH5.5 produces higher leached quantities of selenium than de-ionised water.

9 Conclusions

The results of this research lead to the following conclusions. Increase in fly ash content results in increase in the compressive strength of both excavatable and non-excavatable flowable fill monoliths. For both de-ionised water and pH5.5 leaching,

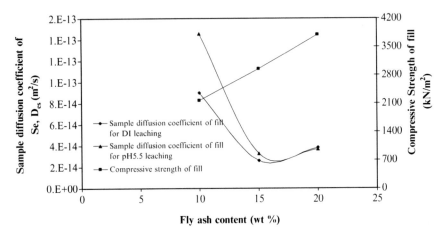

Fig. 8 Test result showing the relationship among fly ash content, selenium diffusion coefficient and compressive strength of non-excavatable fill material

Table 6 Mass of selenium leached from the sample and its leachability index with respect to ash percent and leachant quality

Fly ash percent (%)	Leachant pH	Mass of selenium leached from sample(mg)	Sample leachability index
5% Excavatable fill	DI	0.04	12.2
	pH5.5	0.05	11.9
10% excavatable fill	DI	0.04	12.8
	pH5.5	0.05	12.5
15% excavatable fill	DI	0.04	13.2
	pH5.5	0.04	13.4
20% excavatable fill	DI	0.08	13.0
	pH5.5	0.08	12.9
10% Non-excavatable fill	DI	0.04	13.0
	pH5.5	0.05	12.8
15% Non-excavatable fill	DI	0.03	13.6
	pH5.5	0.04	13.5
20% Non-excavatable fill	DI	0.05	13.4
	pH5.5	0.05	13.4

increase in fly ash content of monoliths tends to suppress the diffusion of arsenic and selenium dramatically until an ash content of about 10% by weight. Although leachant acidity enhances the leaching of both metals, the effects tend to become insignificant as the content of fly ash increases in the monoliths. Perhaps, the buffering effects of fly ash constituents neutralize leachant acidity at higher ash content.

Undoubtedly, increases in fly ash content in the monoliths introduce higher quantities of both As, Se into the monoliths such that higher concentrations of these metals are available for leaching. However, the intensification of pozzolanic activity in the monolith, which should be proportional to fly ash content, may retard the transport of the metals, thereby reducing their leaching rate.

References

ACI Committee 226 1987. Uses of fly ash in concrete. ACI Material Journal ACI, ACI 226-3R-87, 381–409.
ACI Committee 229. 1994. Controlled low-strength materials. Report 229-R, American Concrete Institute, Detroit, Michigan.
Asavapisit, S., Fowler, G., and Cheeseman, C.R. 1997. Solution chemistry during cements hydration in the presence of metal hydroxide wastes. Cement and Concrete Research, 27(8): 1249–1260.
Bai, M., Reogiers, J.C., and Inyang, H.I., 1996. Contaminant transport in non-isothermal fractural media. Journal of Environmental Engineering, ASCE, 122(5): 416–423.
Batchelor, B. 1990. Leach models: Theory and application. Journal of Hazardous Materials, 24: 255–266.
Batchelor, B. 1992. A numerical leaching model for solidified/stabilized wastes. Water Science Technology, 26: 107–115.
Batchelor, B. 1998. Leach models for contaminants immobilized by pH-dependent mechanisms. Environmental Science and Technology, 32: 1721–1726.
Batchelor, B. and Park, J-Y. 1998. Prediction of chemical speciation in stabilized/solidified wastes using a general chemical equilibrium model II: Doped waste contaminants in cement porewaters. Cement and Concrete Research, 29: 99–105.
Cote, P., Bridle, T.R., and Benedek, A. 1986. An approach for evaluating long-term leachability from measurement of intrinsic waste properties. Hazardous and Industrial Solid Waste Testing and Disposal: Sixth Volume. ASTM STP 933, D. Lorenzen, R.A. Conway, L.P. Jackson, A., Hamza, C.L., Perket, and W. J. Lacy, Eds., American Society for Testing and Materials, Philadelphia, PA, pp. 63–78.
Gabr, M.A. and Bowders, J.J. 2000. Controlled low strength material using fly ash and AMD sludge. Journal of Hazardous Materials, 76: 251–263.
Glasser, F.P. 1997. Fundamental aspects of cement solidification and stabilization. Journal of Hazardous Materials, 76: 251–263.
Gougar, M.L.D., Scheetz, B.E., and Roy, D.M. 1996. Ettringite and C-S-H Portland cement phases for waste ion immobilization: A review. Waste Management, 16(4): 295–303.
Hurley, J.P., Nowok, J.W., Bieber, J.A., Dockter, B.A. 1998. Strength development at low temperatures in coal ash deposits. Progress in Energy and Combustion Science, 24: 513–521.
Inyang, H.I. 1992. Energy-related waste materials in geotechnical systems: Durability and environmental considerations. Proceedings of the 2nd International Conference on Environmental Issues and Management in Energy and Mineral Production, Calgary, Canada, pp. 1165–1173.
Inyang, H.I. 2003. Framework for recycling of wastes in construction. Journal of Environmental Engineering, ASCE, 129(10): 887–898.
Inyang, H.I., Ogunro, V.O., and Hooper, F. 2003. Simplified calculation of maximum allowable contaminant concentration in waste-amended construction materials. Resources, Conservation and Recycling, 29(1): 19–32.
Janardhanam, R., Burns, F., and Peindl, R.D. 1992. Mix design for flowable fill fly-ash backfill material. Journal of Materials in Civil Engineering, ASCE, 4(3): 252–263.
Kim, I. and Batchelor, B. 2001. Empirical partitioning of leach model for solidified waste/stabilized wastes. Journal of Environmental Engineering, ASCE, 127(3): 188–195.
Naik, T.R., Singh, S.S., and Ramme, B.W. 2001. Performance and leaching assessment of flowable slurry. Journal of Environmental Engineering, ASCE, 127(4): 359–368.

Ogunro, V.O. and Inyang, H. I. 2004. Relating batch and column diffusion coefficients for leachable contaminants in particulate waste materials. Journal of Environmental Engineering, ASCE, 129(10): 930–942.

Park, C.-K. 2000. Hydration and solidification of hazardous wastes containing heavy metals using modified cementitious materials. Cement and Concrete Research, 30: 429–435.

Poon, C.S. and Chen, Z.Q. 1999. A flow-trough leaching model for monolithic chemically stabilized/solidified hazardous waste. Air and Waste Management Association, 49: 569–575.

Reddi, N.L., Inyang, H.I. 2000. Geoenvironmental Engineering Principles and Applications. Marcel Dekker, New York.

Reijnders, L. 2005. Disposal, uses and treatments of combustion ashes: A review. Resources, Conservation and Recycling, 43: 313–336.

Rosetti, V.A. and Medici, F. 1995. Inertization of toxic metals in cement matrices: Effects of hydration, setting and hardening. Cement and Concrete Research, 25(6): 1147–1152.

Sahu, B.K. and Piyo, P.M. 2002. Improvement in strength characteristics of white Kalahari sands by fly ash. 2nd International Conference on Construction in Developing Countries: Challenges Facing the Construction Industry in Developing Countries, Gabarone, Botswana.

Siddique, R. 2003. Effect of fine aggregate replacement with class F fly ash on the mechanical properties of concrete. Cement and Concrete Research, 33: 539–547.

Sorvari, J. 2003. By-products in earth construction: Environmental assessments. Journal of Environmental Engineering, ASCE, 129(10):899–909.

Subramaniam, K.V., Gromotka, R., Shah, S.P., Obla, K., and Hill, R. 2005. Influence of ultrafine fly ash on the early age response and shrinkage cracking potential of concrete. Journal of Materials in Civil Engineering, ASCE, 17(1): 45–63.

Van der Sloot, H.A. 2002. Characterization of the leaching behaviour of concrete mortars and of cement-stabilized wastes with different waste loading for long term environmental assessment. Waste Management, 22: 181–186.

Part VI
Environmental Monitoring

Geoelectrical Resistivity Imaging in Environmental Studies

A.P. Aizebeokhai

Abstract The presence of contaminants in the environment requires a precise characterization of the nature and extent of contamination for effective remediation. Conventional environmental monitoring has focused largely on point sampling, which involves intrusive processes such as grid drilling. This approach is expensive and provides information only on effects at the sample sites, and hence may not be a true representation of the complex and subtle subsurface geology associated with environmental investigations. Alternative methods that have been used in environmental studies are geophysical methods such as geoelectrical resistivity techniques. Geoelectrical resistivity imaging is used in estimating the resistivity distributions of the subsurface based on several measurements of discrete voltage and current. This paper evaluates the effectiveness of geoelectrical resistivity imaging in environmental applications.

Keywords Environmental studies · non-invasive techniques · geoelectrical imaging · resistivity

1 Introduction

The presence of contaminants in the environment requires precise characterization of the nature and extent of contamination for effective remediation. Conventional environmental monitoring has focused largely on point sampling, which usually

A.P. Aizebeokhai (✉)
Department of Physics, College of Science and Technology, Covenant University, Ota, Ogun State, Nigeria
e-mail: Philips_a_aizebeokhai@yahoo.co.uk

involves intrusive processes such as drilling. This approach is expensive and provides information only on effects at the sample site (Granato and Smith 1999). Intrusive processes can be very dangerous and may not be a true representation of the complex subsurface geology (Ogilvy et al. 1999). Non-invasive techniques, such as geophysical methods, are alternative methods that have been used in environmental monitoring applications.

Many environmental problems amenable to solution by geophysical methods are related to the protection of groundwater from various sources of contamination. The most important sources of subsurface contaminants are hazardous waste disposal sites, landfills, saline water intrusion, and saline water disposal basins and buried hazardous wastes. Other contributors are hydrocarbon spills commonly from exploration and production sites, and buried storage tanks at gas stations, refineries and industrial plants. The escape of leachates from these contaminant sources can lead to serious environmental problems (Simmons et al. 2002). The presence of contaminants in porous rocks significantly alters the physical properties of the rock formations. The degree of alteration depends on the nature, constituents and concentration of the contaminants, as well as the duration of the contamination. Many contaminants decrease pore water resistivity; thus they can be detected and mapped by geoelectrical resistivity imaging methods. This paper attempts to evaluate the effectiveness of geoelectrical resistivity imaging techniques in environmental applications, which is commonly characterized with heterogeneous, subtle, complex and multi-scale subsurface geology.

Subsurface resistivity is related to several geological parameters such as mineral matrix, fluid content, porosity, permeability and degree of water saturation. Other contributing factors include groundwater salinity, volumetric clay content, cation exchange capacity, temperature of pore water, concentration of dissolved salts and contaminants (Shevnin et al. 2006; Lagmanson 2005; Loke 2001). The distributions of subsurface resistivity can be converted to geological images by integrating the knowledge of typical resistivity values for different subsurface materials with the local geology. Electrical conduction in the subsurface is mainly electrolytic because most mineral grains are insulators. Thus, the conduction of electricity is through the interstitial water (or other fluids) in the pores and fissures. Groundwater that fills the pore spaces of rocks is a natural electrolyte. The range of resistivity values for common earth's materials and chemicals (common subsurface contaminants) are given in Table 1 (Palacky 1987; Sharma 1997; Loke 2001; Lagmanson 2005). Igneous and metamorphic rocks typically have high resistivity values. However, geological processes such as dissolution, faulting, shearing and weathering can significantly increase their porosity and fluid permeability thereby increasing their conductivity. Processes such as hardening by compaction or metamorphism, and precipitation of carbonates or silica reduce the porosity and fluid permeability of rocks and hence reduces the conductivity. Sedimentary rocks are generally more porous and permeable than igneous and metamorphic rocks. The resistivity of sedimentary rocks is highly variable, low and depends on its formation factor (Archie 1942).

Table 1 Resistivity values of common earth's materials and chemicals

Materials	Resistivity (Ωm)
Igneous and metamorphic rocks	
Granite	$5 \times 10^3 - 10^6$
Basalt	$10^3 - 10^6$
Slate	$6 \times 10^2 - 4 \times 10^7$
Marble	$10^2 - 2.5 \times 10^8$
Quartzite	$10^2 - 2 \times 10^8$
Sedimentary rocks	
Sandstone	$8 - 4 \times 10^3$
Shale	$20 - 2 \times 10^3$
Limestone	$50 - 4 \times 10^2$
Soils and water	
Clay	$1 - 100$
Alluvium	$10 - 800$
Groundwater (fresh)	$10 - 100$
Sea water	0.2
Permafrost	$6.5 \times 10^2 - 10^5$
Chemicals	
Iron	9.074×10^{-8}
0.01 M potassium chloride	0.708
0.01 M sodium chloride	0.843
0.01 M acetic acid	6.13
Xylene	6.998×10^{16}

2 Geoelectrical Resistivity Surveys

The goal of geoelectrical resistivity surveys is to determine subsurface resistivity distributions by taking measurements of the apparent resistivity on the ground surface. Estimate of the true resistivity is made from these measurements by carrying out inversion on the observed apparent resistivity values and anomalous conditions or heterogeneities are inferred. Traditional applications of geoelectrical resistivity surveys include groundwater prospecting, mining and geotechnical investigations. More recently, the technique have been applied in mapping of contaminant transport (Newmark et al. 1998), hydraulic barriers (Daily and Ramirez 2000), fracture flow paths (Slater et al. 1997), contaminated zone (Amidu and Olayinka 2006; Olayinka and Olayiwola 2000), seepage pathways in embankment dams (Cho and Yeom 2007) and hydrocarbon contaminant plume (Osella et al. 2002).

The use of resistivity to determine the thicknesses and resistivities of layered media has its origin in the work of Conrad Schlumberger who conducted the first experiment in the fields of Normandy in 1912; and about the same time Wenner developed the same idea in the USA (Kunetz 1966). The technique has become an important tool in environmental and engineering applications. The conventional methods of geoelectrical resistivity surveys have undergone significant changes in the last 10–20 years. The traditional horizontal layering (1D) model of interpretation of geoelectrical resistivity data has been replaced with 2D and 3D models of interpretation in complex and highly heterogeneous media. Field techniques has been transformed from measurements made in separate and independent points, with current electrodes spacing growing in logarithmic scale to measuring systems with multi-electrode array along profiles (Dahlin 2001). Data acquisition was more or less carried manually till the 1980s, and this is tedious and slow. The survey was limited to either delineating the variation of apparent resistivity over a surface or compiling quasi-2D sections from a rather limited numbers of vertical electrical soundings (VES). This, however, is still the case in most developing countries, especially in Africa. The use of multi-electrode systems for data acquisition has led to a dramatic increase in field productivity so that one person rather than three or more can conveniently carry out electrical resistivity survey with limited layout.

2.1 Conventional Geoelectrical Resistivity Surveys (1D Interpretation Model)

Two procedures are adopted in convention geoelectrical resistivity surveys. The first is vertical electrical sounding (VES) or drilling (Keofoed 1979) where the mid-point of the electrode array remains fixed, but the electrode spread is increased about the centre. This yields the vertical variations in the subsurface resistivity distribution about the mid-point of the entire electrode spread. The subsurface is assumed to consist of horizontal layers in which resistivity vary only with depth. The apparent resistivity values are usually plotted in a log-log graph. Thus, the one-dimensional model of interpretation of VES is insensitive to lateral variations in the subsurface resistivity, which might lead to changes in apparent resistivity values. This is often misinterpreted as changes in resistivity with depth; however, useful results have been obtained for geological situations such as depth to bedrock and water table where the 1D model is approximately true.

The second approach employed is the constant separation traversing (CST) or profiling method where electrodes separation remains fixed but the entire array is progressively moved along a straight line. This yields information about lateral variations in the subsurface resistivity and is incapable of detecting vertical variations. Data obtained from profiling are mainly interpreted qualitatively. In environmental investigations, the subsurface geology is usually complex, subtle and multi-scale such that both lateral and vertical variations in the resistivity can be very rapid. Thus, the conventional approaches for geoelectrical resistivity surveys are inadequate for environmental applications.

2.2 Geoelectrical Resistivity Imaging (2D and 3D) Surveys

2D geoelectrical resistivity imaging has been used to map areas with moderately complex geology (Griffiths and Barker 1993; Dahlin and Loke 1998). The resistivity of the two-dimensional model changes both vertically and laterally along the survey line but constant in the direction perpendicular to the survey line. The 2D geoelectrical resistivity imaging can yield useful information that is complementary to those obtained using other geophysical methods in many geological situations. For instance, seismic methods can clearly delineate undulating interfaces but may not be able map discrete bodies such as boulders, cavities and contaminated plumes. Similarly, ground-penetrating radar can give detail information of the subsurface but have limited depth of penetration especially in areas with conductive unconsolidated sediments such as clay soils. 2D geoelectrical resistivity imaging surveys are usually carried out using large numbers of electrodes, 25 or more, connected to a multi-core cable.

To obtain a good 2D electrical resistivity image of the subsurface, the coverage of measurements should be two-dimensional. This is done in a systematic manner so that all possible measurements that will yield the best results from the inversion of the apparent resistivity values are made (Dahlin and Loke 1998). If a system has limited number of electrodes the area covered by the survey can be extended horizontally using a technique known as the roll-along method. This can be achieved by moving the cable past one end of the line by several unit electrode spacing, after completing a sequence of measurements. The observed apparent resistivity values are presented in pictorial form using pseudo-section contouring to obtain an approximate picture of the subsurface resistivity. However, the shape of the contours depends on the type of array used as well as the true subsurface resistivity. The pseudo-section plot is a useful guide for detail quantitative interpretation. Poor apparent resistivity data can easily be identified from the pseudo-section plot. The major limitation of the 2D geoelectrical resistivity imaging is that measurements made with large electrode spacing are often affected by the deeper sections of the subsurface as well as structures at a larger horizontal distance from the survey line. This is most pronounced when the survey line is placed near a steep contact with the line parallel to the contact (Loke 2001).

Geological structures encountered in environmental and engineering investigations are inherently three-dimensional in nature. Images resulting from 2D geoelectrical resistivity surveys often contain spurious features due to 3D effects and this usually leads to misinterpretation of the observed anomalies in terms of magnitude and location. Geometrically complex heterogeneous subsurface can therefore not be adequately characterized with 2D geoelectrical resistivity imaging. Due to out-of-plane resistivity anomalies and violation of the 2D assumption, the 2D resistivity imaging will produce misleading images (Bentley and Gharibi 2004). Hence, a 3D geoelectrical resistivity survey with a 3D interpretation model should give the most accurate and reliable results especially in subtle heterogeneous subsurface.

3D geoelectrical resistivity imaging have been used: to map an epithermal area associated with mineral deposits (Li and Oldenburg 1994), track fluid infiltration in vadose zone (Park 1998), delineate soil contaminated with oil and tar (Chambers et al. 1999), investigate an old quarry site used as landfill (Ogilvy et al. 1999), investigate the integrity of a permeable reactive barriers (Slater and Binley 2003) and study a decommissioned sour gas processing plants as part of its remediation programme (Bentley and Gharibi 2004). The 3D surveys were conducted because the heterogeneity of the sites precluded the use of conventional methods or the 2D geoelectrical resistivity imaging technique. The investigations show that 3D geoelectrical resistivity images are superior to the 2D images or the quasi-3D images produced from 2D inversions. Consequently, the 3D geoelectrical resistivity imaging is a better option to properly map the subsurface and its spatial distributions of petrophysical properties or contaminants in environmental and engineering investigations. However, 3D geoelectrical resistivity imaging is far more expensive than the 2D geoelectrical resistivity imaging. In addition, active researches in the field geometry and inversion code for 3D geoelectrical resistivity imaging are on going. Thus 2D geoelectrical resistivity imaging is still widely used in subsurface resistivity mapping even in complex and highly heterogeneous sites.

2.3 Arrays Used in 2D and 3D Geoelectrical Resistivity Surveys

A number of arrays have been used in 2D and 3D geoelectrical resistivity imaging surveys, each suitable for a particular geological situation. The most commonly used arrays include Wenner, Wenner-Schlumberger, dipole-dipole, pole-pole, pole-dipole and gradient arrays. The pseudosections produce by the different arrays over the same structure can be very different. The choice of an array type depends on the geological structures to be delineated, sensitivity of the resistivity meter, the background noise level, sensitivity of the array to vertical and lateral variations in the subsurface resistivity, depth of investigation, horizontal data coverage and signal strength (Loke 2001). The sensitivity function of an array shows the degree to which variations in resistivity of a section of the subsurface will influence the potential measured by the array. Higher values of the sensitivity function indicate greater influence of the subsurface region on the measurements which is mathematically given by the Frechet derivative (McGillivray and Oldenburg 1990).

Most of the pioneering works in 2D geoelectrical resistivity imaging surveys were carried out using the Wenner array (Griffiths and Turnbull 1985; Griffiths et al. 1990). The Wenner array is relatively sensitive to vertical variations in the subsurface resistivity below the centre of the spread but less sensitive to horizontal variations. It has moderate depth of investigation and its signal strength is inversely proportional to the geometric factor. It has the strongest signal strength but the smallest geometric factors, among the common arrays. The major limitation of Wenner array is its relatively poor horizontal coverage with increased electrode spacing. The dipole-dipole array has low electromagnetic coupling between the current and potential electrodes. It is most sensitive to resistivity variations between the electrodes in each dipole pair, and very sensitive to horizontal variations but relatively insensitive to vertical variations of subsurface resistivities. Thus, dipole-dipole array is useful in mapping vertical structures like dykes and cavities, but poor in mapping horizontal structures such as sills or horizontal layers. The depth of investigation is generally shallower than that of Wenner array but has better horizontal data coverage. The major disadvantage of dipole-dipole array is the decrease in signal strength with increasing distance between the dipole pair.

Wenner-Schlumberger array (Pazdirek and Blaha 1996) is a modified form of the classical Schlumberger array, and is moderately sensitive to both horizontal and vertical structures. The array is a good compromise between the Wenner and dipole-dipole array. Its depth of investigation is about 10 times greater than that of Wenner array for the same current electrodes separation. However, its signal strength is smaller than that of the Wenner array, but higher than that for Schlumberger array. Each deeper data level has two data points less than the previous data level unlike the loss of three data points with each deeper level in Wenner array. Thus, its horizontal coverage is slightly better than that for the Wenner.

The pole-pole array, in practice, consists of one current and one potential electrode with the second current and potential electrodes at an infinite distance. Finding suitable locations for these infinite electrodes so as to satisfy this requirement is sometimes difficult. In addition, pole-pole array is often associated with large amount of telluric noise capable of degrading the quality of the measurements. However, this array has the widest horizontal coverage and the deepest depth of investigation but the poorest resolution. Pole-dipole array is an asymmetrical array with asymmetrical apparent resistivity anomalies in the pseudosections over a symmetrical structure. The second current electrode is placed at an infinite distance. It has relatively good coverage but higher signal strength compared with dipole-dipole array. It is insensitive to telluric noise. Repeating measurements with the electrodes arranged in the reverse order can eliminate the asymmetrical effect. The signal strength of the pole-dipole array is lower than that of Wenner and Wenner-Schlumberger arrays.

3 Conclusions

Geoelectrical resistivity surveys have become an important tool in environmental applications where the subsurface geology is usually complex, subtle, multi-scale and highly heterogeneous such that both lateral and vertical variations in the resistivity can be very rapid. The conventional methods for geoelectrical resistivity surveys, which used one-dimensional model for interpretation, are inadequate for environmental studies. 2D geoelectrical resistivity imaging has been used to map areas with moderately complex geology resistivity values. However, images resulting from 2D geoelectrical resistivity surveys can contain spurious features due to 3D effects and this usually leads to misinterpretation of the observed anomalies in terms of magnitude and location. A 3D geoelectrical resistivity survey with a 3D interpretation model gives the most accurate and reliable results especially in subtle heterogeneous subsurface associated with environmental investigation sites.

References

Amidu, S. A. and Olayinka, A. I. (2006): Environmental assessment of sewage disposal systems using 2D electrical resistivity imaging and geochemical analysis: A case study from Ibadan, Southwestern Nigeria. *Environmental and Engineering Geoscience*, **7**(3), 261–272.

Archie, G. E. (1942): The electrical resistivity logs as an aid in determining some reservoir characteristics. *SPE-AIME Transactions*, **146**, 54–62.

Bentley, L. R. and Gharibi, M. (2004): Two- and three-dimensional electrical resistivity imaging at a heterogeneous remediation site. *Geophysics*, **69**(3), 674–680.

Chambers, J. E., Ogilvy, R. D., Meldrum, P. I. and Nissen, J. (1999): 3D electrical resistivity imaging of buried oil-tar contaminated waste deposits. *European Journal of Environmental and Engineering Geophysics*, **4**, 3–15.

Cho, I. and Yeom, J. (2007): Crossline resistivity tomography for delineation of anomalous seepage pathways in an embankment dam. *Geophysics*, **72**(2), G31–G38.

Dahlin, T. (2001): The development of DC resistivity imaging techniques. *Computer and Geosciences*, **27**, 1019–1029.

Dahlin, T. and Loke, M. H. (1998): Resolution of 2D Wenner resistivity imaging as assessed by numerical modeling. *Journal of Applied Geophysics*, **38**(4), 237–248.

Daily, W. and Ramirez, L. (2000): Electrical imaging of engineered hydraulic barriers. *Geophysics*, **65**, 83–94.

Granato, G. E. and Smith, K. P. (1999): An automated process for monitoring groundwater quality using established sampling protocols. *Ground Water Monitoring and Remediation*, **18**, 81–89.

Griffiths, D. H. and Barker, R. D. (1993): Two dimensional resistivity imaging and modeling in areas of complex geology. *Journal of Applied Geophysics*, **29**, 211–226.

Griffiths, D. H. and Turnbull, J. (1985): A multi-electrode array for resistivity surveying. *First Break*, **3**(7), 16–20.

Griffiths, D. H., Turnbull, J. and Olayinka, A. I. (1990): Two-dimensional resistivity mapping with a complex controlled array. *First Break*, **8**(4), 121–129.

Keofoed, O. (1979): *Geosounding Principles 1: Resistivity Sounding Measurements*. Elsevier, Amsterdam.

Kunetz, G. (1966): *Principles of Direct Current Resistivity Prospecting*. Gebruder Borntraeger, Berlin, 103 pp.

Lagmanson, M. (2005): *Electrical Resistivity Imaging*. Advanced Geosciences, San Antonio, TX.

Li, Y. and Oldenburg, D. W. (1994): Inversion of 3D DC resistivity data using an approximate inverse mapping. *Geophysical Journal International*, **116**, 527–537.

Loke, M. H. (2001): *Electrical Imaging Surveys for Environmental and Engineering Studies: A Practical Guide to 2D and 3D Surveys*. 62 pp. Available at www.geoelectrical.com.

McGillivray, P. R. and Oldenburg, D. W. (1990): Methods for calculating Frechet derivatives and sensitivities for the non-linear inverse problem: A comparative study. *Geophysical Prospecting*, **38**, 499–524.

Newmark, R. L., Daily, W. D., Kyle, K. R. and Rimirez, A. L. (1998): Monitoring DNAPL pumping using integrated geophysical techniques. *Journal of Environmental and Engineering Geophysics*, **3**, 7–13.

Ogilvy, R., Meldrum, P. and Chambers, J. (1999): Imaging of industrial waste deposits and buried quarry geometry by 3D resistivity tomography. *European Journal of Environmental and Engineering Geophysics*, **3**, 103–113.

Olayinka, A. I. and Olayiwola, M. A. (2000): Integrated use of geoelectrical imaging and hydrochemical methods in delineating the limits of polluted surface and groundwater at a landfill site in Ibadan area, southwestern Nigeria. *Journal of Mining and Geology*, **37**(1), 53–58.

Osella, A., de la Vega, M. and Lascano, E. (2002): Characterization of contaminant Plume due to hydrocarbon spill using geoelectrical methods. *Journal of Environmental and Engineering Geophysics*, **7**(2), 78–87.

Palacky, G. J. (1987): Clay mapping using electromagnetic methods. *First Break*, **5**, 295–306.

Park, S. (1998): Fluid migration in the vadose zone from 3D inversion of resistivity monitoring. Geophysics, **63**, 41–51.

Pazdirek, O. and Blaha, V. (1996): Examples of resistivity imaging using ME-100 resistivity field acquisition system. *EAGE 58th Conference and Technical Exhibition extended Abstracts*, Amsterdam.

Sharma, P. V. (1997): *Environmental and Engineering Geophysics*. Cambridge University press, Cambridge.

Shevnin, V., Rodriguez, O. D., Mousatov, A., Hennandez, D. F., Martinez, Z. and Ryjov, A. (2006): Estimation of soil petrophysical parameters from resistivity data: Application to oil contamination site characterization. *Geophysical Journal International*, **45**(3), 179–193.

Simmons, C. T., Pierini, M. L. and Hutson, J. L. (2002): Laboratory investigation of variable-density flow and solute transport in unsaturated-saturated porous media. *Transport in Porous Media*, **47**, 215–244.

Slater, L. and Binley, A. (2003): Evaluation of permeable reactive barrier (PRB) integrity using electrical imaging methods. *Geophysics*, **68**, 911–921.

Slater, L., Binley, A. and Brown, D. (1997): Electrical imaging of fractures using groundwater salinity changes. *Groundwater*, **35**, 436–442.

Nitrogen Management for Maximum Potato Yield, Tuber Quality, and Environmental Conservation

S.Y.C. Essah and J.A. Delgado

Abstract There is a need to continue developing new management practices to reduce nitrogen (N) losses that affect air, soil, and water quality. The potato (*Solanum tuberosum* L.) crop is grown in many developing economies and plays a major role in daily food consumption. Traditionally, farmers over-apply N fertilizer in potato production. Three key components identified to increase N use efficiencies (NUE) are the use of crop varieties, a carefully controlled N application rate, and better synchronization between applied N and potato N uptake. The effects of N management on agronomic efficiencies for red (Sangre) and russet (Canela Russet) cultivars grown during 2003 and 2004 in the San Luis Valley of Colorado, U.S.A were studied. In this region the traditional farmer practices range on average from 180 to 240 kg N fertilizer ha^{-1}. The results showed that optimum N application for maximum tuber yield and quality for Canela Russet was 157 kg N ha^{-1} split as 90 kg N ha^{-1} applied at planting and three fertigations of 22 kg N ha^{-1} applied biweekly after initial tuberization. Maximum tuber yield and quality for the Sangre cultivar was observed with an N application rate of 90 kg N ha^{-1} applied at planting. The results indicate that tuber yield and quality can be affected by low N availability and N over-fertilization. It is important that site specific nitrogen management practices are developed that consider cultivar physiological responses to total nitrogen application as well as to the physiological stage when the nitrogen is applied to maximize yields, tuber quality, and economic returns while reducing N losses to the environment.

Keywords Agronomic N use efficiency · crop quality · environmental conservation · nitrogen management · potato · tuber quality

---

S.Y.C. Essah(✉)
Colorado State University, San Luis Valley Research Center, 0249 East Road 9 N, Center, CO 81125 USA
e-mail: sessah@lamar.colostate.edu

J.A. Delgado
USDA-ARS-Soil Plant Nutrient Research, 2150 Centre Ave, Bldg. D Suite 100, Fort Collins, CO 80526 USA
e-mail: jorge.delgado@ars.usda.gov

1 Introduction

Nitrogen (N) fertilizer is key to maintaining the sustainability of worldwide agroecosystems and the feeding of the continuously increasing world population. However, its mobility, dynamics, and transformations contribute to off-site transport that impact the environment (Delgado 2002). Nitrogen loses from agricultural systems have been reported to impact groundwater NO_3-N levels (Hallberg 1989; Juergens-Gschwind 1989). In many regions tile NO_3-N leaching has been identified as a pathway that impacts surface water quality (Randall and Iragavarapu 1995; Randall et al. 1997). Surface off-site N transport has also been reported as a pathway that impacts water bodies (Bjorneberg et al. 2002). Since N is the most mobile and essential dynamic nutrient, we need to continue developing best management practices (BMP) to increase N use efficiency (NUE) while maximizing agricultural production, economic returns, and environmental sustainability.

In recent years, there has been interest in new methods for producing crops of high quality and yield in an environmentally sustainable manner (Delgado 2002; Delgado et al. 2001). Included in these new methods and changes in agricultural systems for higher NUE practices are sustainable crop rotations, use of cover crops, viable N rates, synchronization of inputs and N uptake sinks, fertilizer type, precision agriculture, precision conservation, nitrification inhibitors, N index, and management zones (Berry et al. 2003, 2005; Delgado and Bausch 2005; Kirchmann et al. 2002; Meisinger and Delgado 2002; Shoji 1999). Torstensson et al. (2006) measured the smallest leaching load and high crop yield by planting rye-grass as a cover crop between two conventional cropping systems compared to an organic system where no mineral fertilizer was applied. Similar positive impacts from the use of cover crops in potato systems showed that cover crops served as filter systems, mining NO_3-N from underground water (Delgado 1998).

Potato is grown in many developing economies, and with proper selection of varieties and management practices, this crop can be cultivated and harvested during most parts of the year to feed the local population, with a surplus for export. It takes about 2–3 weeks after planting for the potato sprout to emerge. The young plant depends initially on nutrients from the seed tuber for growth and development until the root system is established. When farmers apply N fertilizer before or at planting as a starter, until the potato develops its root systems, the applied N fertilizer has the potential to leak into the environment. This is especially true for irrigated sandy soils that have a lower water holding capacity, thus needing a higher frequency of irrigation. Pre-plant N applications for potatoes, especially those under irrigated sandy soils tend to have lower NUE (Shoji et al. 2001). One option available to minimize N losses to the environment and to increase NUE is to split N applications to better synchronize N inputs with the times of higher demand by the potato (Iritani 1978; Westermann et al. 1988; Vitosh and Jacobs 1990; Shoji et al. 2001). Additionally, a large pre-plant N application can delay tuberization and reduce tuber quality (resulting in smaller tubers) (Errebhi et al. 1998; Westermann and Kleinkopf 1985). Our study evaluated the potential use of N rates and timing of applications to reduce N inputs and its effects of tuber yield, quality, and agronomic NUE.

CASE 1: Effect of N Rate on Yield and Agronomic NUE of a Red Cultivar – Sangre

Experimental Procedure: The 2003 and 2004 field experiments were conducted at the San Luis Valley Research Center, Colorado, U.S.A., on a gravelly sandy loam (loamy-skeletal mixed [calcareous], frigid *Aquic Ustorthents*). The experiments were included in a 2-year barley (*Hordeum vulgare* L.) – potato rotation. The experimental design was a randomized complete block with four replications. Plot size was 7.5 m long × 3.7 m wide and had four rows spaced at 0.9 m. The total urea ammonium nitrate (UAN) fertilizer rates applied were 0, 90, 157, 224, and 291 kg N ha^{-1}. Treatments were labeled with N rates as S-N$_0$, S-N$_{90}$, S-N$_{157}$, S-N$_{224}$, and S-N$_{291}$, respectively (Table 1).

Soil samples from the top 0.3 m surface layer were collected with an auger before planting, air dried, and sent to Colorado State University (CSU) Soil Testing Laboratory for analysis. Soil NO$_3$-N was determined using a KCl extraction method, with extracts analyzed by colorimetric Lachat method©.[1] Bulk density was determined as described by USDA-SCS (1988). The initial soil NO$_3$-N available was 74 kg N ha^{-1}. The background NO$_3$-N applied with irrigation water was measured (20 kg NO$_3$-N ha^{-1}). Available N from mineralization of soil organic matter was estimated at 30 kg N ha^{-1} for the potato growing season. Atmospheric wet and dry N precipitation was estimated to be 5 kg N ha^{-1}. Barley crop residue N cycled was 10 kg N ha^{-1} (Delgado et al. 2004).

The 2003 and 2004 yields were measured by harvesting and weighing the center two rows from each plot at maturity. Tubers were mechanically size-sorted into tubers <114 g; 114–454 g (marketable size tubers), 114–284 g (medium size marketable tubers), and 284–454 g (large marketable size tubers). The agronomic efficiency

Table 1 Total N fertilizer applied during the growing season was split into applications at planting, and at 2, 4, and 6 weeks after the start of tuberization for the Sangre cultivar

Treatments	Total N applied in growing season	N applied at planting		N applied after start of tuberization		
			(kg N ha^{-1})	2	4	6
S-N$_0$	0	0		0	0	0
S-N$_{90}$	90	90		0	0	0
S-N$_{157}$	157	90		22	22	22
S-N$_{224}$	224	90		45	45	45
S-N$_{291}$	291	90		67	67	67

---

[1] Names are necessary to report factually of available data; however the USDA or CSU neither guarantees nor warrants the standard of the product, and the use of the name by USDA or CSU implies no approval of the product to the exclusion of others that may be suitable. The mention of commercial names is just to present the facts. USDA and CSU do not guarantee the function for the product, nor do we recommend the use of any product over the use of another.

was evaluated as described by Baligar et al. (2001) using Equation (1). Average production yields and tuber quality (size distribution)

$$\text{Agronomic NUE} = \frac{\text{kg of potato tubers in N treated plot} - \text{kg of potato tubers in control plot}}{\text{kg of N fertilizer applied in N treated plot.}} \quad (1)$$

were analyzed using LSD ($P < 0.05$) (SAS Institute 1988). Regression analysis was used to determine the relationship between tuber yield and total N fertilizer rate.

CASE 2: Effect of Pre-plant N Application Rate on Yield and Agronomic NUE of Cultivar Canela Russet

Experimental Procedure: The soil characteristics of the experimental site, experimental design, and data collection are as described in Case 1. A randomized complete block design with four replications of similar size was used. Total N rate applied during the growing season was 157 kg N ha^{-1}, but with a variability of the amount of N applied at planting and/or during the growing season. The treatments included four pre-plant UAN fertilizer rates of 0, 67, 90, and 112 kg N ha^{-1}. The UAN treatments were labeled with pre-plant N rates as CR-N$_0$, CR-N$_{67}$, CR-N$_{90}$, and CR$_{112}$, respectively (Table 2). The remaining N after pre-plant application was applied in three split applications during the growing season at 2, 4, and 6 weeks after start of tuberization. Data analysis was performed on all yield data as previously described. Since this region only receive an average annual precipitation of 178 mm, each study received a total of 457 mm of irrigation applied with solid set. Water management was similar for all studies.

Case 1 "Sangre" Cultivar Results and Discussion: Yield of tubers larger than 114 g and between 114 to 454 g were higher for the S-N$_{90}$ than the S-N$_0$ treatment (Fig. 1a; $P < 0.05$). Nitrogen rates higher than 90 kg N ha^{-1} did not increase tuber yield or improve tuber size distribution (tuber quality) (Fig. 1a, b). There was no response in total tuber yield for the tubers size of 114 to 284 g, and 284 to 454 g (Fig. 1b). This unique study shows that tuber production and quality will be

Table 2 Total N fertilizer applied during the growing season was split into applications at pre-plant and at 2, 4, and 6 weeks after the start of tuberization for the Canela Rusett cultivar

Treatments	Total N applied in growing season	N applied at planting		N applied after start of tuberization		
			(kg N ha^{-1})	2	4	6
CR-N$_{67}$	157	67		30.0	30.0	30.0
CR-N$_{90}$	157	90		22.3	22.3	22.3
CR-N$_{112}$	157	112		15.0	15.0	15.0

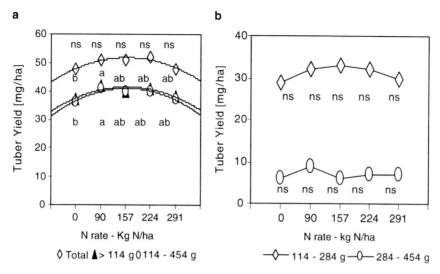

Fig. 1 Relationship between total N rate and tuber yield of Sangre cultivar (1a) and between total N rate and tuber size distribution (1b). Within a compartment rates with different letters are significantly different at P < 0.05

affected by lower N availability observed in the zero N fertilizer treatment. Overfertilization with higher than needed applications will affect yields and yield quality, reducing the economic returns for farmers. Farmers traditionally get a higher economic return from tubers larger than 114 g. If the distribution of larger tubers for the fresh market is reduced, then the economic returns for farmers will be negatively impacted. Farmers must maintain a balance in which plants get enough N to ensure tuber yield and quality without overfertilization that may reduce economic efficiency.

A higher agronomic NUE suggests that most of the N applied was used in crop production, lowering potential of N losses to the environment, while a lower agronomic NUE suggests a higher probability that N is lost to the environment (Baligar et al. 2001). The potential for negative N-related environmental impacts from the overapplication of N fertilizer, causing lower agronomic NUE for N rates above 90 kg N ha^{-1} is shown in Table 3 (P < 0.05).

Case 2 "Canela Russet" Cultivar Results and Discussion: The highest yield, best tuber quality, and highest agronomic NUE was obtained with the CR-N$_{90}$ treatment (Fig. 2a, b; Table 4). When less than the optimum pre-plant 90 kg N ha^{-1} was applied, tuber yield and quality was significantly reduced. Overapplication of initial N fertilizer combined with a lower application of N during the growing season did not improve tuber yield and quality, but rather increased the potential of N loss to the environment, evident by a lower agronomic NUE (Table 4; P < 0.05). Similarly,

Table 3 Effect of N rate on agronomic N use efficiency (NUE) of Sangre cultivar for total yield and tuber size distribution[a]

Treatment	Total yield	>114 g	114–454 g	114–284 g	284–454 g
			Agronomic NUE (kg kg^{-1})		
S-N$_{90}$	42 a	47 a	57 a	30 a	27 a
S-N$_{157}$	19 b	19 b	25 b	24 ab	1 b
S-N$_{224}$	20 b	15 b	17 b	13 b	4 b
S-N$_{291}$	1 c	2 c	5 c	2 c	3 b

[a]Within a column, treatments with different letters are different ($p < 0.05$).

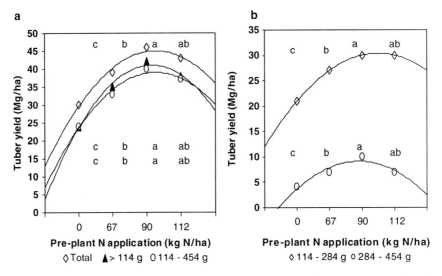

Fig. 2 Relationship between pre-plant N application rate and tuber yield (2a) and tuber size distribution of Canela Russet cultivar (2b). Within a compartment, rates with different letters are significantly different at $P < 0.05$

Table 4 Effect of pre-plant N application rate on agronomic N use efficiency of Canela Russet cultivar for total yield and tuber size distribution[a]

Treatment	Total yield	>114 g	114–454 g	114–284 g	284–454 g
			Agronomic NUE (kg kg^{-1})		
CR-N$_{67}$	132 b	153 b	132 b	88 b	43 b
CR-N$_{90}$	179 a	195 a	179 a	105 a	74 a
CR-N$_{112}$	118 b	120 c	113 c	82 b	31 b

[a]Within a column, treatments with different letters are different ($p < 0.05$).

a lower application of initial N fertilizer combined with a higher application of N during the growing season did not improve tuber yield or quality and also increased the potential of N loss to the environment (Table 4; $P < 0.05$).

2 General Discussion

We found a large range of optimal N fertilizer application rates among cultivars. The optimum range for the Canela Russet cultivar was 157 kg UAN-N ha^{-1} (split as a combination of 90 kg N ha^{-1} at planting with 67 kg N ha^{-1} applied during the growing season), while a lower 90 kg UAN-N ha^{-1} was the optimum for the Sangre cultivar at this site. The total available background N in the system was estimated to be about 139 kg N ha^{-1}. The background N sources were the initial soil NO_3-N, background NO_3-N applied with irrigation water, N from mineralization of soil organic matter and previous crop residue, and atmospheric wet and dry N precipitation. In other words, the best response and tuber quality for the Canela Russet cultivar was obtained with a total of 296 kg N ha^{-1}, a sum combining both N fertilizer applied to the system and background N sources. The best response for the Sangre cultivar was observed with a combined total of 229 kg N ha^{-1}. This study suggest that potato needs large amounts of N since its N recoveries are about 50% in this region (Shoji et al. 2001). This study also shows that tuber yields and quality are sensitive to total N application as well as the amount of N applied during the physiological stage of tuberization.

We recommend that farmers develop a comprehensive N management plan, and that they account for other N sources such as initial soil NO_3-N, NO_3-N in background irrigation water, and N cycling from cover crops or rotation crops. Our results agree with those from Errebhi et al. (1998) and Westermann and Kleinkopf (1985), which showed that larger pre-plant N applications can delay tuberization and reduce tuber quality. We found that N overapplication can also affect tuber quality, especially if excessive N is applied during the growing season.

Since the NO_3-N in background irrigation water was 20 kg NO_3-N ha^{-1}, the plots were already receiving some background N. For the Sangre cultivar, this background N and the N fertilizer applied at planting were enough to maximize tuber yield and quality.

For the Canella Russet cultivar, the maximum yield and quality were obtained by a combination of 90 kg N ha^{-1} at planting and three 22 kg N ha^{-1} applications during the growing season, in addition to the background N. Apparently, the N demand by the Canela Russett cultivar was greater during the growing season, however the higher applications of three 30 kg N ha^{-1} applications during the growing season were too high and negatively affected the tuberization process, diminishing tuber production and quality. These studies show that it is important to develop comprehensive site specific N management recommendations that consider varieties, background N sources, and split N applications to maximize tuber yield and quality while maximizing environmental conservation and economic returns for farmers.

3 Conclusion

Our results clearly show that nitrogen management is the key to maximizing potato yield and tuber quality. Our results are also unique in showing that initial overapplication of N fertilizer not only reduced the agronomic N use efficiency, increasing the potential for N losses to the environment, but can also contribute to lower total tuber yield and quality, lowering economic returns for farmers. It is important that we develop site specific nitrogen management practices that consider cultivars' physiological responses to nitrogen to maximize yields, tuber quality, and economic returns, while reducing N losses to the environment.

References

Baligar, V.C., N.K. Fageria, and Z.L. He. 2001. Nutrient use efficiency in plants. Commun. Soil Sci. Plant Anal. 32:921–950.

Berry, J R., J.A. Delgado, R. Khosla, and F.J. Pierce. 2003. Precision conservation for environmental sustainability. J. Soil Water Conserv. 58:332–339.

Berry, J.K., J.A. Delgado, F.J. Pierce, and R. Khosla. 2005. Applying spatial analysis for precision conservation across the landscape. J. Soil Water Conserv. 60:363–370.

Bjorneberg, D.L., D.T. Westermann, and J.K. Aase. 2002. Nutrient losses in surface irrigation runoff. J. Soil Water Conserv. 57:524–529.

Delgado, J.A. 1998. Sequential NLEAP simulations to examine effect of early and late planted winter cover crops on nitrogen dynamics. J. Soil Water Conserv. 53:241–244.

Delgado, J.A. 2002. Quantifying the loss mechanisms of nitrogen. J. Soil Water Conserv. 57:389–398.

Delgado, J.A. and W.C. Bausch. 2005. Potential use of precision conservation techniques to reduce nitrate leaching in irrigated crops. J. Soil Water Conserv. 60:379–387.

Delgado, J.A., R.R. Riggenbach, R.T. Sparks, M.A. Dillon, L.M. Kawanabe, and R.J. Ristau. 2001. Evaluation of nitrate-nitrogen transport in a potato-barley rotation. Soil Sci. Soc Am. J. 65:878–883.

Delgado, J.A., M.A. Dillon, R.T. Sparks, and R.F. Follett. 2004. Tracing the fate of 15N in a small-grain potato rotation to improve accountability of N budgets. J. Soil Water Conserv. 59:271–276.

Errebhi, M., C.J. Rosen, S.C. Gupta, and D.E. Birong. 1998. Potato yield response and nitrate leaching as influenced by nitrogen management. Agron. J. 90:10–15.

Hallberg, G.R. 1989. Nitrate in ground water in the United States. pp. 35–74, In: R.F. Follett (ed.) Nitrogen Management and Ground Water Protection. Elsevier, New York.

Iritani, W.M. 1978. Seed productivity: Stem numbers and tuber set. Proc. Ann. Washington State Potato Conf. 17:1–4.

Juergens-Gschwind, S. 1989. Ground water nitrates in other developed countries (Europe) – relationships to land use patterns. In: R.F. Follett (ed.) Nitrogen Management and Ground Water Protection. Elsevier, New York, pp.75–138.

Kirchmann, H., A.E.J. Johnston, and L.F. Bergstrom. 2002. Possibilities for reducing nitrate leaching from agricultural land. Ambio 31:404–408.

Meisinger, J.J. and J.A. Delgado. 2002. Principles for managing nitrogen leaching. J. Soil Water Conserv. 57:485–498.

Randall, G.W. and T.K. Iragavarapu. 1995. Impact of long-term tillage systems for continuous corn on nitrate leaching to tile drainage. J. Environ. Qual. 24:360–366.

Randall, G.W., D.R. Huggins, M.P. Russelle, D.J. Fuchs, W.W. Nelson, and J.L. Anderson. 1997. Nitrate losses through subsurface tile drainage in conservation reserve program, alfalfa, and row crop systems. J. Environ. Qual. 26:1240–1247.

Shoji, S. (ed.). 1999. Meister Controlled Release Fertilizers – Properties and Utilization. Konno Printing Co. Sendai, Japan.

Shoji, S., J. Delgado, A. Mosier, and Y. Miura. 2001. Use of controlled release fertilizer and nitrification inhibitors to increase nitrogen use efficiency and to conserve air and water quality. Commun. Soil Sci. Plant Anal. 32:1051–1070.

SAS Institute Inc. (Statistical Analysis System). 1988. SAS/STAT User guide. Ver 6.03. 3rd Edition. SAS Inst., Cary NC.

Torstensson, G., H. Aronsson, and L. Bergstrom. 2006. Nutrient use efficiencies and leaching of organic and conventional cropping systems in Sweden. Agon. J. 98:603–615.

USDA-SCS (United States Department of Agriculture-Soil conservation Service). 1988. National Agronomy Manual. 2nd Edition. Washington, DC.

Vitosh, M.L. and L.W. Jacobs. (1990). Nutrient management to protect water quality. Michigan State Univ. Ext. Bull.: (Water quality series) (25). 6pp.

Westermann, D.T. and G.E. Kleinkopf. 1985. Nitrogen requirements of potatoes. Agron. J. 77:616–621.

Westermann, D.T., G.E. Klienkopf, and L.K. Porter. 1988. Nitrogen fertilizer efficiencies on potatoes. Am. Pot. J. 65:377–386.

In Vitro Analysis of Enhanced Phenanthrene Emulsification and Biodegradation Using Rhamnolipid Biosurfactants and *Acinetobacter calcoaceticus*

N.D. Henry and M. Abazinge

Abstract The ability of biosurfactants and *Acinetobacter calcoaceticus* to enhance the emulsification and biodegradation of phenanthrene was investigated. Phenanthrene is a polycyclic aromatic hydrocarbon (PAH) that may be derived from various sources, for example incomplete combustion of petroleum fuel, thus it occurs ubiquitously throughout the environment. Phenanthrene biodegradation has been reported to be greatly enhanced in the presence of surfactants (Cuny et al. 1999; Chen et al. 2001). It is weakly soluble in water (1.2 mg L^{-1}, 1 atm, 25 °C); therefore, its' biodegradation is strictly limited by its bioavailability (Chen et al. 2001). Emulsification assays were carried out to assess the stability of emulsions formed between phenanthrene and water in the presence of rhamnolipid biosurfactants. An increase in emulsion stability has been shown to equal an increase in bioavailability of a hydrophobic PAH (Dean et al. 2001). Emulsion stability was determined by height of emulsion layer and optical density measurements. Results show phenanthrene and water emulsifications were stabilized for a period up to 10 days at levels ranging from 70–80% with the use of un-encapsulated biosurfactants. Microencapsulated biosurfactants stabilized the emulsion up to 89% for 15 days. Experimental microcosm studies to assess biodegradation rates were carried out over 15 days in 40-mL bioreactors. The reactors were sampled at t = 0, 3, 6, 9, 12, 15 days. Biodegradation rates were determined from measurements of carbon dioxide respiration and phenanthrene concentrations. Results show that on average, more phenanthrene was mineralized (96.4% over 15 days) by bacteria amended with non-encapsulated rhamnolipid biosurfactant (NERhBS).

Keywords Bioremediation · biosurfactants · phenanthrene · microencapsulation

---

N.D. Henry and M. Abazinge (✉)
Environmental Sciences Institute, Florida A&M University

1 Introduction

Polycyclic aromatic hydrocarbons (PAHs) are prevalent contaminants in the environment as a result of fossil fuel combustion and by-product waste from industrial activities (Banat 1995; Dean et al. 2001). Biodegradation by microorganisms is one of the primary methods of removal of PAHs from contaminated sites (Wong et al. 2004). One of the main reasons for prolonged persistence of PAHs in contaminated environments is their low water solubility, which restricts their bioavailability to biodegrading microorganisms (Barkay et al. 1999). Therefore, approaches to enhance PAH biodegradation often attempt to increase their apparent solubility by treatments such as addition of biosurfactants (Piehler and Paerl 1996; Dean et al. 2001).

Biosurfactants are biologically synthesized surface-active agents produced by many various microorganisms and represent a wide range of chemicals and molecular structures (Desai and Banat 1997; Barkay et al. 1999). They posses properties comparable to chemically produced surfactants and can be used for various applications many of them in environmental management (Maslin and Maier 2001; Ron and Rosenberg, 2001). In general, biosurfactants could enhance the apparent solubility of PAHs by micellar formation, which commences at the critical micelle concentration (CMC) and then solubility is proportional to surfactant concentration (Wong et al. 2004). However, biodegradation of PAHs is not always correspondingly enhanced by surfactants (Wong et al. 2004). Some researchers have found that addition of surfactants enhanced PAH biodegradation (Piehler and Paerl 1989; Barkay et al. 1999; Dean et al. 2001), whereas others reported no effect or inhibition by surfactants (Laha and Luthy 1991; Tiehm 1994; Volkering et al. 1995; Boonchan et al. 1998; Wong et al. 2004). The conflicting results may be due to varied interactions amongst microbial communities, types of PAHs targeted for biodegradation, or the type of surfactant used in the approach.

The goal of the work reported here was to test the efficiency of a microparticle system of biosurfactants, previously described in Henry et al. (2005), using emulsification assays and traditional biodegradation protocols. The use described here is only one potential application of the microparticle system. Rhamnolipid biosurfactants have been used as effective biopesticides on agricultural, horticultural and turf sites to eradicate certain pathogenic fungi (EPA 2004). This application of the microparticle system would be potentially effective in developing countries because it is inexpensive to formulate and it does not require clean-up of secondary waste, as none is produced. The objectives of the work are to assess the ability of encapsulated and non-encapsulated rhamnolipid biosurfactants to emulsify phenanthrene (PHE), a PAH; and to compare the capabilities of *Acinetobacter calcoaceticus* to degrade PHE in the presence of encapsulated and non-encapsulated rhamnolipid biosurfactants.

2 Materials and Methods

Chemicals: Phenanthrene (purity >98%; Aldrich Chemical Company, St. Louis, MO) was used in this study as the target PAH for biodegradation.

Biosurfactant: Rhamnolipid biosurfactants produced by *Pseudomonas aeruginosa* (Jeneil Biosurfactant Co., Saukville, WI) were used because of their known ability to solubilize and emulsify hydrophobic organic compounds (Al-Tahhan et al. 2000). This study involved the use of microencapsulation as a means to deliver biosurfactants to target PAHs. Microencapsulation involves enclosing a desired substance within a biodegradable polymer, forming a microparticle. The desired substance is then released over time as the polymer degrades. The development of the microparticle system used here has been described by Henry et al. (2005).

Acinetobacter calcoaceticus (ATCC 31012): *Acinetobacter calcoaceticus* was obtained from the American Type Culture Collection (ATCC) as a pure culture. Strain ATCC 31012 (*A. calcoaceticus*-RAG-1) was used in this study. Freeze-dried bacteria were rehydrated with 0.5 mL of nutrient broth and mixed well. Aseptically, the total mixture of the vial was transferred to a test tube of nutrient broth medium (5–6 mL). The culture was incubated at 30 °C and allowed to grow for about 4 days, depending on the bacteria response.

Determination of emulsifying activity: A micro-modification of the emulsification assay described by Navon-Venezia et al. (1995) and Cooper and Goldberg (1997) was used to measure emulsifying activity. Samples to be tested were prepared by introducing 2.5 mL of each rhamnolipid biosurfactant (RhBS) solution and PHE into 10 mL glass test tubes. The samples were incubated at room temperature for 1 h. The height of the emulsion layer between the two aqueous phases was determined after the 1 h settling period. Changes in the emulsion layer were noted until the emulsion stabilized.

In order to assess the effectiveness of the microparticle system, the emulsification assay was further modified to account for release of the biosurfactant from the microparticles. Ten milligrams of microparticles were added to a 10 mL test tube containing 4 mL of TM buffer (20 mM Tris-HCl buffer [pH 7.0], 10 mM $MgSO_4$), and then 2.5 mL of PHE was added. The samples were incubated at room temperature for 24 h before any observations were made to allow time for biosurfactant to be released from the microparticles. The changes in height of the emulsion layer between the two aqueous phases were determined after the 1 h settling period and the initial 24 h incubation period, as described above.

Biodegradation microcosm study: Aqueous degradation tests were performed to estimate PHE degradation rates by *A. calcoaceticus*, grown in nutrient broth, supplemented with minimal salts at 30 °C, in the presence of non-encapsulated rhamnolipid biosurfactants (NERhBS) and encapsulated rhamnolipid biosurfactants (ERhBS). The bacterial culture medium, which consisted of 0.5 g K_2HPO_4, 1.0 g of NH_4Cl, 2.0 g of Na_2SO_4, 2.0 g of KNO_3, 0.2 g of $MgSO_4 \cdot 7 H_2O$, and 0.002 g $FeSO_4 \cdot 7 H_2O$ per liter of distilled water, was prepared in accordance with the procedure described by Wong et al. (2004).

Aqueous PHE concentrations were analyzed by high performance liquid chromatography (HPLC), equipped with a UV detector (254 nm). Samples (50 µL) were injected into a C_{18} reverse phase column (4.6 × 250 mm) and eluted with an isocratic mobile phase consisting of 80% acetonitrile and 20% water (v/v) delivered at a flow rate of 1.0 mL min^{-1}. Eluted peaks were monitored by UV absorption at 254 nm.

3 Results

Phenanthrene emulsification by rhamnolipid biosurfactant: Figure 1 summarizes the time required for RhBS to form a stable emulsion with PHE. NERhBS formed a stable emulsion more quickly than ERhBS; however the ERhBS held a more stable emulsion for a longer period of time. The NERhBS held a stable emulsion for up to 10 days, however after that time period, the sample vials began to mold so the results were hindered. A stable emulsion was observed for up to 360 h with the ERhBS, which was significantly higher ($p < 0.05$) than that observed with the NERhBS.

In order to quantify the volume of PHE emulsified by RhBS, the Emulsification Index (EI), determined based on the equation described by Bodour et al. (2004),

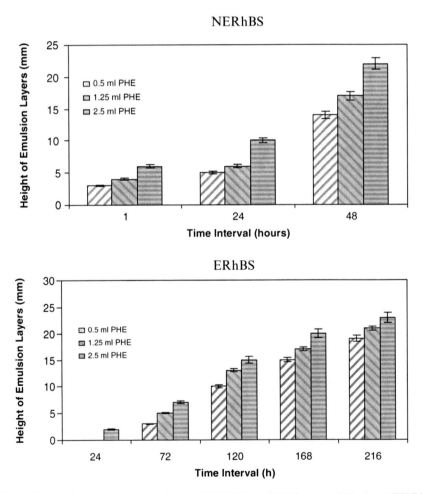

Fig. 1 Changes in height of emulsion layers of NERhBS and ERhBS up to stabilization. NERhBS achieved a stable emulsion after only 48 h, while ERhBS achieved the same at approximately 216 h

was used. The RhBS was a strong and stable emulsifier. PHE concentrations of 5 mg L^{-1} exhibited EI of up to 89% by ERhBS with biosurfactant concentration of 45 mg L^{-1} after 360 h (Table 1). There was no significant difference ($p > 0.05$) among the EI in the ERhBS treatments, showing that biosurfactant concentration had no influence on the results. Similar results were shown in the NERhBS treatments, wherein biosurfactants with concentrations of 30 and 45 mg L^{-1} had no significant difference ($p > 0.05$) in the emulsion stability. After 168 h, the NERhBS test vials became turbid and the presence of mold formation was obvious. A decrease in the height of the emulsion layer was also detected.

Phenanthrene biodegradation in the presence of biosurfactants and *A. calcoaceticus*: Biodegradation of PHE in the presence of NERhBS, ERhBS and A. calcoaceticus was evaluated by determining the change in PHE concentration in solution phase (Fig. 2) over a 15 day incubation period. A. calcoaceticus was able to utilize PHE as a carbon source and degrade it accordingly. The PHE biodegradation percentage values (mean ± SD%) after 6 days for NERhBS + AC, ERhBS + AC and AC treatments were 58 ± 3%, 40 ± 2% and 29 ± 2%, respectively. At the end of the incubation period, 96 ± 4%

Table 1 Emulsion activity

Emulsifying activity of biosurfactant with phenanthrene

Biosurfactant concentration	Emulsification index NERhBS					Emulsification index ERhBS				
	E_{24}	E_{48}	E_{120}	E_{168}	E_{264}	E_{24}	E_{48}	E_{120}	E_{168}	E_{360}
30 mg L^{-1}	80	80	80	80	75	0	46	74	85	85
45 mg L^{-1}	80	80	80	80	78	0	35	75	89	89

The emulsion index (E = vol. of emulsion layer × total vol^{-1}) × 100. Values represent mean percentages of emulsion of the oil layer in the test tubes. EI values for ERhBS increase over time because biosurfactants are consistently released from microparticles during the emulsification assay.

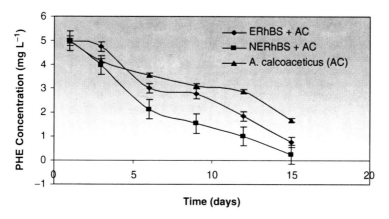

Fig. 2 Biodegradation of phenanthrene (mg L^{-1}) in the microcosm study. Data points represent means of determinations from triplicate experimental microcosms. Vertical bars correspond to 1 standard deviation about the mean

of the initial concentration of PHE had been degraded by the treatment with NERhBS + AC, which was significantly higher ($p < 0.05$) than the other treatments. It is interesting to note that the growth medium color changed depending on the treatment. This varied from colorless in the AC treatment, deep orange-yellow with NERhBS, to light orange-yellow with ERhBS. These changes in color are a result of the appearance of phenanthrene metabolites (Mueller et al.1990; Lal and Kanna 1996; Cuny et al. 1999).

4 Conclusions

The results of this study show that biodegradation activity of PHE-degrading bacteria is influenced by the bioavailability of the phenanthrene in the aqueous phase. By demonstrating that rhamnolipid biosurfactants enhance the emulsification of phenanthrene, it can inferred that the biosurfactants facilitate the transport of PHE into microbial cells, thus increasing its biodegradation. This positive effect on biodegradation is not a general response to the addition of biosurfactants, as some biosurfactants may inhibit biodegradation (Laha and Luthy 1991; Cuny et al. 1999).

This study also investigated the effectiveness of a previously developed microparticle system. Microparticles were observed to be a promising method of delivery of biosurfactants as results yielded by ERhBS + AC were considerably similar to those yielded by NERhBS + AC treatments. Further investigation of this system is underway. Once further testing is completed, this system could serve as an economically and environmentally friendly alternative for clean-up of residual PAH waste. This system could also be used as a method to deliver needed nutrients to indigenous bacteria, to increase their natural abilities to degrade such wastes.

References

Al-Tahhan, R.A., Sandrin, T.R., Bodour, A.A., and Maier, R.M. 2000. *Applied and Environmental Microbiology.* 66(8):3262–3268.
Banat, I.M. 1995. *Acta Biotechnology.* 15:251–267.
Barkay, T., Navon-Venezia, S., Ron, E.Z., and Rosenberg, E. 1999. *Applied and Environmental Microbiology.* 65(6):2697–2702.
Bodour, A.A., Guerrero-Barajas, C., Jiorle, B.V., Malcomson, M.E., Paull, A.K., Somogyi, A., Trinh, L.N., Bates, R.B., and Maier, R.M. 2004. *Applied and Environmental Microbiology.* 70(1):114–120.
Boonchan, S., Britz, M., and Stanley, G.A. 1998. *Biotechnology and Bioengineering.* 59:482–494.
Chen, G., Strevelt, K. A., and Varegas, A. 2001. *Biodegradation.* 12:433–442.
Cooper, D. and Goldberg, B.G. 1997. *Microbiology.* 53:224–229.
Cuny, P., Faucet, M., Acqualviva, J., Bertrand, C., and Gilewicz, M. 1999. *Letters in Applied Microbiology.* 29:242–245.
Dean, S.M., Jin, Y., Cha, D.K., Wilson, S.V., and Radosevich, M. 2001. *Journal of Environmental Quality.* 30:1126–1133.
Desai, J.D. and Banat, I.M. 1997. *Microbiology and Molecular Biology Reviews.* 61:47–64.

EPA. 2004. *Biopesticides Registration Document.* http://epa.gov/pesticides/biopesticides/ingredients/tech_docs/brad_110029.pdf

Henry, N.D., Abazinge, M.D., Johnson, E., and Jackson, T. 2005. *Bioremediation Journal.* 9(3–4):121–128.

Laha, S. and Luthy, R.G. 1991. *Environmental Science and Technology.* 25:1920–1930.

Maslin, P. and Maier, R.M. 2001. *Bioremediation Journal.* 4(4):295–308.

Mueller, J.G., Chapman, P.J., Blattman, B.O., and Pritchard, P.H. 1990. *Applied Microbiology and Biotechnology.* 56:1079–1086.

Navon-Venezia, S., Zosim, Z., Gottlieb, A., Legmann, R., Carmell, S., Ron, E.Z., and Rosenberg, E. 1995. *Applied and Environmental Microbiology.* 61(9):3240–3244.

Piehler, M.F. and Paerl, H.W. 1996. *Biodegradation.* 7(3):239–247.

Ron, E.Z. and Rosenberg, E. 2001. *Environmental Microbiology.* 3(4):229–236.

Tiehm, A. 1994. *Applied and Environmental Microbiology.* 60:258–263.

Volkering, F., Breure, A.M., van Andel, J.G., and Rolkens, W.H. 1995. *Applied and Environmental Microbiology.* 61:1699–1705.

Wong, J.W.C., Fang, M., Zhao, Z., and Xing, B. 2004. *Journal of Environmental Quality.* 33:2015–2025.

Digital Elevation Models and GIS for Watershed Modelling and Flood Prediction – A Case Study of Accra Ghana

D.D. Konadu and C. Fosu

Abstract Geographical Information Systems (GIS) and Digital Elevation Models (DEM) can be used to perform many geospatial and hydrological modelling including drainage and watershed delineation, flood prediction and physical development studies of urban and rural settlements. This paper explores the use of contour data and planimetric features extracted from topographic maps to derive digital elevation models (DEMs) for watershed delineation and flood impact analysis (for emergency preparedness) of part of Accra, Ghana in a GIS environment.

In the study two categories of DEMs were developed with 5 m contour and planimetric topographic data; bare earth DEM and built environment DEM. These derived DEMs were used as terrain inputs for performing spatial analysis and obtaining derivative products. The generated DEMs were used to delineate drainage patterns and watershed of the study area using ArcGIS desktop and its ArcHydro extension tool from Environmental Systems Research Institute (ESRI).

A vector-based approach was used to derive inundation areas at various flood levels. The DEM of built-up areas was used as inputs for determining properties which will be inundated in a flood event and subsequently generating flood inundation maps. The resulting inundation maps show that about 80% areas which have perennially experienced extensive flooding in the city falls within the predicted flood extent. This approach can therefore provide a simplified means of predicting the extent of inundation during flood events for emergency action especially in less developed economies where sophisticated technologies and expertise are hard to come by.

Keywords DEM · GIS · drainage modelling · watershed · flood modelling · ArcGIS · ArcHydro

---

D.D. Konadu (✉) and C. Fosu
Department of Geomatic Engineering, Kwame Nkrumah University of Science and Technology (KNUST), Kumasi, Ghana.
e-mail: konadu@gmail.com; fosucol@hotmail.com

1 Introduction

Accra, the administrative and commercial capital of Ghana has serious problems related to urban flooding as in many cities the world over. This situation is highlighted during the rainy season – between the months of May and July each year – when residences in parts of the city experience ankle to knee deep inundations. Daily activities become virtually paralysed and heavy traffic jams crop up due to stagnant water on the streets and other motorable passages in parts of the city. This scenario creates large infrastructure problems for parts of the city and a huge economic loss in production as well as large damages to existing property, goods and even loss of human lives.

Geographic Information System (GIS) has proven to be very resourceful in dealing with many natural disasters in terms of modelling, prediction, damage assessment and evacuation exercises. GIS can assist in all stages of flood disaster management: prediction, preparation, and prevention, mitigation, and post disaster activities especially when it is integrated with Digital Elevation Models. Digital Elevation Models (DEMs) have been proved to be a valuable tool for the topographic parameterisation of hydrological models which are the basis for any flood modelling process. The existence of digital topographic data in the form of contours (Elevation data) and both man-made and natural features presents an affordable source of data for generating DEM and the subsequent use of the derived DEM in hydrological studies. In this paper a vector base GIS and DEM have been used to delineate watershed boundaries and predict areas of possible inundation during a flood event in the city of Accra using ArcGIS 9.0® software package with its ArcHydro® Tool from ESRI.

The main objectives of the study were:

- Develop DEM from digital topographic data
- Delineate watershed and drainage patterns of the study area from the derived DEM
- Model flood extent and areas of possible inundation using vector-base approach from the derived DEM

1.1 The Study Area

The study area, Accra, (Map shown in Fig. 1) is the administrative and commercial capital of Ghana. It is characterized by low relief topography with occasional hill and an average elevation of 20m above mean sea level. It is bound in the north by the Akwapim ranges from where most of the natural streams that drain the city take their source. The area falls within the anomalous dry equatorial climate region and experiences double maxima rainfall and a prolonged dry season with occasional dry harmattan condition. Rainfall in this area has two peak periods, from May to August and from October to November, with an annual rainfall ranging from 780 to1,200mm (Nyarko 2002). It is during these periods that the city experiences serious inundation.

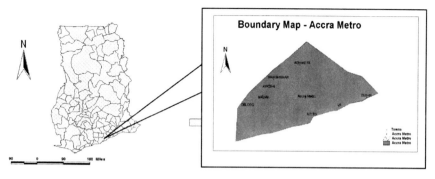

Fig. 1 Map of Accra

2 DEM Data Modelling and Development

A digital elevation model (DEM) is defined as "any digital representation of the continuous variation of relief over space," (Burrough 1986). The preference of data sources and terrain data sampling technique is vital to the quality of the resulting DEM. There are four main methods of data acquisition for the generation of DEMs. These include: Data acquired by space-borne platforms (Satellite Remote Sensing), Air-borne platforms (Photogrammetry and Laser Altimetry), Digitising of contours from already existing topographical maps and Terrestrial Survey techniques (e.g. Spirit levelling, GPS).

2.1 Data Source

In this paper, digital contours and planimetric features (including building footprints, streams, roads etc.) obtained from a 1:2,500 topographic maps of the study have been used. The contours were at an interval of 5 m and had been created from aerial photographs using photogrammetric procedures. This data is very cheap in terms of cost and has an appreciable accuracy. The contours were further interpolated for the generation of the DEM.

2.2 DEM Development

In order to construct a comprehensive DEM it is necessary to establish the topological relations as well as an interpolation model to approximate the surface behaviour (Weibel and Heller 1991). A variety of DEM data structures have being in use over time (Peuker 1978; Mark 1979). However, today, majority of DEMs conform to one

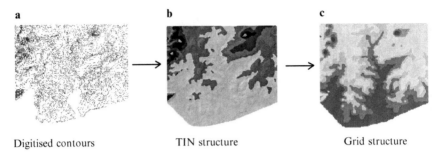

Fig. 2 Various DEM representation

of two data structures: a Triangulated Irregular Network (TIN) (Peuker et al. 1978) or a Regular Grid (or elevation matrix) which is also a format GIS supports (Chen and Kolditz 2005). In hydrologic modelling the first step is to develop this surface.

Watershed and drainage patterns are phenomena which are directly related to the actual ground and must be delineated based on the actual ground elevation so should flood simulation. The digital contours which gives a representation of the bare earth elevation was used as the basis for the creation of a TIN surface in ArcGIS using the 3D Analyst. The output TIN was then converted to a Grid Elevation model. The figure (Fig. 2) above shows the result of the DEM generated from the digital contours covering the study area.

3 Watershed and Drainage Modelling

An effective and proactive means of analyzing any flood event is by determining exactly where water entering a given area will flow; including the general direction and the magnitude of flow. DEMs provide good terrain representation and are the basis for automatic watersheds delineation in GIS technology platforms. The grid DEM generated above has been used for this purpose in this paper.

3.1 Watershed Delineation and Stream Network Generation

Using the ArcHydro Tools, catchment delineation of the watershed and the natural drainage patterns (stream network) of the study area, Accra was generated. Raster analysis is performed to generate data on flow direction, flow accumulation, stream definition, stream segmentation, and watershed delineation. These data are then used to develop a vector representation of catchments and drainage lines (Fig. 3).

4 Flood Simulation

This study has used a vector-based approach in simulating flood extent based on the derived drainage lines, their depth and capacity to hold rainfall run-off. This approach is a very basic flood simulation model which only requires the extents of flood levels, without any information on movement and volume of water. The model uses contours to represent particular flood levels originating from drainage lines. These are selected based on the contours derived from the DEM which have been further interpolated to smaller contour intervals. A catchment has been selected from the delineated watershed of the study area and used as a case study for the simulation. The generalised methods used for the flood simulation follows below.

4.1 Determination of Cross-sections Across Drainage Line

Cross-sections were drawn across the drainage line at locations where there is a sharp change in elevation. The average of the bottom elevation of two successive cross-sections were determined and used to represent the average elevation of the drainage line between them. A total of six cross-sections were determined from the beginning of the drainage line to the end with an average bottom elevation difference between successive cross-sections of about 5 m (See Table 1 below). Fig. 4 shows one of the cross-section positions along the drainage line of the catchment under study.

4.2 Flood Contour Derivation

The mean bottom elevation of consecutive cross-sections determined from the profiles represents the average minimum elevation between the cross-sections above which features in the catchment will be inundated during a flood event, depending on their elevation. With this elevation measure as a guiding factor, 0.5, 1, 1.5 and 2 m were selected as flood water levels. Flood level contours for each of the flood water levels were computed and derived for each of the consecutive cross-sections and then selected from the contours derived at 0.5 m interval and saved as new layers in ArcGIS (Fig. 5). Below is a table showing flood level contours derived from flood water levels for each of the consecutive cross-sections (Table 1).

Table 1 Cross-sections and mean bottom elevation

Cross-section	Bottom elevation	Mean bottom elevation of consecutive cross-section
A-A'	31.5	28.75
B-B'	26.0	24.75
C-C'	23.5	20.50
D-D'	17.5	13.75
E-E'	10.0	7.00
F-F'	4.0	

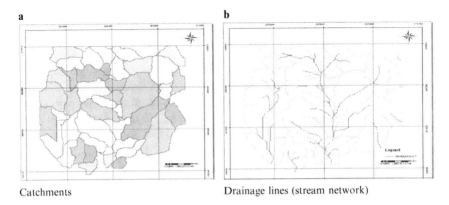

Catchments   Drainage lines (stream network)

Fig. 3 Maps of delineated catchments and drainage lines of study area (Threshold 500)

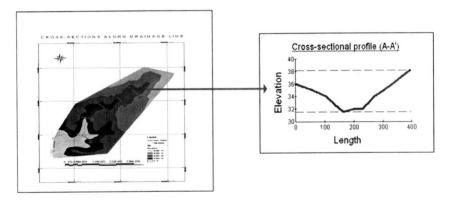

Fig. 4 A cross-section of the drainage line

4.3 Concept Validation and Data Cleaning

The DEM reveals variation in elevation along the drainage line, ranging between 31.5 and 4 m above datum. This suggests that inundations during flood events will vary based on the bottom elevation of the drainage line and that of the adjacent areas. It is therefore rational to section the drainage line with cross-sections and then use the mean bottom elevation between consecutive cross-sections to represent the mean elevation along the drainage line between these cross-sections. The resulting contours represented the flood extents at a given flood level. Many of the areas selected were far and non-adjacent to the drainage line i.e. areas that would not get logically flooded from the drainage line at a given flood level, based on the terrain model and these areas (contour lines) were erased.

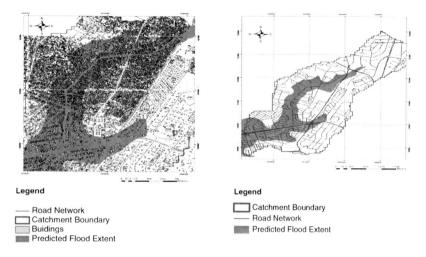

Fig. 5 Maps showing predicted flood extent for 1 m flood water level and properties which may be affected

Table 2 Derived flood level contours

Cross-sections	Mean bottom elevation	Flood level contours at specific flood water levels			
		0.5 m	1.0 m	1.5 m	2.0 m
A-A'/B-B'	**28.75**	29.25	29.75	30.25	30.75
B-B'/C-C'	**24.75**	25.25	25.75	26.25	26.75
C-C'/D-D'	**20.50**	21.0	21.5	22.0	22.5
D-D'/E-E'	**13.75**	14.25	14.75	15.25	15.75
E-E'/F-F'	**7.00**	7.50	8.00	8.50	9.00

5 Results and Analysis of Flood and Map Derivatives

The flood level contours derived for the selected flood water levels indicate areas that face possible inundation in the event of any flood at the specified water levels. The selected flood level contours when converted into polygons and overlaid on the land use and land cover maps of the study area shows elements which would be affected during the event of such flood water levels. In this particular study the major affected elements include residential and commercial establishments, social amenities and other infrastructure such as transportation routes and transit points. This is due to that fact that the area under study is a built up area and hence the land cover will inevitably be constituted of the above mentioned elements.

From the above, the following flood hazard maps have been derived; firstly, general flood hazard maps of a composite land cover/land use showing all elements liable to the predicted floods and secondly maps showing only transportation routes (Roads and streets) that may be affected by the predicted floods. The affected roads give an indication of the traffic congestion that may be experienced on such routes during flood events. This serves as means of managing traffic as well as determining fastest routes for evacuation and rescue operations during flood events. The map derivatives are shown in the figures below.

6 Conclusion

For efficient urban and floodplain and management as well as physical environmental development control and town planning purposes, it is imperative to map various flood prone zones as means for setting development guidelines and instituting emergency response modus operandis. It can therefore be concluded that mapping the extent of flood has overt uses in illustrating the area affected by a particular historical flood or a modeled flood of a given probability of occurrence. Once the possible flood levels have been derived, maps for each flood level can be effortlessly produced with GIS. GIS tools have also shown to have the ability to readily perform different land use based overlay analysis in a planning context, and produce maps of resultant analysis. When overlaid on property or an infrastructure database, it could be analyzed which property or infrastructure is immediately at risk. The visual representations which GIS affords have shown in this case to definitely add value to the results of the numerical modeling for informed planning and hazard management.

It should however be noted that this approach has been used as a prelude to performing hydraulic simulation in a distributed hydrological model for flood modelling in order to assess the type of analysis which can be performed in a vector GIS for flood hazard mapping and planning by disaster management organizations and city authorities with less complexities and cost.

References

Burrough, P. A. (1986) Principles of Geographical Information Systems for Land Resources Assessment, Clarendon Press, Oxford.
Chen, C. and Kolditz, O. (2005) DEM-based structural modeling and TIN technology for more accurate terrain in GIS, ZAG Publisher, Tübingen.
Mark, D. M. (1979) Phenomenon-bases data structuring and digital elevation modeling. Geo-processing 1: 27–33.
Miller, C. L. and Laflamme, R. A. (1958) The digital terrain model – theory and application. Photogrammetric Engineering 24 (3): 433–442.
Nyarko, E. K. (2002) Application of a rational model in GIS for flood risk assessment in Accra, Ghana. Journal of Spatial Hydrology 2:1.
Peuker, T. K. (1978) Data structures for digital elevation models: Discussion and comparison. In Haward Papers on GIS, 5, 1–5.
Steve Kopp, E. S. R. I., Dean Djokic, E. S. R. I., and Al Rea, U. S. G. S. (2005) Introduction to GIS and hydrology. Proceedings of ESRI International Pre conference Seminars, 2005, www.crwr.utexas.edu/gis/gishydro05/Introduction/Presentation/IntroGISHydro05.ppt.
Weibel, R. and Heller, H. (1991) Digital terrain modelling, Geographical Information Systems: Principles and applications, Longmann Group, UK.

Trace Metal Pollution Study on Cassava Flour's Roadside Drying Technique in Nigeria

E.O. Obanijesu and J.O. Olajide

Abstract Cassava flour, generally consumed in Africa as food, is a major source of carbohydrate. Its common drying technique in Nigeria is sun drying for cost optimization whereby the flour (in powder form) is spread by the roadside for moisture content reduction process. This research was carried out at five major traffic highways in Nigeria to study the level of trace element pollution introduced through this drying method, identifying the sources of the pollutants mainly as automobile exhaust emission (major) and street dust (minor). At each site, ten samples (from the four corners and the center) were collected, mixed, digested and analyzed using Graphite Furnace Atomic Absorption Spectroscopy (FAAS) technique to determine the concentration of ten elements (Fe, Cd, As, Pb, Ni, Co, Cu, Cr, Mn and Zn). Analysis of certified standard reference material IAEA-V-10 Hay (Powder) was carried out to ensure accuracy and precision of the technique. Except for zinc, all samples have comparatively high concentrations. Specifically, Fe, As, Pb, Cu and Cd have concentrations as high as 7.2, 5.70, 17.16, 4.57 and 0.39 g/70 kg respectively as against the maximum human uptake limits of 0.01, 0.014, 0.08, 0.11 and 0.03 g/70 kg respectively.

The results show that even though, cassava flour is a rich source of the essential and beneficial minerals required for healthy living, it's drying technique exposes it to the excessive intake of some of these trace metals which could be hazardous to human health. Alternative drying techniques are recommended.

Keywords Trace metals · Drying technique · Cassava flour · FAAS · Health impacts

---

E.O. Obanijesu (✉)
Chemical Engineering Department, Ladoke Akintola University of Technology, Ogbomoso, Nigeria
e-mail: emmanuel257@yahoo.com

J.O. Olajide
Food Science and Engineering Department, Ladoke Akintola University of Technology, Ogbomoso, Nigeria
e-mail: ralfola@yahoo.co.uk

1 Introduction

Cassava flour (popularly known as "Lafun") as a food in Nigeria is a rich source of carbohydrate which contains essential and beneficial mineral needed for body morphological processes. Carbohydrates, often referred to as the major fuel of the tissues (Robert et al. 2000), release energy needed by the body to function properly in its daily activity. The ancient means of preserving this food source is sun-drying and it is still practice in Nigeria whereby the flour is spread by the roadside in order to reduce the cost of drying. The dried foodstuff can then be stored for a long period of time without deteriorating. However, through this drying technique, the foodstuff is subject to air pollution through the trace metal released from the exhaust pipes of passing-by vehicles. Air pollution is the transfer of harmful amounts of natural and synthetic materials into the atmosphere as a direct or indirect consequence of human activities (Vesilind et al. 1993).

Because the emissions of these pollutants occur near ground level, they are not diluted and dispersed as effectively as pollutants emitted from chimneys. The emissions of these pollutants are dependent principally upon the type (diesel or petrol), the quality (Table 1) and the quantity of fuel consumed, the combustion technology employed and the mode in which the road vehicle is driven. Further studies have shown that the extent of contamination of the roadside food also depends on the volume of traffic and nearness to the highway (WHO 1992; Brewer 1997).

The presence of these fugitive particles increase the inorganic component of the cassava flour such as the poisonous group like Lead, Arsenic, Cadmium (Aribike and Akinpelu 2000) and the beneficial group like the nutritive minerals e.g. zinc, calcium, manganese, copper, iron and Nickel (Robert et al. 2000). However, though, some of these inorganic elements are essential to man, their presence in excessive amount in food taken into the body may cause morphological abnormalities, reduce growth, increased mortality rate and mutagenic effect in human (Prasad 1996).

These metals, when taken into the body through food contamination cause damages to human health (Table 2). A greater part of the damage caused is irreversible (Ogunkola and Agboola 2004) thereby leaving the situation more harmful.

Table 1 Concentration of metals in common brands of fuel sold in Nigeria

Super brand of petrol	Levels of metals							
	Na	K	Cr	Mn	Fe	Cu	Zn	Pb
National	7.1	12.5	12.5	4.7	25.3	2.0	4.0	615
A.P	11.5	9.5	19.0	6.5	14.0	3.0	3.5	605
Texaco	4.0	16.5	22.0	4.0	8.0	3.0	3.0	695
Mobil	3.5	19.5	19.0	5.0	9.0	3.0	4.0	792
Aviation gas	7.0	13.0	19.0	4.2	25.5	2.5	6.5	915
Total	3.5	15.5	27.0	4.5	8.5	4.0	5.0	655

Source: (Shalangwa 2004)

Table 2 Effects of excessive intake of trace metals

Trace metal	Toxic effects on human health
Manganese	Parkinson disease results into symptoms which are slowness, poverty of movement, rigidity and postural instability (Nelson and Cox 2000)
Zinc	Gastro intestinal irritation and vomiting occurs, dehydration, nausea, muscular in coordination if even and cough (Robert et al. 2000)
Iron	Tissue damages, stiffness, pains in ankles wrist, knee and finger joints. It also causes greyish skin pigmentation (Nelson and Cox 2000)
Copper	Genetic disease, skin and mucous irritation, liver disease and haemolytic anaemia (Nelson and Cox 2000)
Cobalt	Vomiting, nausea, vision blurring and heart problems. It could lead to thyroid damage, sterility, hair loss, diarrhoea and eventually death (Robert et al. 2000)
Chromium	Dermatitis, ulceration of the skin and perforation of the nasal septum, chronic catarrh and its carcinogen (ATSDR 2006)
Nickel	Causes cancer of the lungs, nose, larynx and prostrate, it leads to heat disorders, birth defects, asthma and allergic reactions such as skin rashes (Trombetto et al. 2005)
Cadmium	Poisoning occurs frequently with zinc. The end-result is painful and rheumatic in nature (Nelson and Cox 2000). High level of this poisonous metal damages the lungs severely and causes death
Mercury	Damages the brain, causes trouble breathing, birth defect, pneumonia, gum problem, hallucination, memory loss, tremor of hand, tongue and eyelids (NJSDH 2006)
Arsenic	Destroys the blood vessels and causes skin cancer (ON 2000)
Lead	Highly toxic element, its intake leads to lead poisoning which induces changes in porphyrin metabolism and produces clinical symptoms resembling acute intermittent which causes anaemia for it accumulates in the bone marrow where red blood corpuscles are formed (Obioh et al. 1998; Padgett and Corash 1998)

In recognition of the various sources of these pollutants into the petroleum products (Table 2), the internationally recommended dietary range (Table 3) coupled with the consequences of their presence above the upper limits (Table 4), this research work is carried out to study the quantity and effects of these trace metals on cassava flour spread along the roadside obtained from some towns in Nigeria. Samples were taken from five different cites widely spread with different vehicular density within Nigeria (Table 5). Recommendations are also made based on the obtained results.

2 Methodology

2.1 Sample Collection and Preparation

Ten samples were collected at each spot from the corners and centre on a bright sunny day from cassava flour spread to dry. At each city, samples were collected from five different locations and properly mixed to represent a city. To avoid further

Table 3 The source of trace metals into the automobile exhaust

Pollutant	Sources
Lead	Lead alkyl in leaded gasoline as antiknock agent to increase octane number
Zinc	Wear from tyres or brake from automobiles and vehicle exhaust and as a result of zinc containing in its lubricating oil
Manganese	Organic forms of manganese present in automotive gas oil existing as a gasoline additive
Cadmium	Mineral combined with other elements such as oxygen, chlorine or sulphur
Nickel	Usually present as catalyst during refining operations such as catalytic cracking
Cobalt	Deposits from dead plants and animal and also from soil or earth crust
Arsenic	From soil humus and earth crust

Table 4 US recommended dietary allowances

Trace metal	Recommended dietary range/adequate level (mg/day)	Tolerable upper intake level (mg/day)
Iron[a] (Fe)	6–15	40–50
Cobalt[a] (Co)	3–5	6–10
Copper[a] (Cu)	2–3	4–10
Chromium[a] (Cr)	0.1–1	2–3
Manganese[a] (Mn)	2–5	6–11
Zinc[a] (Zn)	12–15	10–40
Cadmium[b]	0.00	0.00
Arsenic[b]	0.00	0.00
Lead[b]	0.00	0.00

[a]NAS (1998).
[b]Robert et al. (2000).

Table 5 Table showing collection with traffic classification

Sample	Town	State	Traffic classification
1	Lafia	Nasarawa	High traffic density fast movement
2	Owo	Ondo	Moderately high traffic density slow movement
3	Fiditi	Oyo	High traffic density slow movement
4	Ibilo	Kogi	Moderately low traffic density slow movement
5	Sekona	Osun	Traffic density fast movement
6	Control	–	No traffic

contamination during sampling, transport and storage, the cassava flour samples were kept in labelled polyethylene bags.

2.2 Digestion

Each sample was ground to powder to increase the reaction surface area followed by screening to ensure homogeneity. Two grams of each sample was placed in a round bottom flask and 10 ml of concentration Nitric acid was added. The solution was well shaken to ensure even distribution and the solution was heated on an electro mantle consistently between 95–100 °C as observed by the thermometer. More solution of 10 ml concentrated nitric acid was added to the solution after every 15 min until there were no more brown fumes generated from the solution indicating complete digestion of the food sample. The solution was allowed to cool after which 10 ml of concentrated hydrochloric acid was added followed by filtration into the beaker using funnel and filter paper. Distilled water was added to the filtered solution to make up 50 ml solution. This procedure was repeated for each sample and the resulting solutions sent to a laboratory for Atomic Absorption Spectrophotometry (AAS) analysis.

3 Results and Discussion

The analysis results are presented in Table 6 while graphically represented in Fig. 1. Generally speaking, trace metal contents in all the samples are above the recommended dietary level except for the control which was not exposed to such situation. This confirms that the passing–by vehicles contribute enormously to this metallic poisoning.

The level of pollution in sample 3 (Fiditi) is high because there is high traffic density but slow movement on the account of the roughness of the road, the resulting situation creates a rapid vehicular build up which reaches a peak at about 8:00 and 18:00 h. In a situation of high traffic density or 'go-slow' it is most unlikely that smokes from automobile exhaust accumulate on the sample.

Similar trend is observed in sample 1 (Lafia) which also has high traffic density but smooth road. The pollution was not as much as that of Fiditi however for there was no hold-up to encourage accumulation.

In sample 2 (Owo), high pollution value was also recorded because of the moderately high traffic, slow movement and convergence at this area with it's build up and consequent pollution of the atmosphere.

Table 6 Trace metal concentrations of the samples

Trace element	Sample					
	1	2	3	4	5	6
Iron (Fe)	7.20	8.00	12.80	3.10	2.30	0.01
Cadmium (Cd)	0.386	0.28	0.43	0.06	0.31	0.00
Arsenic (As)	5.70	7.20	9.10	7.90	0.41	0.00
Lead (Pb)	17.16	23.41	25.06	11.40	9.86	0.00
Nickel (Ni)	0.40	0.40	0.50	0.90	0.20	0.20
Cobalt (Co)	0.02	0.02	0.07	0.01	0.00	0.00
Copper (Cu)	4.57	4.85	10.00	3.31	2.52	0.01
Chromium (Cr)	0.04	0.02	0.04	0.02	0.00	0.00
Manganese (Mn)	0.32	0.16	0.33	0.19	0.13	0.07
Zinc (Zn)	0.02	0.03	0.04	0.00	0.01	0.00

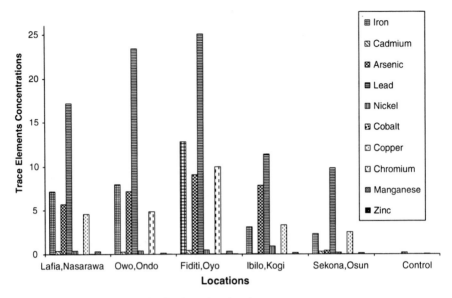

Fig. 1 Trace metal concentration levels against locations

The low values recorded for sample 5 (Sekona) has to do with low traffic density and fast movement because of the large aerial space with low traffic density and for sample 4 (Ibilo) the low value obtained is due to the moderately low traffic density and slow movement of automobile.

4 Conclusion

The results of the study show that concentration levels of some trace metals in cassava flour which is a staple food in Nigeria is very high in three of the study areas and thus may have harmful effect if consumed by humans. The source of these excessive trace metals in the cassava flour could be directly linked to vehicular movement. Areas with high traffic density or 'go-slow' have high concentration levels of some trace metals, whereas, areas with low traffic density have low concentration levels of the trace metals analyzed.

5 Recommendation

Based on this result, it is advisable to discourage people from applying this technique for cassava drying. This could be achieved through public education (e.g. seminars) by making people be aware of the consequences of there actions. While an alternative but cheap drying technique is researched into, fabricated and recommended for their use in replacement. Furthermore, preventive measures such as governmental environmental policies should be put in place and implemented to ensure strict compliance.

References

Aribike, D. S. and Akinpelu, A. (2000). Lead deposition in Nigeria Chemistry, Engineering Journal, November, 1–9

ATSDR (2006). Chromium Toxicity Physiologic Effects, Department of Health and Human Services, Agency for Toxic Substances and Disease Registry, Retrieved November 10, 2006, http://www.atsdr.cdc.gov/HEC/CSEM/chromium/physiologic_effects.html

Brewer, P. (1997). Vehicles as a source of heavy metal contamination in the environment, Unpublished M.Sc. thesis, University of Reading, Berkshire, pp. 1–87

NAS (1998). Dietary Reference Intakes, National Academy of Sciences, Retrieved March 18, 2006, http://www.nal.usda.gov/etext/000/05.html

Nelson, L. N. and Cox, M. M. (2000). Lehninger Principles of Biochemistry, 4th edition, W.H. Freeman and Company, New York, pp. 31–64

NJSDH (2006). The Health Effects of Mercury, Division of Occupational and Environmental Health, New Jersey State Department of Health, USA http://www.pp.okstate.edu/ehs/training/Mercury.htm

Obioh, I. B., Oluwole A. F., and Akeredolu, F. A. (1998). Atmospheric Lead Emissions and Source Strengths in Nigeria, 1998 inventory, Pollution Research Group, Department of Physics, Obafemi Awolowo University Ile-Ife, Nigeria, pp. 271–272

Ogunkola, S. A. and Agboola, O. B. (2004). Environmental impact of waste and its consequences, Chemical Engineering Journal, June, 44

ON (2000). Arsenic and CCA Pressure-Treated Wood: Toxic Effects of Arsenic, Origen Networks, Retrieved November 1, 2006, http://www.origen.net/tox.html

Padgett, B. and Corash, L. (1998). Blood lead concentration in remote Himalayan population, Science, 210, 1135–1136

Prasad, A. S. (1996). Deficiency of Zinc in Men, and Its Toxicity, in Trace Elements in Human Health and Disease, A.S. Prasad and D. Oberleas (eds), Academic, New York, Vol. 1, pp. 20–24

Robert, K. M., Daryl, K. G., Peter, A. M., and Victor, W. R. (2000). Harper's Biochemistry, 25th edition, McGraw Hill, New York, pp. 658–670

Shalangwa, D. K. (2004). The determination of total lead fall our from the atmosphere on certain exposed Nigerian Food, Unpublished B.Sc. thesis, Department of Food Science, University of Maiduguri, Maiduguri, Nigeria, pp. 1–58

Trombetto, D., Mondello, M. R., Cimino, F., Cristani, M., Pergozzi, S., and Saija, A. (2005). Toxic effect of nickel in an in vitro model of human oral epithelium, Journal of Toxicology Letters, 159, 219–225

Vesilind, P. A., Pierce, J. J., and Weiner, R. F. (1993). Environmental Pollution and Control, 3rd edition, Heinemann, London, pp. 120–138

WHO (1992). Urban Air Pollution in Mega Cities of the World, World Health Organization, Blackwell, Oxford, pp. 1–15

Impact of Industrial Activities on the Physico-chemistry and Mycoflora of the New Calabar River in Nigeria

O. Obire and W.N. Barade

Abstract A total of 196 samples were collected at an oil-servicing company (station A) and a control station of the New Calabar River during a 7 months investigation. The samples were analyzed for physico-chemical parameters and mycoflora as to assess the impact of the company activities on the river. Generally, values of temperature, pH, BOD_5, total organic carbon (TOC), chloride, sulphate, oil and grease were higher in station A than in the control. Analysis of variance using paired t-test showed significant difference in temperature, pH, transparency, DO, BOD_5 and in total fungal count at $P \geq 0.01$ and in TOC at $P \geq 0.05$ between the stations. Heavy metal concentration for cadmium, chromium, lead, nickel and zinc were higher in station A which also showed a high tendency of eutrophication. Mean total fungal counts were up to $\times 10^2$.cfu/ml and the frequency of fungi isolated were *Aspergillus* (8.12%), *Byssochlamys* (1.98%), *Candida* (14.31%), *Cephalosporium* (6.73%), *Cladosporium* (6.09%), *Fusarium* (7.42%), *Mucor* (4.06%), *Penicillum* (9.99%), *Rhizopus* (8.07%), *Saccharomyces* (17.73%), *Sporobolomyces* (10.15%) and *Trichoderma* (5.35%). These fungal genera contain species that are potential pathogens. The high values of BOD_5 and heavy metals and low values of DO and total fungal counts in station A is attributed to the activities of the oil servicing company.

Keywords Chemicals · DO · BOD_5 · heavy metals · fungi · pathogens

---

O. Obire (✉) and W.N. Barade
Department of Applied and Environmental Biology, Rivers State University of Science & Technology, P. M. B. 5080, Port Harcourt, Nigeria
e-mail: omokaro515@yahoo.com

1 Introduction

The new Calabar River is a black water type located in Rivers State of Nigeria (Erondu and Chinda 1991). It lies on the eastern arm of the Niger Delta and empties into some creeks and coastal lagoon bordering the Atlantic Ocean. The River is one of the major rivers within the Niger Delta region of Nigeria. It is used as routes and harbor for marine transport boats and barges from oil and oil servicing companies. Notable among these companies are Tidex Nigeria Limited, Wilbros Nigeria Limited, Horizon Fibers limited, West Africa oil field services Limited, and Trans Coaster Limited.

Considering the arrays of companies sited within this area the river receives contaminants and wastes which include auto and marine fuels, heavy oils, spent lubricants and other petroleum products, untreated sewage, human and animal faeces and various kinds of industrial waste. In spite of its importance there is general dearth of information on its water quality (Erondu and Chinda 1991).

Considering the above, it is important to carry out physicochemical and mycological studies of the New Calabar River as to ascertain its water quality. Fungi are chosen as the biological end-point following a general neglect of fungal studies over the years. Apart from that, fungi are also very important in biogeochemical cycle of nutrients and contribute extensively to the atmospheric content of some gases such as CO, CO_2, SO_2 and halomethane (Harper and Hamilton 1988). Microbial populations are nutrient dependent. Hence it is necessary to study changes in fungal frequencies with the associated mineral nutrients in the River. The research is therefore designed to investigate some physico-chemical parameters including the concentrations of some major anions (Cl^-, SO_4^{2-}, and PO_4^{3-}) of the New Calabar River to determine the level of eutrophication based on biological oxygen demand (BOD) and dissolved oxygen (DO) content; to isolate, enumerate and identify the fungi present and their relative frequencies; and to subject the data obtained to statistical analyses. This is aimed at ascertaining the impact of industrial activities on the water quality of the New Calabar river.

2 Materials and Methods

2.1 Description of Sampling Stations and Collection of Water Samples

Following the preliminary survey of the New Calabar River by boat cruise, the Choba-Aluu stretch of the river was chosen as the sampling area for this study. The oil servicing sampling station had the presence of marine boats, tug boats, house boats, etc. and the presence of oil films on the surface of stems and roots of mangrove trees and other vegetation in this station. The control station is located beyond the point where the inhabitants of the Umuihuechi village carry out their domestic activities and is 'believed' to be free from any domestic or industrial activities.

Seven water samples were collected from a depth of about 25–30 cm from each sampling station for physico-chemical analyses, Dissolved Oxygen (DO) and biological oxygen demand (BOD_5) determinations, heavy metal analyses and mycological (fungal) analyses. Samples were collected bi-weekly for a period of 7 months.

2.2 Determination of Physicochemical Parameters and Heavy Metals

The following physico-chemical parameters – water temperature, transparency, hydrogen ion concentration, (pH), flow velocity, chloride, sulphate and phosphate, total oil and grease, total organic carbon, DO and BOD_5, orthophosphate, sulphate, chloride, total organic carbon (TOC), and oil and grease were determined according to standard methods of APHA (1992). Measurement of BOD_5 was by use of automatic digital pH meter (Model METTLE DELTA-340) made in England. The decline in DO was measured using a Polarographic Oxygen Meter (YSI-model 54ARC Ohio USA). Determination of heavy metals concentrations of the water samples was done by use of Atomic Absorption Flame Emission Spectro-Photometer (AAFES), Model-SHIMADZU-AA-6300. The analyzed heavy metals are copper, chromium, iron, nickel, zinc, cadmium and lead.

2.3 Fungal Cultivation, Enumeration, Characterization and Identification

Czapek Dox Agar was prepared in accordance with the modification of Czapek solution by Smith (1971) in agar. A 0.1 ml aliquot of 10^{-1} dilution was plated onto agar plates and incubated at 27.5°C in an inverted position for 5 days. The discrete colonies that developed were counted and the mean of replicate plates were recorded. Fungal cultures were observed while still on plates and after wet mount in lacto-phenol on slides under the compound microscope. Observed characteristics were recorded and compared with the established identification key of Malloch (1997).

3 Results

The average range and mean in parenthesis of the physico-chemical parameters determined and fungal counts of water samples of the stations of the New Calabar River are as shown in Table 1.

The average heavy metal distribution in the stations of the New Calabar River is as shown in Table 2.

Table 1 Average range and mean in parenthesis of physicochemical parameters (mg/l) and fungal counts of stations of the New Calabar River

Parameter	Oil servicing company station	Control station
Temperature (°C)	24.8–31.4 (27.0)	23.7–27.6 (25.2)
pH unit	5.07–5.85 (5.5)	4.84–5.31 (5.1)
Transparency (cm)	98–151 (133.3)	160–200 (189.0)
DO	4.4–5.7 (4.8)	6.8–11.9 (8.4)
BOD_5	26–30.5 (27.8)	4.3–8.6 (6.4)
Total organic carbon (TOC)	23.5–34.7 (29.0)	12.4–14.2 (13.0)
Oil and grease	0.26–0.75 (0.47)	Not detected
Chloride	4.5–7.8 (6.63)	3.8–4.2 (4.0)
Sulphate	8.4–10.4 (9.5)	6.4–8.6 (7.8)
Phosphate	0.16–0.24 (0.20)	0.18–0.31 (0.24)
Fungal count ($\times 10^2$)	2–8 (5.3)	7–14 (11.3)

Table 2 Average heavy metal distribution of the New Calabar River

Heavy metal ($\mu g\ l^{-1}$)	Cadmium (Cd)	Copper (Cu)	Chromium (Cr)	Iron (Fe)	Lead (Pb)	Nickel (Ni)	Zinc (Zn)
Oil servicing company station	127.2	41.4	128.5	124.5	75.7	16.7	16.8
Control station	44.2	63.2	38.6	197.6	29.2	10.7	14.5

Fungal genera and frequency of isolation from the both stations of the New Calabar river were; *Aspergillus* (8.12%), *Byssochlamys* (1.98%), *Candida* (14.31%), *Cephalosporium* (6.73%), *Cladosporium* (6.09%), *Fusarium* (7.42%), *Mucor* (4.06%), *Penicillum* (9.99%), *Rhizopus* (8.07%), *Saccharomyces* (17.73%), *Sporobolomyces* (10.15%) and *Trichoderma* (5.35%).

4 Discussion

The present study has revealed some physico-chemical constituents including heavy metals, and mycoflora (fungi) of the New Calabar River. Generally, values of temperature, pH, BOD_5, total organic carbon (TOC), chloride, sulphate, oil and grease were higher in the oil servicing company station (station A) than in the control. Analysis of variance using paired t-test showed significant difference in temperature, pH, transparency, DO, BOD_5 and in total fungal count at $P \geq 0.01$ and in TOC at $P \geq 0.05$ between the stations.

The higher temperature recorded in station A is due to the direct discharge of warm wastewaters from machinery and automobiles of the oil-servicing company into the river.

The phosphate values obtained in this investigation were generally low ($0.21\,mg\,l^{-1}$). They are however within the range expected of black waters. The low chloride concentration of the New Calabar is an indication of the reduced influence of nutrient laden seawater on this zone. This accounts for the transition from oligohaline at the company station to fresh water in the control station. High transparency recorded in the investigation (an average transparency 133.3 and 189.0 cm Secchi disc) is within the range of 1.3–2.3 m recommended for classification of River types as black water (Whitton 1975).

The concentration of dissolved oxygen is a major limiting factor to the mineralization of organic matter in a river. In this study station A recorded a lower DO content which must have accounted for the lower fungal counts recorded in this station.

Concentrations of some heavy metals cadmium, chromium, lead, nickel and zinc were higher in station A. The values of heavy metals obtained are considered high because they were the filtered value. The high value of cadmium, chromium and lead is associated with the industrial activities in the oil servicing company which include the use of anti corrosion paints and paints of various kinds. The high concentration of iron ($>100\,\mu g\,l^{-1}$) in both stations is significant because low pH (3–4) and low DO concentrations may cause iron toxicity (Brain 1993). The persistent high levels of sulphates may also lead to increased release of soluble reduced metal species (e.g. Fe, Zn, and Cu), from sediment into surface water (Gilmour 1992) thereby increasing their availability. The pH of both stations of the New Calabar River is acidic. However, the sulphate concentration of the station A is higher than in the control. Sulphuric acid is believed to be a major strong acid component of deposition in most industrialized regions. Effects of acidification of natural waters include changes in biological species composition and densities and changes in biogeochemical cycling of a number of elements (Gilmour 1992). In this study, the mean total fungal counts (fungal population) were lower in station A than in the control. However, *Aspergillus, Cephalosporium, Cladosporium, Fusarium, Mucor, Rhizopus*, and *Trichoderma* occurred more in station A, while *Byssochlamys, Candida, Penicillum, Saccharomyces* and *Sporobolomyces* species occurred more in the control.

Most of these fungal genera are known to contain species that are potential pathogens or opportunistic pathogens. The main hazardous species belong to *Aspergillus, Penicillum, Cladosporium, Mucor*, and *Fusarium* which have been implicated in being causative agents in asthma, hypersensitivity pneumonitis and pulmonary mycosis (Ponton et al. 2000). The high occurrence of *candida* in both stations is of considerable concern as the genera can cause candidiasis of the skin prevalent in persons on broad-spectrum antibiotic therapy. Members of the genus can also produce endocarditis, septicemia, protracted urinary tract infections, kidney and lung infections, esophagitis, oral thrush, diaper rash infections in infants, vaginal infection in women and other soft tissues infections. Fungi are also implicated in Dermatomycoses, a superficial fungal infection that penetrates only the epidermis, hair, or nails. Such infections include athlete's foot, jock itch, and ringworm (Romano et al. 1998; Deshpande and Koppikar 1999).

Fusarium species are common plant pathogens and causative agents of superficial and systemic infections in humans (Mayayo et al. 1999). *Fusarium* spp. produces mycotoxins. Ingestion of grains contaminated with these toxins may give rise to allergic symptoms or be carcinogenic in long-term consumption (Schaafsman et al. 1998). It is worthy of note that the New Calabar River serves as a source of water for many domestic purposes such as food processing of various grains including maize used in producing palp (*akamu*). The presence of these potential fungal pathogens is of considerable concern as regards public health as these waters are used for domestic purposes, recreation and aesthetics.

5 Conclusion

Statistical analysis of the physicochemical constituents and of fungal counts of this investigation revealed significant differences between the stations. The high values of BOD_5 and heavy metals and low values of DO and total fungal counts in the company station is an indication that the industrial activities of the oil servicing company have effect on the water quality and on the fungal population of the New Calabar River. Heavy metal toxicity may be induced if the levels are left unchecked. The presence of potential fungal pathogens is of considerable concern as regards public health. The findings justified the need to constantly monitor the water quality and changes in fungal populations and types of fungi of the New Calabar River as to ascertain the cause and implications of such changes. There is the urgent need for companies sited along the river to treat their wastewater or effluent as to improve its quality before discharge into the river. Potable water should be provided for the communities to prevent the use of the river water for domestic, recreation, and other purposes. The New Calabar River requires specific remediation effort to prevent it from further deterioration and improve its quality.

References

APHA (American Public Health Association) (1992). *Standard Methods for the Examination of Water and Waste-Water*, 18th ed. APHA, Washington, DC.
Brain, M. (1993). *Ecology of Fresh Waters*. Blackwell, London, pp. 10–59.
Deshpande, S.D. and Koppikar, G.V. (1999). A study of mycotic keratitis in Mumbia. *Indian J. Pathol. Microbiol.* 42:81–87.

Erondu, E.S. and Chinda, A.C. (1991). *Variations in the Physico-chemical Features and Phytoplankton of the New Calabar River at Aluu, Rivers State Nigeria*. Nigerian Institute for Oceanography and Marine research (NIOMR) Tech. Paper No. 75.

Gilmour, C.C. (1992). Effects of acid deposition on microbial processes in natural waters. In: Ralph, M. (ed) *Environmental Microbiology*. Wiley, New York, pp. 33–35.

Harper, D.B. and Hamilton, J.T.G. (1988). Biosynthesis of chloromethane in *Phellinus pomaceus*. *J. Gen. Microbiol*. 134:2831–2839.

Malloch, D. (1997). *Moulds Isolation, Cultivation and Identification*. Department of Botany, University of Toronto, Toronto.

Mayayo, E., Pujol, I., and Guarro, J. (1999). Experimental pathogenicity of four opportunist *Fusarium* species in a murine model. *J. Med. Microbiol*. 48:363–366.

Ponton, J., Ruchel, R., Clemons, K.V., Coleman, D.C., Grillot, R., Guarro, J., Aldebertt, D., Ambroise-Thomas, P., Cano, J., Carrillo-Munoz, A.J., Gene, J., Pinel, C., Stevens, D.A., and Sullivan, D.J. (2000). Emerging pathogens. *Med. Mycol*. 38:225–236.

Romano, C., Miracco, C., and Difonzo, E.M. (1998). Skin and nail infection due to *Fusarium oxysporum* in Tuscany, Italy. *Mycoses* 41:433–437.

Schaafsmann, A.W., Nicol, R.W., Savard, M.E., Sinha, R.C., Reid, L.M., and Rottinghaus, G. (1998). Analysis of Fusarium toxins in maize and wheat using thin layer chromatography. *Mycopathologia* 142:107–113.

Smith, G. (1971). *An Introduction to Industrial Mycology*. Edward Arnold Publishers, London, p. 277.

Whitton, B.A. (1975). *River Ecology, Guidelines to Water Quality*. Blackwell, Oxford, pp. 141–800.

Plants as Environmental Biosensors: Non-invasive Monitoring Techniques

A.G. Volkov, M.I. Volkova-Gugeshashvili, and Albert J. Osei

Abstract Plants are continuously exposed to a wide variety of perturbations including variation of temperature and/or light, mechanical forces, gravity, air and soil pollution, drought, deficiency or surplus of nutrients, attacks by insects and pathogens, etc. It is essential for all plants to have survival sensory mechanisms against such perturbations. As a consequence, plants generate various types of intracellular and intercellular electrical signals mostly in the form of action potentials or variation potentials in response to these environmental changes. However, over a long period, only certain plants with rapid and highly noticeable responses to environmental stresses have received much attention from plant scientists. Of particular interest to our recent studies on ultra fast action potential in green plants, we discuss in this review the possibility of utilizing green plants as fast biosensors for molecular recognition of the direction of light, monitoring the environment, and detecting the insect attacks as well as the effects of pesticides and defoliants.

Keywords Bioelectrochemical signaling · biosensors · action potential · phototropism · acid rain

1 Introduction

A biosensor is defined as a device that either detects, records, and transmits information related to a physiological change/process in a biological system, or uses biological materials to monitor the presence of various chemicals in a substance. A variety of

---

A.G. Volkov (✉), M.I. Volkova-Gugeshashvilla
Department of Chemistry, Oakwood University, 7000 Adventist Blvd., Huntsville, AL 35896, USA
e-mail: gvolkov@oakwood.edu

Albert J. Osei
Department of Mathematics and Computer Science, Oakwood University, 7000 Adventist Blvd., Huntsville, AL 35896, USA
e-mail: osei@oakwood.edu

E.K. Yanful (ed.), *Appropriate Technologies for Environmental Protection in the Developing World*,
© Springer Science+Business Media B.V. 2009

plant and animal-tissues have been incorporated into various electrochemical transducers to detect and quantify a range of biologically important analytes including drugs, hormones, toxicants, neurotransmitters and amino acids. A detailed discussion on biosensors utilized in chemical or biological analysis is beyond the scope of this review. Based on our investigations on fast bioelectrochemical signaling events in green plants and similar examples reported in the literature by other plant scientists, we discuss here the evidence supporting the foundation for utilizing the entire green plant as a fast biosensor for monitoring the environmental perturbations in the close vicinity of a living plant.

Nerve cells in animals and phloem cells in plants share one fundamental property: they possess excitable membranes through which electrical excitations can propagate in the form of action potentials (Ksenzhek and Volkov 1998; Bose 1925; Volkov et al. 2000; Mwesigwa and Volkov 2001a, b). Plants generate bioelectrochemical signals that resemble nerve impulses, and are present in plants at all evolutionary levels. Prior to the morphological differentiation of nervous tissues, the inducement of nonexcitability after excitation and the summation of subthreshold irritations were developed in the vegetative and animal kingdoms in protoplasmatic structures.

The cells, tissues, and organs of plants transmit electrochemical impulses over short and long distances. It is conceivable that action potentials are the mediators for intercellular and intracellular communication in response to environmental irritants (Mwesigwa et al. 2000; Shvetsova et al. 2001; Volkov 2006a, b, c; Brown and Volkov 2006). Action potential is a momentary change in electrical potential on the surface of a cell that takes place when it is stimulated, especially by the transmission of an impulse (Brown and Volkov 2006).

Initially, plants respond to irritants at the site of stimulation; however, excitation waves can be distributed across the membranes throughout the entire plant. Bioelectrical impulses travel from the root to the stem and vice versa. Chemical treatment, intensity of the irritation, mechanical wounding, previous excitations, temperature, and other irritants influence the speed of propagation (Ksenzhek and Volkov 1998; Volkov 2006a, b, c).

Conductive bundles of vegetative organisms sustain the flow of material and trigger the conduction of bioelectrical impulses. This feature supports the harmonization of biological processes involved in the fundamental activity of vegetative organisms.

The conduction of bioelectrochemical excitation is a rapid method of long distance signal transmission between plant tissues and organs. Plants quickly respond to changes in luminous intensity, osmotic pressure, temperature, cutting, mechanical stimulation, water availability, wounding, and chemical compounds such as herbicides, plant growth stimulants, salts, and water. Once initiated, electrical impulses can propagate to adjacent excitable cells. The change in transmembrane potential creates a wave of depolarization or action potential, which affects the adjoining resting membrane (Brown and Volkov 2006).

Electrical potentials have been measured in our laboratory at the tissue and whole plant level by using the experimental set-up described in Fig. 1. Measurements were taken inside a Faraday cage mounted on a vibration-stabilized table. An *IBM*-compatible microcomputer with multi I/O plug-in data acquisition board NI 6052E DAQ (*National Instruments*) was interfaced through a NI SC-2040 Simultaneous Sample and Hold (*National Instruments*). The multifunction NI 6052E data acquisition board provides high resolution and a wide gain range and supports continuous, high-speed data acquisition. Single channels can be sampled at any gain up to 333 kSamples/s. The digitized data includes negligible time skew (less than 50 ns) between channels. Measuring signals were recorded as ASCII files using *LabView (National Instruments)* software. Nonpolarizable reversible Ag/AgCl electrodes were used to measure the electrical signals. The temperature was held constant since these electrodes are sensitive to the temperature. Ag/AgCl electrodes were prepared from Teflon coated silver wire (*A-M Systems, Inc.*). Plants were irradiated in directions A or B at different wavelengths using narrow band pass interference filters from *GS Edmund Scientific* (Barrington, NJ) with a central wavelength tolerance of ±1 nm.

2 Plants as Biosensors for Monitoring the Acid Rain

Acid rain is the most serious environmental problem and has impact on agriculture, forestry, and human health (Shvetsova et al. 2002; Volkov et al. 2002). Chemical reactions involving aerosol particles in the atmosphere are derived from the interaction of

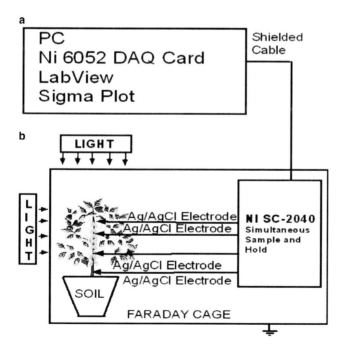

Fig. 1 Experimental set-up for measuring electrical signals in green plants

gaseous species with the liquid water. These reactions are associated with aerosol particles and dissolved electrolytes. For example, the generation of HONO from nitrogen oxides takes place at the air/water interface of seawater aerosols or in clouds. Clouds convert between 50% and 80% of SO_2 to H_2SO_4. This process contributes to the formation of acid rain. Acid rain exerts a variety of influences on the environment by greatly increasing the solubility of different compounds, thus directly or indirectly affecting many forms of life.

Acid rain has a pH below 5.6. Sulfuric acid and nitric acid are the two predominant acids in acid rain. Approximately, 70% of the acid content in acid rain is sulfuric acid, with nitric acid contributing to the rest 30%. Spraying the soybean plant with an aqueous solution of H_2SO_4 in the pH region from 5.0 to 5.6 does not induce action potentials. However, action potentials were generated in soybean either by spraying the leaves of the plant (1 mL) or deposition of 10 ∝L drops of aqueous solution of

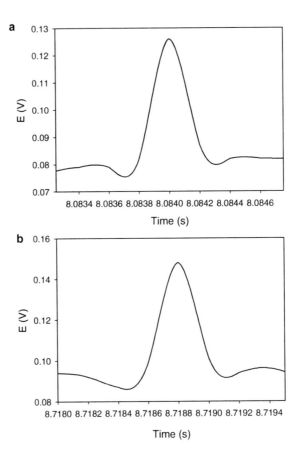

Fig. 2 Potential difference between two Ag/AgCl electrodes in the stem of the soybean measured 7 (**a**) and 100 (**b**) h after adding 25 mL of 1 mM $Al(NO_3)_3$ to the soil. The soil pH after sterilization was 7. Distance between Ag/AgCl electrodes was 5 cm. Volume of soil was 0.5 L. The soil around the plant was treated with water every day. Room temperature was 22°C. Humidity was 45–50%

H_2SO_4 or HNO_3 in the pH region from 0 to 4.9 on leaves. The duration of single action potentials after spraying the plant with HNO_3 and H_2SO_4 was 0.2 and 0.02 s, respectively (Shvetsova et al. 2002; Volkov et al. 2002).

The evolution of plants occurred in the presence of many mineral nutrients even in high to potentially toxic concentrations. Of those nutrients, aluminum in particular is the most abundant metal (8%) in the Earth's crust. Aluminum concentrations in mineral soil solutions are usually well below 50 ∝M at pH values higher than 5.5, but rise sharply at lower pH. Subsequently, plant roots have been continuously exposed to potential toxic concentrations in the soil environment. On the other hand, aluminum is not considered an essential element, but many plants usually contain from 0.1 to 500 ppm and the addition of small amounts of aluminum to a nutrient solution may promote plant growth.

An important advantage of plants at mildly acidic, neutral, or alkaline pH values is that most phytotoxic forms of Al are relatively insoluble in soil. However, at pH 5 and below, Al may accumulate to toxic concentrations that prevent root growth and plant functions. Thus aluminum compounds become more soluble in acidified soils. In accordance with soil science, soil acts as a buffering system when the pH of soil slowly increases to a neutral value after soil acidification due to reaction of protons with metal oxides such as Al which is normally insoluble at a neutral pH. Therefore, we continued our study on the effects of Al salts on electrical signaling in soybean. Figure 2 show that Al ions induce fast action potentials in soybean.

3 Electrical Signals Induced by Pesticides

Pesticides 2, 3, 4, 5, 6-pentachlorophenol (PCP), 2, 4-Dinitrophenol (DNP), carbonylcyanide m-chlorophenylhydrazone (CCCP), and carbonylcyanide-4-trifluoromethoxyphenylhydrazone (FCCP) act as insecticides and fungicides. PCP is the primary source of dioxins found in the environment. This pollutant is a defoliant and herbicide. PCP is utilized in termite control, wood preservation, seed treatment, and snail control. The pesticide DNP is used to manufacture dye and wood preservative. DNP is often found in pesticide runoff water. The electrochemical effects of CCCP, PCP, DNP, and FCCP have been evaluated on soybean plants (Mwesigwa et al. 2000; Shvetsova et al. 2001; Volkov et al. 2000, 2001; Labady et al. 2002; Mwesigwa and Volkov 2001a, b).

CCCP decreased the variation potentials of soybean from 80–90 mV to 0 mV after 20 h. CCCP induced fast action potentials in soybean with amplitude of 60 mV (Labady et al. 2002). The maximum speed of propagation was 25 m/s. Exudation is a manifestation of the positive root pressure in the xylem. After treatment with CCCP, the exudation from cut stems of the soybean remains the same. Therefore, the addition of CCCP did not cause a change in the pressure, although it may influence the zeta potential due to depolarization (Labady et al. 2002).

The addition of aqueous solution of PCP also causes the variation potential in soybeans to stabilize at 0 mV after 48 h. Rapid action potentials are induced. These action potentials last for 2 ms, and have amplitudes of 60 mV. The speed of propagation is 12 m/s; after 48 h, the speed increased to 30 m/s.

DNP induces fast action potentials and decreases the variation potential to zero in soybeans (Mwesigwa et al. 2000). The addition of aqueous DNP to the soil induces fast action potentials in soybeans. After treatment with an aqueous solution of DNP, the variation potential, measured between two Ag/AgCl electrodes in a stem of soybean, slowly decreases from 80–90 mV (negative in a root, positive on the top of the soybean) to 0 during a 48-h time frame. The duration of single action potentials, 24 h after treatment by DNP, varies from 3 to 0.02 s. The amplitude of action potentials is about 60 mV. The maximum speed of action potential propagation is 1 m/s. After 2 days, the variation potential stabilized at 0. Fast action potentials were generated in a soybean, with amplitude of about 60 mV, 0.02 s duration time, and a speed of 2 m/s. Fromm and Spanswick studied the inhibiting effects of DNP on the excitability of willow by recording the resting potential in the phloem cells (Fromm and Spanswick 1993). In willow, 10^{-4} M DNP rapidly depolarized the membrane potential by about 50 mV.

The FCCP also induced action potentials in soybean (Shvetsova et al. 2001). The maximum speed of these action potentials within 20 h after the treatment was 10 m/s. After 100 h, the action potentials were still being produced. The amplitude of 60 mV remained constant. The duration was 0.3 ms, and the speed of propagation was 40 m/s (Shvetsova et al. 2001).

Constant release of hazardous metal pollutants into the environment has become a global problem. Contamination of soil, ground and surface waters with such pollutants can negatively affect all levels of an ecosystem, and thus, the clean-up of contaminated soils and waters is one of the most important challenges the environmental scientists face today.

4 Insect-Induced Electrochemical Signals in Potato Plants

Volkov and Haack were the first to afford a unique opportunity to investigate the role of electrical signals induced by insects in long-distance communication in plants (Haack and Volkov 1995a, b).

Action and resting potentials were measured in potato plants (*Solanum tuberosum* L.) in the presence of leaf-feeding larvae of the Colorado potato beetle (*Leptinotarsa decemlineata* (Say); Coleoptera: Chrysomelidae). When the larvae were allowed to consume upper leaves of the potato plants, after 6–10 h, action potentials with amplitudes of 40 ± 10 mV were recorded every 2 ± 0.5 h during a 2-day test period. The resting potential decreased from 30 mV to a steady state level of 0 ± 5 mV. The action potential induced by the Colorado potato beetle in potato plants propagates slowly and hence, the speed of propagation can be measured with two Ag/AgCl electrodes. The action potential propagates from plant leaves with Colorado potato beetles down the stem, and to the potato tuber (Volkov and Haack 1995). The speed of propagation of the action potential does not depend on the location of a working electrode in the stem of the plant or tuber, or the distance between the working and reference electrodes (Volkov and Haack 1995).

5 Molecular Recognition of the Direction of Light by Green Plants

Phototropism is one of the best-known plant tropic responses. A positive phototropic response is characterized by a bending or turning toward the source of light. When plants bend or turn away from the source of light, the phototropic response is considered negative. A phototropic response is a sequence of the four following processes: reception of the directional light signal, signal transduction, transformation of the signal to a physiological response, and the production of directional growth response. Phototropin is a blue light (360–500 nm) flavoprotein photoreceptor responsible for phototropism and chloroplast orientation.

Inside the Faraday cage the soybean plant was irradiated in the direction A (Fig. 1) with white light for 2 days with a 12:12 h light: dark photoperiod prior to the conduction of experiments. Action potentials are not generated when the lights are turned off and on. Changing the direction of irradiation from direction A to direction B generates action potentials in soybean approximately after 1–2 min. These action potentials depend on the wavelength of irradiation light. Irradiation at wavelengths 400–500 nm induces fast action potentials in soybean with duration time of about 0.3 ms; conversely, the irradiation of soybean in the direction B at wavelengths between 500 and 630 nm fails to generate action potentials. Irradiation between 500 and 700 nm does not induce phototropism. Irradiation of soybean by blue light induces positive phototropism (Volkov et al. 2004).

6 Conclusion and Future Perspective

Green plants interfaced with a computer through data acquisition systems can be used as fast biosensors for monitoring the environment, detecting effects of pollutants, pesticides, defoliants, predicting and monitoring climate changes, and in agriculture, directing and fast controlling of conditions influencing the harvest.

Acknowledgements This work has been supported by NASA grant NAG8-1888.

References

Bose, J.C. 1925. Transmission of stimuli in plants. Nature, 115, 457 pp.
Brown, C.L. and Volkov, A.G. 2006. Electrochemistry of plant life. In: Plant Electrophysiology (A.G Volkov, ed.), Springer, Berlin/New York.

Fromm, J. and Spanswick, R. 1993. Characteristics of action potential in willow (Salix viminalis L). Exp. Bot., **14**, 119–125.
Haack, R.A. and Volkov, A.G. 1995a. Bioelectrochemical signals in potato plants. Russ. J. Plant Physio. **42**, 17–23.
Haack, R.A. and Volkov, A.G. 1995b. Insect induced bioelectrochemical signals in potato plants. Bioelectrochem. Bioenerg. **35**, 55–60.
Ksenzhek, O.S. and Volkov, A.G. 1998. Plant Energetics. Academic, San Diego, CA.
Labady, A., Thomas, D'J., Shvetsova, T. and Volkov, A.G. 2002. Plant electrophysiology: Excitation waves and effects of CCCP on electrical signaling in soybean. Bioelectrochemistry **57**, 47–53.
Mwesigwa, J. and Volkov, A.G. 2001a. In: Liquid Interfaces in Chemical, Biological and Pharmaceutical Applications (A.G Volkov, ed.), Marcel Dekker, New York.
Mwesigwa, J. and Volkov, A.G. 2001b. Electrochemistry of soybean: Effects of uncouplers, pollutants, and pesticides. J. Electroanal. Chem. **496**, 153–157.
Mwesigwa, J., Collins, D.J. and Volkov, A.G. 2000. Electrochemical signaling in green plants: Effects of 2, 4-dinitrophenol on resting and action potentials in soybean. Bioelectrochemistry **51**, 201–205.
Shvetsova, T., Mwesigwa, J. and Volkov, A.G. 2001. Plant electrophysiology: FCCP induces fast electric signaling in soybean. Plant Sci. **161**, 901–909.
Shvetsova, T., Mwesigwa, J., Labady, A., Kelly, S., Thomas, D'J., Lewis, K. and Volkov, A.G. 2002. Soybean electrophysiology: Effects of acid rain. Plant Sci. **162**, 723–731.
Volkov, A.G. 2006a. Green plants: Electrochemical interfaces. J Electroanal. Chem. **483**, 150–156.
Volkov, A.G. 2006b. Electrophysiology and phototropism. In: Communication in Plants. Neuronal Aspects of Plant Life (F. Baluška, S. Mancuso, and D. Volkman, eds.), Springer, New York, pp. 351–367.
Volkov, A.G. (ed.). 2006c. Plant Electrophysiology, Springer, New York/Berlin.
Volkov, A.G., Collins, D.J. and Mwesigwa, J. 2000. Plant electrophysiology: Pentachlorophenol induces fast action potentials in soybean. Plant Sci. **153**, 185–190.
Volkov, A.G., Labady, A., Thomas, D. J. and Schvetsova, T. 2001. Green plants as environmental biosensors: electrochemical effects of Carbonyl cyanide 3-chlorophenylhydrazone on soybean. Analyt. Sci. **17**, i359–i362.
Volkov, A.G., Mwesigwa, J., Jovanov, E., Labady, A., Thomas, D'J., Lewis, K. and Shvetsova, T. 2002. Acid Rain Induces Action Potentials in Green Plants. Proceedings, IV International Workshop on Biosignal Interpretation BSI2002, Eds. S. Cerutti, M. Akay, L. T. Manardi, S. Sato and C. Zywietz, Polytechnic University Press, Milaro, Italy, 513–517.
Volkov, A.G., Dunkley, T.C., Morgan, S.A., Ruff, D., Boyce, Y. and Labady, A. 2004. Bioelectrochemical signaling in green plants induced by photosensory systems. Bioelectrochemistry **63**, 91–94.

Index

A
AAS, 337
Acid mine drainage (AMD), 199, 229, 247
Acid rain, 351
Acidic effluent, 200
Action potential, 350
Activated carbon, 115
Adsorption, 118
Africa, 26
Agronomic NUE, 309
ArcGIS, 328
ArcHydro, 328
ARPU, 110
Aquifer, 240

B
Bagasse ash, 272
Barrier, 258
Bioavailability, 322
Biodiesel, 76
Biodiversity, 94
Bioelectrochemical signaling, 350
Biofilm, 259
Biogas, 95, 145
Biosensors, 351
Biosurfactants, 318
BOD5, 343

C
California bearing ratio, 274
Cancer, 60
Carbon absorbents, 117
Cassava flour, 334
Catalytic hydrocracking, 76
Chocolate production, 37
Cleaner production, 143
Coal fly ash, 248

Coal mining, 237
Cocoa supply chain, 37
Co-combustion, 85
Compaction, 274
Composite thin film, 182
Covers, 207
Cow dung, 97
Crop yield, 308
Cuprous oxide, 182

D
Decentralised systems, 165
Deforestation, 26
DEM, 326
Desalination, 124
Developing countries, 152
Digester, 96
DO, 342
Drainage modelling, 328
Drying technique, 333–339
Durability, 272
Dynamic control, 219

E
Energy, 17
Environmental conservation, 307–314
Environmental impacts, 36
Environmental studies, 297–303
Equipment safety, 226

F
Fertilizer, 17
Flood modelling, 326
Flowable fill, 285
Fly ash, 282

Forests, 26
Fungi, 342

G
Geoelectrical resistivity, 298–303
GIS, 326
Greenhouse gas, 248
Groundwater quality, 171
Gypsum-saturated, 237

H
Heavy metals, 343
Herbal vegetables, 45
Human health, 334
Hydrate formation climate change, 106

I
Integrated waste management concept, 83–90
Irrigation, 236

J
Jatropha curcas, 75–81

L
Land degradation, 27
Lateritic soil, 271–279
LCA, 36
Leaching, 281–293
Lime, 247–254

M
Malawi, 169–178
Market, 47–50
Medicinal plants, 45
Medicine, 43–50
Membranes, 126–128
Microbial activity, 200
Microbial risk assessment, 53–58
Microencapsulation, 319
Mineral admixture, 17
Mining, 206
Modern, 139–150
Moringa oleifera, 171
Municipal solid waste (MSW), 85

N
Nanites, 69
Nanofiltration and reverse osmosis, 124
Nanotechnology, 59–70
Natural Gas, 102
Natural wetland, 199–201
Neutralization, 229–234
Nitrogen management, 307–314
Non-invasive techniques, 298
NORM, 3

P
Pathogen retention, 155
Pathogens, 345
Phenanthrene, 318
Photo electrode, 182
Photoelectrocatalysis titanium dioxide, 184
Phototropism, 355
Photovoltaic, 124
Polymers, 258
Potato, 308
Poultry manure, 96

Q
Quantum dots, 60

R
Radiation, 4
Radioactivity, 3
Reactive mine tailings, 248
Reclamation, 206
Renewable energy, 165
Resistivity, 298
Resources, 47
Reversing, 30
Rice straw, 16
Risk management, 57
RLR, 231

S
Salts, 211
Shallow wells, 170
Silica, 16
Situational awareness, 219
Slag, 231
Solar energy, 124

Solid recovered fuels (SRF), 84
Strategies, 26
Strength, 16
Submerged offshore pipeline, 101
Submerged ultrafiltration pretreatment, 123
Subsurface resistivity, 298
Sustainable, 26

T
Thermo-gravimetric analysis, 88
Trace metals, 335
Traditional healers, 44
Truck vibrations, 222
Tuber quality, 308

U
Ultrafiltration, 152–160

Unconfined compressive strength, 277
United Nations, 67

V
Vegetable leaves, 3–6
Vegetation, 206

W
Waste management, 119, 257
Waste, 141
Watershed scale, 206
Water supply, 152
Water, 140, 236
Watershed, 325–332

X
X-ray diffraction, 232